Study Guide for Starr and

BIOLOGY
THE UNITY AND DIVERSITY OF LIFE

Study Guide for Starr and Taggart's

BIOLOGY
THE UNITY AND DIVERSITY OF LIFE

FIFTH EDITION

JANE B. TAYLOR
Northern Virginia Community College

Wadsworth Publishing Company
Belmont, California
A Division of Wadsworth, Inc.

Biology Editor: Jack C. Carey
Editorial Assistant: Susan Belmessieri
Production Editor: Jerilyn Emori
Managing Designer: Carolyn Deacy
Print Buyer: Randy Hurst
Permissions: Robert Kauser
Interior Design: Detta Penna
Copy Editor: Betty Duncan-Todd
Compositor: Graphic Typesetting Service, Los Angeles, California

Credits: Illustrations from C. Starr and R. Taggart, *Biology,* Fourth and Fifth Editions, Wadsworth, 1987 and 1989: Root tip sketch by Marian Reeve (p. 165); buttercup photo by Chuck Brown (p. 165); cell diagrams after Weier et al., *Botany,* Sixth Edition, Wiley, 1982 (p. 165); brain and digestive system photos from C. Yokochi and J. Rohen, *Photographic Anatomy of the Human Body,* Second Edition, Igaku-Shoin Ltd., 1979 (p. 223 and p. 285). Other illustrations originally from C. Starr and R. Taggart, *Biology,* Fourth and Fifth Editions, Wadsworth, 1987 and 1989 (pp. 51, 81, 91, 129, 161, 163, 180, 197, 201, 233, 240, 242, 247, 250, 256, 261, 277, 285, 295, 311, 313, 355, 358, 364, 369).

Printed in the United States of America   49

2 3 4 5 6 7 8 9 10—93 92 91 90 89

ISBN 0-534-09181-4

CONTENTS

PREFACE

This study guide is like a tutor; it increases the efficiency of your study periods; it condenses the major ideas of your text; it asks you to do a series of specific tasks to demonstrate your ability to recall key concepts and terms and relate them to life; it tests you on your understanding of the factual material and indicates what you might wish to reexamine or clarify; and it gives you a preliminary estimate of your next test score based on specific material. Most important, though, the study guide and text together help you make informed decisions about matters that affect your own well-being and the well-being of your environment. In the years to come our survival will increasingly depend on administrative and managerial decisions based on an informed biological background.

HOW TO USE THIS STUDY GUIDE

After this preface, you will find an outline that will show you how the study guide is organized and will help you use it efficiently. Each chapter begins with an outline of the topics discussed, to provide an overview of what follows. The content of each text chapter is then broken up into sections, which are labeled 1-I, 1-II, and so on. Each section has many parts. The *Summary* stresses important concepts and indicates the text pages covered; it is followed immediately by a list of *Key Terms*. To help you memorize the terms so that you can improve your grades, flash cards (with a term or name on one side and its definition on the reverse side) are extremely useful and well worth the time and effort.

A series of learning *Objectives* follows each *Key Terms* section. These are tasks that you should be able to accomplish if you have understood the assigned reading in the text. There are generally three levels of difficulty: Some objectives require that you memorize the terms and concepts; some require that you understand the material; and others require that you apply your understanding of the terms and concepts to different situations.

So that you can immediately evaluate your mastery of the terms and concepts, each *Objectives* section is followed by *Self-Quiz Questions*, which are answered in the back of the study guide. Any wrong answers will show you which portions of the text you need to reexamine.

Immediately following the last section of each chapter are two *Chapter Tests*. The first consists of multiple-choice (and some matching) questions; the answers to these are also in the back. The second invites you to try your hand at applying the major concepts to situations in which there is not necessarily a single pat answer (so none is provided in the back of the study guide). Your text generally will provide enough clues to get you started on an answer, but these sections are intended to stimulate your thought and provoke group discussions.

Finally, *Crossword Puzzles* are included following Chapters 2, 9, 22, 29, 42, and 49, as another way to help you match terms with definitions. Answers to the puzzles are in the answer section.

I would like to thank Maria S. Taylor and Cynthia J. Phelps for their care in keying the manuscript onto computer disks and helping solve some vexing computer problems. Special thanks go to the staff at Wadsworth and GTS, especially Jack Carey, Sue Belmessieri, Jerilyn Emori, and Betty Duncan-Todd, for their help and support.

Structure of the Study Guide

The outline below indicates how each chapter in this study guide is structured.

1

Chapter Title ————————→ **ON THE UNITY AND DIVERSITY OF LIFE**

Chapter Outline ————————→ | **Chapter Outline** |

General Objectives ————————→ A List of Tasks to Be Accomplished

Section Title ————————→ **1-I ORIGINS AND ORGANIZATION**

These categories occur for each section within a chapter ————————

→ **Summary**

→ **Key Terms**

→ **Objectives**

→ **Self-Quiz Questions**

True-False

Matching

Fill-in-the-Blanks

Chapter Tests ————————→ **UNDERSTANDING AND INTERPRETING KEY CONCEPTS**

INTEGRATING AND APPLYING KEY CONCEPTS

1

ON THE UNITY AND DIVERSITY OF LIFE

ORIGINS AND ORGANIZATION	DIVERSITY IN FORM AND FUNCTION
UNITY IN BASIC LIFE PROCESSES	The Tropical Reef
Metabolism	The Savanna
Growth, Development, and Reproduction	A Definition of Diversity
Homeostasis	ENERGY FLOW AND THE CYCLING OF RESOURCES
DNA: Storehouse of Constancy and Change	PERSPECTIVE

General Objectives

1. List features that distinguish living organisms from dead organisms. Then distinguish a dead organism from a rock.
2. Explain what is meant by the term *diversity*, and speculate about what caused the great diversity of life forms on Earth.
3. Describe the general pattern of energy flow through Earth's life forms, and explain how Earth's resources are used again and again (cycled).

1-I
(pp. 3–5)

ORIGINS AND ORGANIZATION

Summary

Living organisms tend to have specific form, function, and behavior. They tend to have specific types of atoms and molecules, characteristic patterns of organization, and particular ways of obtaining and systematically using energy and materials. Organisms are constructed of one or more cells. They can generally adjust to many long-term and short-term environmental changes. Living forms can reproduce themselves and are committed to characteristic programs of growth and development. And organisms interact on several different levels of organization in nature.

Key Terms

NOTE: The following underscored terms are especially important and are often boldfaced or italicized in the text. You should be able to define and explain all of them.

life	cell	energy
organism	adaptive potential	bacterium

liposomes	molecule	population
virus	organelle	community
bacteriophage	tissue	ecosystem
subatomic particle	organ	biosphere
atom		

Objectives

After reading the section and thinking about the contents, you should be able to:

1. Compare the basic structures of (a) a frog and a bacterium and (b) a virus and a rock.
2. List some specific functions or examples of behavior carried out by a frog, a bacterium, a virus, and a rock.
3. Define and contrast *energy* and *materials*. Explain how each is related to a living organism and to work.
4. Arrange in order from smallest to largest the levels of organization that occur in nature. Define each as you list it.

Self-Quiz Questions

True-False

If false, explain why.

___ (1) A bacterium can reproduce itself if it has the appropriate materials, but a virus cannot.

___ (2) Food storage, reproduction, and locomotion are all examples of work that cells do.

Matching

Choose the most appropriate answer to match with each of the following terms.

(3) ___ atom

(4) ___ cell

(5) ___ community

(6) ___ ecosystem

(7) ___ molecule

(8) ___ organelle

(9) ___ population

(10) ___ subatomic particle

(11) ___ tissue

A. A proton, neutron, or electron
B. A well-defined structure within a cell, performing a particular function
C. The smallest unit of life
D. Two or more atoms bonded together
E. All of the populations interacting in a given area
F. The smallest unit of a pure substance that has the properties of that substance
G. A community interacting with its nonliving environment
H. A group of individuals of the same species in a particular place at a particular time
I. A group of cells that work together to carry out a particular function

UNITY IN BASIC LIFE PROCESSES
    Metabolism
    Growth, Development, and Reproduction
    Homeostasis
    DNA: Storehouse of Constancy and Change

Summary

All forms of life show metabolic activity; they extract and transform energy from their environment and use it for manipulating materials in ways that ensure their own maintenance, growth, and reproduction. Each living organism is a continuum of patterns that unfold during its life cycle. The patterns unfold in the same way for all organisms of its kind and correspond to specific aspects of the environment.

All forms of life depend on homeostatic controls. These maintain internal conditions within some tolerable range even when external conditions change. They also govern new kinds of adjustments in the internal state as the life cycle proceeds. DNA is a storehouse of patterns for all heritable traits in living things. Mutations introduce variations in patterns, and the environment—both internal and external—determines whether any changes in the patterns permit the organism to survive.

Key Terms

energy transfer	development	variations
ATP	larva	inheritance
photosynthesis	pupa	DNA
aerobic respiration	homeostasis	mutations
reproduction	dynamic homeostasis	trait
metabolism		

Objectives

1. Explain what is meant by *energy transformation* and give an example.
2. State which energy sources you can tap each day and identify the forms into which you transform that energy for both long-term storage and immediate use.
3. State which energy is trapped by green plants and identify the forms in which plants store energy.
4. Explain what rocks would be able to do (that they cannot do now) if they carried out metabolism.
5. List in order the stages of the moth life cycle; then state why you think the larval form does not eat the same food as the adult.
6. List some examples of homeostatic activities in living organisms.
7. Define *inheritance* and indicate what is passed from parent to offspring that issues the instructions resulting in particular developmental and maintenance patterns.
8. Explain how variations can occur in offspring and why variations are important for the long-term survival of a population.
9. Imagine a community of different populations in which mutations never occur. Predict what will probably happen in these populations over thousands of years as the environment changes.

Self-Quiz Questions

True-False

If false, explain why.

____ (1) Converting light into heat is an example of an energy transformation.

____ (2) Trapping light energy in the chemical bonds of sugar is an example of an energy transformation.

____ (3) To stay alive, an organism must obtain energy from someplace else.

____ (4) Green plants absorb sugar, water, and minerals from the soil. The sugar absorbed in this manner is used to do cellular work.

____ (5) Larvae that consume foods not eaten by the adults can coexist in the same region without competing with the adults for the same food supply.

____ (6) The moth adult is equipped with jaws that enable it to obtain enough energy to fly.

____ (7) Instructions that result in particular developmental and maintenance patterns being followed are encoded in the ATP molecule.

____ (8) A mutation is never advantageous to the organism in which it occurs.

____ (9) One of the major groups of bacteria lacks homeostatic controls.

____ (10) Sweating is part of a homeostatic control system that keeps body temperature more or less constant in mammals.

Fill-in-the-Blanks

In humans, (11) _____ serves as the main long-term storehouse of energy, and (12) _____ is used as a more immediate energy source to do metabolic work. (13) _____ is the capacity for acquiring and using energy for stockpiling, tearing down, building up, and eliminating materials in controlled ways. The four stages in the moth life cycle, in sequence, are (14) _____, (15) _____, (16) _____, and (17) _____. (18) _____ is the capacity to maintain internal conditions within some tolerable range, even when external conditions vary. (19) _____ are changes that occur in the kind, structure, sequence, or number of DNA's component parts.

DIVERSITY IN FORM AND FUNCTION
    The Tropical Reef
    The Savanna
    A Definition of Diversity
ENERGY FLOW AND THE CYCLING OF RESOURCES
PERSPECTIVE

Summary

The array of different organisms on Earth is the total of variations that have proved adaptive over time in obtaining available resources, in tolerating diseases and toxic substances, and in escaping from predators and parasites.

    Almost all existing forms of life depend directly or indirectly on one another for materials and energy. There is a one-way flow of energy from the sun, through producer organisms, and on through consumers and decomposers.

Key Terms

diversity	geologic record	producer
coral	predator	consumer
sea anemone	scavenger	herbivores
savanna	ecology	carnivores
ungulate	cycling of minerals	decomposers
natural selection	microorganisms	evolution

Objectives

1. Explain why there are so many different kinds of fish and coral on a tropical reef.
2. Explain why there are so many hoofed, herbivorous mammals on the African savanna.
3. Explain how huge, ponderous elephants depend on and benefit from the efforts of tiny dung beetles.
4. Diagram how raw materials, producers, consumers, and decomposers are interrelated in the flow of energy and the cycling of materials through an ecosystem.
5. Consult Figure 1.12 in the main text and explain why energy cannot be totally recycled even though matter can.

Self-Quiz Questions

True-False

If false, explain why.

_____ (1) There are more different species of organisms associated with a tropical reef than there are living off the coast of Greenland because the mutation rate is much higher in the coral reef ecosystem.

_____ (2) There are more different species of organisms associated with the savanna than with the arctic tundra because there is a greater variety and abundance of food in the savanna.

_____ (3) Organisms that are well adapted to prevailing ecological conditions obtain a larger share of the resources and thus tend to survive longer and reproduce more often than less-adapted organisms.

___ (4) Dung beetles act as consumers that help decomposers recycle some of the nitrogen and minerals in elephant fecal material.

___ (5) A carnivore is a consumer, but a herbivore is a producer.

___ (6) Each time an energy transformation occurs, some energy is lost, usually as heat.

___ (7) The theory of evolution by natural selection explains how there can be so much diversity among the Earth's organisms, but it does not explain why organisms share so many of the same characteristics.

Matching   Choose the most appropriate answer for each.

(8) ___ coral

(9) ___ diversity

(10) ___ predator

(11) ___ savanna

(12) ___ scavenger

(13) ___ sea anemone

(14) ___ sponge

(15) ___ ungulate

(16) ___ consumer

(17) ___ decomposer

(18) ___ producer

A. Tiny marine organisms with tentacles and cylindrical bodies encased in calcareous skeletons
B. A small marine organism with weapon-studded tentacles and cylindrical body
C. Has pores opened toward the oncoming, food-laden currents
D. A hoofed, plant-eating animal
E. All of the variations in form and behavior that have accumulated in different types of organisms
F. Grassland of tropical or subtropical regions, punctuated with scattered trees and shrubs
G. Feeds on dead animal flesh or other decaying matter
H. An organism that lives by preying on other organisms
I. Traps solar energy in its tissues
J. Organism that eats other organisms
K. Organism that breaks down the tissues of other organisms and releases the raw materials contained therein

Fill-in-the-Blanks   Solar energy is first trapped in the chemical bonds of substances in the tissues of (19) _____; then it is passed to plant-eating organisms called (20) _____, the tissues of which are in part consumed by flesh-eating organisms called (21) _____. These last are in turn consumed by other flesh-eaters, which, when they die, are broken down by (22) _____ such as bacteria and fungi into (23) _____ such as CO_2, O_2, nitrogen, and minerals.

CHAPTER TEST   **UNDERSTANDING AND INTERPRETING KEY CONCEPTS**

___ (1) About 12 to 24 hours after the last meal, a person's blood sugar level normally varies from about 60 to 90 milligrams per 100 milliliters of blood, though it may attain 130 mg/100 ml after meals high in carbohydrates. That the blood sugar level is maintained within a fairly

narrow range despite uneven intake of sugar is due to the body's ability to carry out _____.
(a) adaptation
(b) inheritance
(c) metabolism
(d) homeostasis

___ (2) As an eel migrates from saltwater to freshwater, the salt concentration in its environment decreases from as much as 35 parts of salt per 1,000 parts of seawater to less than 1 part of salt per 1,000 parts of fresh water. The eel stays in the freshwater environment for many weeks because of its body's ability to carry out _____.
(a) adaptation
(b) inheritance
(c) metabolism
(d) homeostasis

___ (3) A boy is color blind just as his grandfather was, even though his mother had normal vision. This situation is the result of _____.
(a) adaptation
(b) inheritance
(c) metabolism
(d) homeostasis

___ (4) The digestion of food, the production of ATP by respiration, the construction of the body's proteins, cellular reproduction by cell division, and the contraction of a muscle are all part of _____.
(a) adaptation
(b) inheritance
(c) metabolism
(d) homeostasis

___ (5) A lion can mate with a tiger, but any offspring produced will be infertile and thus will not be able to produce offspring. This situation suggests that lions and tigers *cannot* be part of the same _____.
(a) ecosystem
(b) community
(c) population
(d) family

___ (6) The story about dung beetles and elephants was intended to make the point that _____.
(a) animals of such different sizes as dung beetles and elephants carry on metabolism
(b) mutations introduce variations in the heritable patterns of organisms
(c) each living organism is a continuum of patterns that unfold during its life cycle, and the patterns unfold in the same way for all of its kind
(d) almost all existing forms of life directly or indirectly depend on one another for materials and energy

___ (7) Which of the following does *not* involve using energy to do work?
(a) Atoms being bound together to form molecules
(b) The division of one cell into two cells
(c) The digestion of food
(d) None of these

___ (8) Which of the following is *not* true?
   (a) According to the definition of an organ, the human hand is an organ.
   (b) Bacterial cells have a nucleus and several different kinds of organelles.
   (c) The fundamental nature of matter and energy determine the structure and organization of both the living and nonliving world.
   (d) According to the definition of a molecule, carbon dioxide (CO_2) is a molecule.

CHAPTER TEST

INTEGRATING AND APPLYING KEY CONCEPTS

(1) Humans have the ability to maintain body temperature very close to 37° C.
   (a) What conditions would tend to make the body temperature drop?
   (b) What measures do you think your body takes to raise body temperature when it drops?
   (c) What conditions would cause body temperature to rise?
   (d) What measures do you think your body takes to lower body temperature when it rises?
(2) (a) What is required for an animal to become well adapted to a particular environment?
   (b) Do you consider yourself to be biologically well adapted to your environment?
   (c) Are there any ways that you could improve your ability to adjust to desert life?
   (d) To arctic life?
   (e) To life in outer space?

2

METHODS AND ORGANIZING CONCEPTS IN BIOLOGY

General Objectives

1. List as many steps of the scientific approach to understanding a problem as you can.
2. Explain how people came to believe that the populations of organisms that inhabit Earth have changed through time.
3. Understand as well as you can the ideas and evidence that biologists use to explain how life might have changed through time.

2-I
(pp. 19-22)

ON SCIENTIFIC PRINCIPLES
   Commentary: Testing the Hypothesis Through Experiments

Summary

The scientific method of approaching questions is a commitment to systematic observation and testing. Various different tools and experimental designs are used to record observations, test hypotheses, and draw conclusions. Most experiments compare a *control group* (used as a standard by which to judge results) with the *experimental group*, and only one variable at a time is tested quantitatively. Discipline, objectivity, suspended judgment, testing, and repeatability of experimental results are the bases of experimental principles. Occasionally, accidents and intuition contribute to the experimental process.

Key Terms

principle	objectivity	randomization
hypothesis, -ses	subjective	sampling error
induction	testable	significant
deduction	independent variable	"law"
suspended judgment	experimental group	
theory	control group	

Objectives

1. Contrast inductive reasoning with deductive reasoning.
2. Outline the principal steps generally used in the scientific method of investigating a problem.
3. Explain how observations differ from conclusions and how a hypothesis differs from a theory.
4. Explain why one or more control groups are used in an experiment.
5. Explain why most biologists regard the modern theory of evolution by natural selection as "scientific" and the idea that life has always existed in the forms we see today as "not scientific."

Self-Quiz Questions

True-False

If false, explain why.

___ (1) Deduction is a form of reasoning that proceeds from scientific observations to the development of a generalized concept.

___ (2) Quantitative studies generally use instruments or tools to obtain precise measurements.

___ (3) A control group helps to establish how far a variable deviates from standard values.

___ (4) When any test group is not equivalent to a natural population, the test group is said to be randomized.

Sequence

Arrange the following steps of the scientific method in correct chronological sequence from first to last:

(5) ___   A. Carry out the tests. Repeat as often as necessary to find out whether results consistently are as predicted.

(6) ___   B. Use trained judgment in selecting and summarizing the relevant preliminary observations from what could be nearly

(7) ___   infinite observational trivia.

C. Report objectively on the results of the tests and the

(8) ___   conclusions drawn from them.

D. Review all available preliminary observations. Be sure to note

(9) ___   the range of conditions under which they have been made.

E. Devise ways to test whether the explanation is valid. Think

(10) ___   through how different but related conditions might affect the outcome. Be sure the test you devise addresses these so-called variables.

F. Work out a hypothesis that seems in line with the observations.

Labeling

Assume that you have to determine what object is inside a sealed, opaque box. Your only tools to test the contents are a bar magnet and a triple-beam balance. Label each of the following with an O (for observation) or a C (for conclusion).

(11) ___ The object has two flat surfaces.

(12) ___ The object is composed of nonmagnetic metal.

(13) ___ The object is not a quarter, a half-dollar, or a silver dollar.

(14) ___ The object weighs x grams.

(15) ___ The object is a penny.

2-II
(pp. 23–26)

EMERGENCE OF EVOLUTIONARY THOUGHT
    Linnean System of Classification of Organisms
    Challenges to the Theory of Unchanging Life
    Lamarck's Theory of Evolution

Summary

2,300 years ago, Aristotle categorized organisms in a hierarchy according to the view that life proceeded gradually from lifeless matter to ever more complex forms of animal life—the Great Chain of Being. In the 1700s, Linnaeus developed the binomial system of nomenclature, which assigned a two-part (genus + species) name to each recognizably distinct kind of organism, established a concise hierarchical system, and reinforced the prevailing idea that each species is unique and unchanging. By the late 1700s and early 1800s, the discoveries of similar structures, vestigial (that is, have no apparent function) structures, and index fossils in Earth's geological strata caused biologists to question the idea that all forms of life had persisted unchanged since Earth's creation. Works published by Cuvier and Lamarck proposed that organisms have changed over time (have evolved), but extensive testing failed to support their theories of *catastrophism* and the *inheritance of acquired characteristics*.

Key Terms

Aristotle	family	Buffon
Scala Naturae	order	Cuvier
species	class	catastrophism
Linnaeus	phylum	Lamarck
binomial system of	division	inheritance of acquired
nomenclature	kingdom	characteristics
genus	fossil	vestigial structures
specific epithet	era	

Objectives

1. State the contributions of Aristotle, Linnaeus, Buffon, Cuvier, and Lamarck to our current views of the origins of Earth's species.
2. Arrange, in order from greater to fewer organisms included, the following categories of classification: class, family, genus, kingdom, order, phylum, species.
3. Explain why the global explorations of the sixteenth century created problems for biologists and how the binomial system of nomenclature not only helped solve some of those problems but also encouraged biologists who spoke different languages to compare information and solve problems.
4. Define and contrast similar and vestigial structures.

5. Explain how the discovery of similar and vestigial structures caused biologists to reappraise their views on the unchanging nature of species.
6. Explain how the geological record of fossils in rock layers contradicted the idea that species are fixed and unchanging.
7. Connect the idea of use and disuse of organs with the theory of inheritance of acquired characteristics.

Self-Quiz Questions

Fill-in-the-Blanks

The theory of inheritance of acquired characteristics is primarily linked with the name of (1) _____. (2) _____ believed that there had been several "centers of creation," that the origins of species had been spread out geographically, and that species might have become changed through time.

Sequence

Arrange in correct hierarchical order with the largest, most inclusive category first and the smallest, most exclusive category last:

(3) ___   A. Class
              B. Family
(4) ___   C. Genus
(5) ___   D. Kingdom
              E. Order
(6) ___   F. Phylum
(7) ___   G. Species

(8) ___

(9) ___

True-False

If false, explain why.

___ (10) The most inclusive (largest) taxonomic category is the phylum.

___ (11) Linnaeus was one of the first biologists to believe that species evolve substantially as time passes.

___ (12) There are more different species in a class than in an order.

___ (13) The binomial system developed by Linnaeus was a phylogenetic system based on the evolutionary relationships of organisms.

___ (14) Similar structures are body parts that have no apparent role in the functioning of the organism.

___ (15) The organisms that were present in the earliest rock layers are essentially the same as those in the most recent (uppermost) strata.

___ (16) According to Lamarck, parts of the body that have been stimulated by constant use grow larger than normal, and the greater development of those body parts is somehow transmitted to the offspring.

___ (17) Cuvier developed the theory that an organism's free will acts together with environmental influences to excite "vital fluids" that enlarge specific body parts.

Choose the one most appropriate answer.

____ (18) Which of these men believed that life has evolved *and* that new species are formed? (a) Aristotle (b) Linnaeus (c) Cuvier (d) Lamarck (e) None of these

---

2-III
(pp. 26–32)

EMERGENCE OF THE PRINCIPLE OF EVOLUTION
Naturalist Inclinations of the Young Darwin
Voyage of the *Beagle*
Darwin and Wallace: The Theory Takes Form

Summary

Spurred by the ideas of Lyell and Malthus and by his experiences on the voyage of the *Beagle*, Darwin developed the theory of evolution by natural selection; he was later joined by Wallace in suggesting that species do change over time. Heritable variations occur among members of a species. Each population produces more offspring than can survive to reproduce. Bearers of the traits that improve chances for surviving and reproducing under prevailing environmental conditions tend to produce more descendants than other members of the population; this differential reproduction ensures that the most adaptive ("fit") traits will show up more frequently in the next generation.

Key Terms

Darwin	Galápagos Islands	artificial selection
Lyell	*Essay on the Principle of Population*	differential reproduction
Principles of Geology		Wallace
uniformitarianism	natural selection	*Archaeopteryx*
Malthus	variety	"missing links"

Objectives

1. Distinguish between *population* and *species*.
2. List in sequential order the statements that compose the Darwin-Wallace theory of evolution by natural selection.
3. Indicate how the writings of Lyell, Hutton, and Malthus helped shape Darwin's early formulation of his theory.
4. Outline briefly how new species can evolve.
5. Explain how scientists determine which individuals should be grouped into the same species.
6. Describe Wallace's role in the formulation of the Darwin-Wallace theory.

Self-Quiz Questions

True-False

If false, explain why.

____ (1) If two or more individuals from different populations can interbreed and produce offspring that can survive and reproduce, those individuals should be grouped in the same species.

____ (2) If two organisms are in the same species, they also must be a part of the same population.

___ (3) When resources become scarce, competition for similar resources promotes greater specialization among the competitors.

___ (4) Malthus believed that the food supply of a population increases faster than the population increases.

<table>
<tr><td>

2-IV
(pp. 32–33)

</td><td>

AN EVOLUTIONARY VIEW OF DIVERSITY
PERSPECTIVE
SUMMARY

</td></tr>
</table>

Summary

A phylogenetic system categorizes organisms by using data from the fossil record, comparative anatomy, genetics, biochemistry, reproductive biology, behavior, ecology, geology, and geography to construct the major lines of evolutionary descent. Whittaker's five-kingdom system classifies organisms according to their structure and their means of obtaining nutrients and energy.

Key Terms

R. Whittaker	Animalia	eubacteria
phylogenetic system of classification	autotroph	cyanobacteria (blue-green algae)
	heterotroph	
Monera	prokaryotic	diatoms
Protista	eukaryotic	parasite
Plantae	archaebacteria	chordate
Fungi		

Objectives

1. Explain what is meant by a phylogenetic system of classification and how it differs from the way organisms were originally classified by Linnaeus.
2. Characterize the kingdoms in Whittaker's five-kingdom system and distinguish each from the other four kingdoms. To do this, you may have to consult Table 2.2.

Self-Quiz Questions

True-False

If false, explain why.

___ (1) Phylogenetic classification is a process by which organisms inherit acquired characteristics.

___ (2) Modern classification systems try to categorize organisms phylogenetically whenever the necessary background information is available.

___ (3) All prokaryotes are monerans.

CHAPTER TEST

UNDERSTANDING AND INTERPRETING KEY CONCEPTS

___ (1) To eliminate from consideration the influence of uncontrolled variables during experimentation, one should _____.

(a) increase the sampling error as much as possible and use suspended judgment
(b) establish a control group identical to the experimental group except for the variable being tested
(c) use inductive reasoning to construct a hypothesis
(d) make sure the experiments are repeatable

___ (2) A hypothesis should *not* be accepted as valid if _____.
(a) the sample studied is determined to be representative of the entire group
(b) a variety of different tools and experimental designs yield similar observations and results
(c) other investigators can obtain similar results when they conduct the experiment under similar conditions
(d) several different experiments, each without a control group, systematically eliminate each of the variables except one

___ (3) The principal point of the Darwin-Wallace theory of evolution by natural selection was that _____.
(a) long-term heritable changes in organisms are caused by use and disuse
(b) those mutations that adapt an organism to a given environment somehow always arise in the greatest frequency in the organisms that occupy that environment
(c) mutations are caused by all sorts of environmental influences
(d) survival of characteristics in a population depends on competition between organisms, especially between members of the same species

___ (4) Cuvier, an anatomist and paleontologist, proposed that _____.
(a) all present-day organisms have descended, with adaptations, from one—or possibly a few—original organisms
(b) Earth's history has been marked by several periods when destruction of populations was widespread and that, after one such period, the Earth was repopulated by the survivors
(c) evolutionary changes in organisms are caused by use and disuse
(d) although evolution is responsible for all the changes that happen to species, God created the original members of each species

___ (5) A control group _____ .
(a) is not subjected to experimental errors
(b) is exposed to experimental treatments
(c) is maintained under strict laboratory conditions
(d) is treated exactly the same as the experimental group
(e) varies from the experimental group only in the factor being tested by the experiment

___ (6) Statistical tests are designed to determine if differences between experimental and control groups are _____.
(a) valid
(b) significant
(c) qualitatively different
(d) null
(e) quantitatively different

___ (7) The least inclusive of the taxonomic categories listed is _____.
(a) family
(b) phylum

(c) class

(d) order

(e) genus

___ (8) The oldest fossils _____ .

    (a) demonstrate the widest distribution

    (b) represent the most highly evolved plants and animals

    (c) generally are found buried deepest in the ground

    (d) are found in Africa

    (e) are primitive marine invertebrates

___ (9) Which of the following theories was important to Darwin in the formulation of his theory of evolution?

    (a) Catastrophism

    (b) Inheritance of acquired characteristics

    (c) Uniformitarianism

    (d) Continental drift

    (e) Special creationism

___ (10) Malthus proposed that _____.

    (a) advanced organisms evolve from simple forms

    (b) populations tend to outgrow their food supplies

    (c) organisms inherit acquired characteristics

    (d) the Earth was more than 6000 years old

    (e) humans produce new forms of domesticated plants and animals through artificial selection

CHAPTER TEST

INTEGRATING AND APPLYING KEY CONCEPTS

(1) Suppose you're an ecologist working in an African game preserve that includes elephants. You want to discover the precise migratory habits of the elephants so that they can encounter humans less frequently and thus have better chances to survive.

    (a) What sources would you go to in order to review all available observations?

    (b) Which hypotheses might you try to test?

    (c) What tools might you use in testing your hypotheses?

    (d) Name three variables that might affect the migratory movements of elephants.

    (e) How could you isolate each of these variables and study the effects of just one variable at a time so as to deduce its contribution to elephant migratory behavior?

(2) What sorts of topics are usually regarded by scientists as untestable by the kinds of methods that scientists generally use?

(3) Do you think that all humans on Earth today should be grouped in the same species?

Crossword
Number One

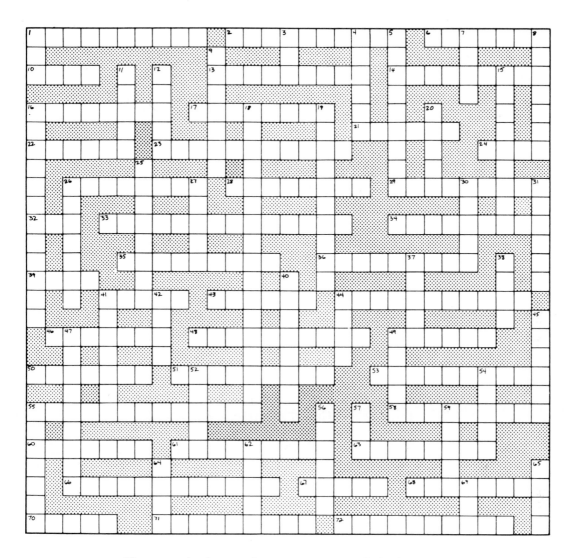

The terms in the puzzle are not necessarily biology terms.

Across

1. a state of order based on submission to rules and authority
2. an educated guess
6. proposed that characteristics acquired during a lifetime could be inherited by offspring
10. smallest unit of a pure substance that has the properties of that substance
13. subatomic _____; very small parts
14. choosing or choice
16. the largest classification category of plants
17. not inherited
21. includes blue-green algae, bacteria, and their allies
22. a taxonomic group of related genera
23. the process of genetic transmission of characters or characteristics
24. a filmy layer of extraneous or impure matter that forms on the surface of a liquid or body of water

26. an organism that can synthesize its own food from appropriate inorganic substances
28. an organism that must depend on other organisms for its food
29. reasoning that proceeds from the general to the specific
32. age; geological span of time
33. change from one form to another
34. all of the regions on earth that support self-sustaining and self-regulating ecological systems
35. variety; different kinds
36. change in the frequency that a gene occurs in a population
39. Persia, in older days, is _____ now
41. French impressionist painter, 1840–1926
43. coformulator of the theory of natural selection
44. a state of physiological equilibrium produced by a balance of functions and of chemical composition within an organism
46. coformulator of the theory of evolution by natural selection
48. the physical and chemical processes that maintain life
49. a group of closely related classes of organisms
50. a term that characterizes the nature of a person or thing
51. occurring or persisting as a rudimentary or degenerate structure
53. corresponding in structure and evolutionary origin
55. the theory that geological processes such as mountain building, flooding, earthquakes, etc., occur through time in violent, disruptive surges rather than at a constant rate
58. discrete structure that has a specific function within a cell
60. *Scala* _____, by Aristotle
61. a fundamental doctrine on which new concepts are based or from which they are drawn
63. taxonomic group that includes unicellular or colonial eukaryotes
66. relating to being tested
67. a taxonomic group of closely related families
68. two or more atoms bonded together
70. wrote *Principles of Geology*
71. a biological community interacting with its nonliving environment
72. capable of being done again

Down

1. the molecular blueprint for life
3. Japanese sash
4. organ _____
5. held in abeyance; maintained in an undecided state
7. medieval war club with a spiked or flanged head; an aromatic spice
8. most inclusive (largest) category of classification
9. least inclusive category of classification of organisms
11. oneness; commonality
12. a group of organisms that absorb food they have digested outside their bodies
15. reasoning that proceeds from the specific to the more general
16. not proceeding at the same rate
18. the theory that the erosive action of wind and water and the gradual uplifting of mountain ranges have occurred at fairly constant rates through time
19. an organism that breaks down the remains of dead organisms
20. a variety or subspecies of organism
25. a heritable alteration of the genes or chromosomes of an organism
26. the kingdom that includes multicellular photosynthetic organisms
27. to direct to a source for help or information

30. a castrated rooster
31. the capacity to do work
37. a hard bonelike structure in a jaw or skeleton used to seize, hold, or chew
38. one of two offspring born at the same birth
40. having two names or a two-part name
41. said that populations increase faster than their food supply
42. a large, wide-mouthed pitcher or jug
44. an organism that harbors and provides nourishment for a parasite
45. a group of cells that cooperate to carry out a particular function
47. Greek philosopher (384–322 B.C.) who attempted to categorize all organisms according to a "great chain of being"
49. resort city in Utah
52. a volatile, highly flammable liquid that consists of two hydrocarbon groups linked by an oxygen atom
54. _____ Garbo
55. used as a standard for judging the results of an experiment
56. a coherent set of ideas that form a general frame of reference for further studies in the field of inquiry
57. the energy "currency" of a cell; a temporary energy-storage molecule
59. _____ Adams, American photographer of natural landscapes
62. a taxonomic group of closely related orders
64. the opposite of death
65. the smallest unit of a living organism that is capable of independent function
69. fade away, diminish

3

CHEMICAL FOUNDATIONS FOR CELLS

ORGANIZATION OF MATTER
    Atoms and Ions
    Isotopes
    How Electrons Are Arranged in Atoms
BONDS BETWEEN ATOMS
    The Nature of Chemical Bonds
    Ionic Bonding
    Covalent Bonding
    Hydrogen Bonding
    Hydrophobic Interactions
    Bond Energies

ACIDS, BASES, AND SALTS
    Acids and Bases
    The pH Scale
    Buffers
    Dissolved Salts
WATER MOLECULES AND CELL ORGANIZATION
    Hydrogen Bonding in Liquid Water
    Solvent Properties of Water
    Water and the Organization Underlying the Living State
SUMMARY

General Objectives

1. Understand how protons, electrons, and neutrons are arranged into atoms and ions.
2. Explain how the distribution of electrons in an atom or ion determines the number and kinds of chemical bonds that can be formed.
3. Know the various types of chemical bonds, the circumstances under which each forms, and the relative strengths of each type.
4. Understand the essential chemistry of water and of some common substances dissolved in it.

3-I
(pp. 36–38)

ORGANIZATION OF MATTER
    Atoms and Ions
    Isotopes

Summary

All events in the living world depend on the organization and behavior of atoms and molecules. All energy transfers and transformations within and between living things occur at the level of atoms and molecules. The shapes and behavior of cells are determined by molecules contained within them.

    The structure of each atom in a given element makes it different from atoms of other elements. The number and arrangement of its subatomic particles (protons, neutrons, and electrons) dictate the behavior of the atoms of a specific element. Uncharged neutrons in the nucleus, together with the protons, account for the *mass (weight) number* of an element. The mass number of an atom is equal to its atomic weight, a term useful when comparing types of atoms. The number of protons alone dictates the *atomic number* of the element and helps, indirectly, to establish how that atom reacts with other atoms. How the electrons are arranged outside the nucleus determines the ways in which the atom reacts with other atoms and the kinds of ions the atom forms.

The number of positively charged protons in an *atom* always equals the number of negatively charged electrons; thus, atoms have no net charge. In an *ion*, the number of protons is not equal to the number of electrons; consequently, ions are electrically charged. All atoms of an element have the same number of protons but can vary slightly in the number of neutrons they have. The variant forms are called *isotopes*.

Key Terms			
element	electron	atomic weight
compound	molecule	ion
atom	electric charge	radioactive isotopes
proton	atomic nucleus	isotopes
neutron	atomic number	tracers
mixture	mass number	

Objectives

1. Define *energy* and explain what it means for a living organism.
2. Explain how an organism that could obtain energy from different sources (a food generalist) might have an advantage over organisms that obtain energy from only one source (food specialists).
3. Indicate the relationships between *atoms* and *elements* by comparing the definition of each term.
4. Compare the definitions of *atom* and *molecule*.
5. Define *electric charge* and state how (a) two identically charged particles react to each other and (b) two oppositely charged particles react to each other.
6. List and describe the three types of subatomic particles in terms of their location (inside or outside the nucleus) and charge.
7. Define and distinguish *atomic number* and *mass number*.
8. Show how an atom of carbon differs from atoms of oxygen, hydrogen, and nitrogen.
9. Explain what isotopes of an element are and how differences in the number of neutrons and in the mass numbers distinguish each isotope.
10. Define *ion* and explain how an atom can be converted into an ion.
11. Name the four most abundant elements in the human body. (See Table 3.1.)

Self-Quiz Questions

True-False

If false, explain why.

___ (1) Conversion of light energy into chemical energy, as occurs in plants, is an energy transformation.

___ (2) The number of protons in the nucleus of an atom is equal to the number of electrons outside the nucleus.

___ (3) The number of protons in a nucleus helps, indirectly, to determine how that atom reacts with other atoms.

___ (4) If two particles bear the same electric charge, they are attracted to each other.

___ (5) Carbon 12 and Carbon 14 are isomers of each other.

___ (6) A sodium atom can be converted into a sodium ion if the sodium atom can lose an electron.

Matching   Choose the one most appropriate answer for each.

(7) ___ atom

(8) ___ atomic number

(9) ___ electric charge

(10) ___ electron

(11) ___ element

(12) ___ energy

(13) ___ ion

(14) ___ isotope

(15) ___ mass number

(16) ___ matter

(17) ___ molecule

(18) ___ neutron

(19) ___ proton

(20) ___ pure

A. A charged atom
B. An uncharged subatomic particle
C. The number of protons in the nucleus of one atom of an element
D. The smallest neutral unit of an element that shows the chemical and physical properties of that element
E. That which occupies space and has mass
F. One or more atoms linked together by one or more chemical bonds
G. Contributes the capacity to do work—that is, to move a certain amount of material across a specific distance
H. A positively charged subatomic particle
I. Composed entirely of only certain atoms or molecules
J. The sum of protons + neutrons
K. A negatively charged subatomic particle
L. Ninety-two different types occur in nature
M. A form of an element, the atoms of which contain a different number of neutrons from other forms of the same element
N. Pushes away similar particles but attracts oppositely charged particles

---

3-II
(pp. 38–40)

ORGANIZATION OF MATTER (cont.)
How Electrons Are Arranged in Atoms

Summary

An *orbital* is a specific region of space outside the nucleus in a particular *energy level*, where an electron is most likely to be located. An orbital can contain no more than two electrons. Negatively charged electrons, arranged in orbitals outside the nucleus, tend to get as close as possible to the positively charged protons in the nucleus and stay as far away as possible from each other. As soon as the inner orbitals are filled with two electrons, orbitals of the next highest energy level are the next to fill. Elements composed of atoms that have two electrons in each occupied orbital tend to be chemically nonreactive substances; atoms that have one or more orbitals containing only one electron tend to react chemically with other such atoms.

Electrons occupying the orbitals in the highest occupied (outermost) energy level of an atom are the ones that interact with other atoms to form either a molecule or some other group of associated atoms. Orbitals have specific shapes, which determine the shapes of the molecules that the orbitals in part compose. An electron can absorb certain amounts of incoming energy and as a result spend more of its time farther out from the pull of the nucleus. If the outside energy source is removed, the excited electron eventually gives off the extra energy it has absorbed and returns to whichever orbital closest to the nucleus can accommodate it.

Key Terms

orbital	1s orbital	electron dot model
energy level	orbital model	absorb
nonreactive	shell model	release
reactive		

Objectives

1. Define *orbital* and indicate how electrons are related to orbitals.
2. Describe the relationships among energy levels, orbitals, and electrons. Explain what happens when an orbital is filled with two electrons.
3. Examine Table 3.2 (in the main text) and state which of the elements listed tend to react chemically with other substances. Which elements listed do not tend to react chemically?
4. Describe the situation that enables atoms to become reactive.
5. Explain how the shapes of the orbitals can influence the specific shapes of molecules that are formed by several atoms becoming linked together.
6. Distinguish the orbital model, the shell model, and the electron dot model of electron distribution from each other.
7. Explain what happens to the behavior of one or more of an atom's electrons (a) when the atom absorbs energy and (b) when the atom releases energy.

Self-Quiz Questions

True-False

If false, explain why.

____ (1) In an atom, the volume of space that can accommodate (at most) two electrons is called an *energy level*. called orbital

____ (2) When an atom absorbs energy, electrons that are part of that atom move faster and tend to spend more of their time farther from the nucleus than they did before the energy was absorbed.

____ (3) For an atom to react with another atom, all orbitals of that atom must already contain two electrons.

BONDS BETWEEN ATOMS
   The Nature of Chemical Bonds
   Ionic Bonding
   Covalent Bonding
   Hydrogen Bonding

Summary

In a chemical bond, one or more orbitals of one atom become linked with one or more orbitals of another atom (or atoms), establishing an energy relationship between the participating atoms that holds the atoms at certain distances from each other and causes the group of atoms to assume a specific three-dimensional shape.

In cells, molecules are held together by strong covalent bonds that involve a sharing of electron pairs, and by weaker interactions (ionic bonds and hydrogen bonds) between atoms or ions having opposite charge.

In an *ionic bond*, a positive and a negative ion are linked by the mutual attraction of opposite charges. A *covalent bond* is formed when two different atoms share a pair of electrons. If the pair is shared equally, the covalent bond is nonpolar; if shared unequally, the bond is polar. Water is a molecule with polar covalent bonds.

In a *hydrogen bond*, an electronegative atom weakly attracts a hydrogen atom that is covalently bonded to a different atom. Hydrogen bonds help stabilize many large biological molecules, and they give liquid water some structure.

Key Terms

chemical bond	product	covalent bond
energy relationship	law of conservation of	nonpolar covalent bond
formula	mass	polar covalent bond
chemical equation	mole	electronegative
reactant	ionic bond	hydrogen bond

Objectives

1. Define and use in a sentence the term *chemical bond*.
2. Explain how a chemical bond can be formed between two atoms.
3. Identify which bonds are strong and which bonds are weak.
4. Describe covalent, ionic, and hydrogen bonds fully, so as to distinguish each from the others.
5. Contrast nonpolar and polar covalent bonds.
6. Explain why two single atoms, each with a lone electron in one of its outermost orbitals, would tend to become bonded. Explain what is gained by forming a bond.
7. State one way that living organisms use hydrogen bonding.

Self-Quiz Questions

True-False

If false, explain why.

____ (1) A chemical bond is formed when two atoms share a pair of electrons, and a chemical bond is formed by atoms losing or gaining one or more electrons.

___ (2) An ionic bond is stronger than a covalent bond.

___ (3) Covalent bonds typically occur within molecules; weaker bonds usually form between two or more different molecules but can also form within the same molecule.

___ (4) The atoms in table salt (sodium chloride) are linked together with covalent bonds.

___ (5) In order to form a covalent bond, two orbitals from different atoms, each containing one electron, must overlap so that a pair of electrons may be shared between the two nuclei.

___ (6) Chemical bonds form between atoms because the two atoms bonded together are more stable than the atoms are when they are separate.

Short-Answer Essay

(7) How are double and triple bonds formed within molecules?

3-IV
(pp. 42–44)

BONDS BETWEEN ATOMS (cont.)
    Hydrophobic Interactions
    Bond Energies
ACIDS, BASES, AND SALTS
    Acids and Bases
    The pH Scale
    Buffers
    Dissolved Salts

Summary

Certain substances dissolve in water because they can become dispersed among and form weak bonds with the polar water molecules. Such substances are *hydrophilic*. Substances that cannot form weak bonds with water molecules are *hydrophobic*. Hydrophobic molecules (such as oil) tend to cluster in a watery environment.

Covalent bonds formed between atoms of carbon, nitrogen, oxygen, and hydrogen typically contain bond energies of between 80 and 110 kilocalories per mole. [A mole is a specific number (6.023×10^{23}) of atoms or molecules of any substance.] Under conditions that generally prevail in cells, ionic bonds are much weaker (generally about 5 kcal/mole) than covalent bonds.

Water can dissociate into *hydroxide ions* and *hydronium ions*. A substance whose molecules release hydrogen ions in solution is known as an acid. Any substance that combines with hydrogen ions in solution is a base. The greater the hydrogen ion concentration, the lower the pH value. pH values are a shorthand measure of the degree to which a solution is acidic, basic, or neutral. All biochemical reactions that usually take place in water are sensitive to changes in pH. It is fortunate that the pH of pure water is 7, or neutral, because that contributes to the stability of biochemical processes in cells. Each kind of cell is adapted to a particular pH range, usually not far from neutrality.

Inside cells, *buffer molecules* help maintain pH values within rather narrow tolerance ranges. Buffer molecules combine with or release hydrogen ions in response to changes in cellular pH. Cell functioning also depends on essential ions such as K^+, Na^+, Ca^{++}, Mg^{++}, and Cl^-, NO_3^- (nitrate), $PO_4^{\equiv}$ (phosphate), and HCO_3^- (bicarbonate).

Key Terms

Objectives

1. Explain what "to dissolve" means.
2. Distinguish between *hydrophilic* and *hydrophobic* substances.
3. Give an example of something that is influenced by hydrophobic interactions.
4. Compare covalent, ionic, and hydrogen bonds in terms of their typical bond energies.
5. Explain how water dissociates into its ions, by writing a chemical reaction like the one shown on page 43 of the main text. State whether you think the reaction is reversible.
6. Describe the pH scale, specifying the acidic range, the basic range, and the point of neutrality.
7. Distinguish between an acid and a base. Give three examples of acids and three examples of bases. (See Figure 3.8 in the main text.)
8. Explain why the pH of substances is important to living organisms.
9. Define the term *salt* and state how a salt is related to ions.
10. List four essential ions.

Self-Quiz Questions

True-False

If false, explain why.

___ (1) A kilocalorie is equal to 6.023 atoms of a substance.

___ (2) The shapes of lipid-based molecules are influenced by their hydrophobic interactions with water molecules.

___ (3) Hydrophilic substances generally dissolve in water.

___ (4) Ionic bonds and hydrogen bonds contain approximately similar bond energies.

Fill-in-the-Blanks

Two examples of essential ions are (5) _____ and (6) _____.

H_2CO_3 $\left(\text{H-O-C} \begin{smallmatrix} \nearrow O \\ \\ \searrow O-H \end{smallmatrix} \right)$ dissociates to form H$^+$ plus (7) _____, the

(8) _____ ion. Ionic substances formed by the reaction between an acid and a base are called (9) _____. Sometimes polar molecules (10) _____; that is, they separate into two (or more) ions. Nonpolar regions of large molecules have no net charge and show little tendency to form hydrogen bonds with water; they are said to be (11) _____ (water-dreading).

Choose the most appropriate answer for each.

(12) ___ acid

(13) ___ base

(14) ___ essential ion

(15) ___ hydrogen ion

(16) ___ hydroxyl ion

(17) ___ neutral

(18) ___ salt

A. OH^-
B. Na^+, Cl^-
C. H^+
D. NaCl
E. Combines with hydrogen ions in solution
F. Releases hydrogen ions in solution
G. 7

3-V
(pp. 45–48)

WATER MOLECULES AND CELL ORGANIZATION
   Hydrogen Bonding in Liquid Water
   Solvent Properties of Water
   Water and the Organization Underlying the Living State
SUMMARY

Summary

Water makes up about 75 to 85 percent of an active cell's weight. None of life's activities can proceed without water because (1) water stabilizes the cellular temperature so that chemical reactions can proceed in an orderly fashion, (2) water's cohesive nature resists penetration and allows whole columns of water to be moved about, and (3) water dissolves ionic substances and most polar molecules.

Water molecules have negatively and positively charged ends. The polar nature of water encourages spheres of hydration to be formed around ions, preventing ions from reacting with each other and forcing them to remain dispersed in the cytoplasm rather than concentrating in some part of the cell. Thus, water molecules help keep vital ions available for reactions throughout the cell. Large molecules may have charged regions that tend to attract either the negative or the positive ends of water molecules (that is, to be hydrophilic) and other regions that bear no net charge (in other words, are hydrophobic).

The properties of water profoundly influence the organization and behavior of substances that make up the cellular environment.

Key Terms

cohesion	calorie	lipid bilayer
stabilization	evaporation	solutes
specific heat	heat of fusion	spheres of hydration
heat of vaporization	solvent	dissolved
temperature	surface tension	

Objectives

1. Explain how the water that constitutes the bulk of most organisms helps those organisms tolerate diverse variable environmental conditions.
2. Explain why none of the activities associated with the term *living* can occur without water.
3. Distinguish between *specific heat* and *heat of vaporization*.
4. Give two specific examples of organisms that depend directly on the cohesive properties of water and describe the nature of that dependence. Speculate about what those organisms would do if water molecules were not cohesive.

5. Explain why the fact that water is an outstanding solvent is of such importance to living organisms. State how you think life would be limited if water could dissolve only a few different substances.

6. Explain the role that hydrogen bond formation plays in some of the physical and chemical properties of water.

Self-Quiz Questions

Fill-in-the-Blanks

(1) _____ is the amount of heat energy a single gram of a substance can absorb before its temperature increases by 1° C. As water absorbs heat and the molecules move more quickly, (2) _____ _____ between neighboring water molecules are more readily broken down and reformed. The amount of heat energy that must be absorbed before a single gram of liquid is converted to gaseous form is the (3) _____ _____ _____ of that substance. When molecules at the surface of a liquid begin moving fast enough to escape into the atmosphere, the process is called (4) _____. The surface of the liquid left behind drops in temperature because much of its energy has been (5) _____ by the escaping molecules and converted into their energy of (6) _____. In pure water, each molecule is attracted to and hydrogen-bonded with other water molecules; such an attraction between like molecules is called (7) _____. Water is an outstanding (8) _____ because it dissolves many different types of (9) _____.

True-False

If false, explain why.

___ (10) Cytoplasm is a formless substance with little structure.

___ (11) Water molecules tend to cluster around each positively charged ion with their positive ends pointing toward the ion, thus forming a sphere of hydration.

Short-Answer Essays

(12) Explain why proteins remain dispersed in watery systems and why particles in a suspension tend to fall out.

(13) Explain how lipid bilayers, the essential structure of all cell membranes, are formed. Use the terms *hydrophobic tails* and *hydrophilic heads* in your answer.

CHAPTER TEST   **UNDERSTANDING AND INTERPRETING KEY CONCEPTS**

___ (1) A molecule is ___*a*___.
- (a) a combination of two or more atoms
- (b) less stable than its constituent atoms separated
- (c) electrically charged
- (d) a carrier of one or more extra neutrons

___ (2) A hydrogen bond is _____.
- (a) a sharing of a pair of electrons between a hydrogen and an oxygen nucleus
- (b) a sharing of a pair of electrons between a hydrogen nucleus and either an oxygen or a nitrogen nucleus
- (c) an attractive force that involves a hydrogen atom and an oxygen or a nitrogen atom that are either in two different molecules or within the same molecule
- (d) none of the above

___ (3) A mixture of sugar and water is an example of a(n) _____.
- (a) compound
- (b) solution
- (c) suspension
- (d) colloid
- (e) ion

___ (4) Radioactive isotopes have _____.
- (a) excess electrons
- (b) excess protons
- (c) excess neutrons
- (d) insufficient neutrons
- (e) insufficient protons

___ (5) The shapes of large molecules are controlled by ___*c*___ bonds.
- (a) hydrogen
- (b) ionic
- (c) covalent
- (d) inert
- (e) single

___ (6) A pH solution of 10 is _____ times as basic as a pH of 7.
- (a) 2
- (b) 3
- (c) 10
- (d) 100
- (e) 1000

___ (7) In a lipid bilayer, _____ tails point inward and form a region that excludes water.
- (a) acidic
- (b) basic
- (c) hydrophilic
- (d) hydrophobic
- (e) none of the above

INTEGRATING AND APPLYING KEY CONCEPTS

Explain what would happen if water were a nonpolar molecule instead of a polar molecule. Would water be a good solvent for the same kinds of substances? Would the nonpolar molecule's specific heat likely be higher or lower than that of water? Surface tension? Cohesive nature? Ability to form hydrogen bonds? Is it likely that the nonpolar molecules could form unbroken columns of liquid? What implications would that hold for trees?

4

CARBON COMPOUNDS IN CELLS

General Objectives

1. Understand how small organic molecules can be assembled into large macromolecules by condensation. Understand how large macromolecules can be broken apart into their basic subunits by hydrolysis.
2. Memorize the functional groups presented and know the properties they confer when attached to other molecules.
3. Know the general structure of a monosaccharide with six carbon atoms, glycerol, a fatty acid, an amino acid, and a nucleotide.
4. Know the macromolecules into which these essential building blocks can be assembled by condensation.
5. Know where these carbon compounds tend to be located in cells or organelles and the activities in which they participate.

4-I
(pp. 49–51)

THE ROLE OF CARBON IN CELL STRUCTURE AND FUNCTION
    Families of Small Organic Molecules
    Properties Conferred by Functional Groups
    Condensation and Hydrolysis

Summary

Oxygen, hydrogen, and carbon compose 93 percent of your body's weight; water accounts for much of the first two elements, but carbon is the most important structural element in the body because a carbon atom can form as many as four covalent bonds with carbon as well as other elements. Chains or rings of carbon atoms form the main structure of strandlike, globular, or sheetlike molecules in cells, some of which contain millions of atoms.

Four families of small organic compounds—simple sugars, fatty acids, amino acids, and nucleotides—serve as energy sources and as building blocks for the macromolecules present in cells. *Functional groups* confer special properties to organic molecules. Macromolecules such as polysaccharides, fats, proteins, and nucleic acids are assembled by *condensation*, which is the covalent linkage of small molecules. The condensation process is guided by specific *enzymes* and may also involve forming water as a product. Macromolecules generally are split apart into two or more parts by reaction with water; this process of *hydrolysis*, too, is guided by specific enzymes.

Key Terms

organic	ethyl group, —C_2CH_3	enzyme
inorganic	alcohols	condensation
simple sugar	hydroxyl group	polymer
macromolecule	carbonyl group	monomer
functional groups	carboxyl group	hydrolysis
hydrocarbons	amino group	
methyl group, —CH_3	phosphate group	

Objectives

1. Explain why carbon atoms are part of so many substances in living organisms.
2. Distinguish between the terms *organic* and *inorganic*.
3. Illustrate, by drawing their carbon skeletons, a linear chain, a branched chain, and a ring structure.
4. (a) Distinguish the *hydroxyl group* from the *carboxyl group* (Table 4.1).
   (b) Describe the amino and phosphate groups.
5. Show how two amino acids can be joined together by removing a water molecule.
6. Look closely at pictures of amylose, cellulose, triglyceride, and nucleic acid; identify the basic building blocks of each and note how these blocks are joined.

Self-Quiz Questions

Fill-in-the-Blanks

A(n) (1) _____ group contains an oxygen atom and a hydrogen atom.

A(n) (2) _____ group contains two oxygen atoms, a carbon atom, and a hydrogen atom. A(n) (3) _____ group can contain four oxygen atoms, two hydrogen atoms, and a phosphorus atom. A(n) (4) _____ group may contain one nitrogen atom and two hydrogen atoms.

True-False

If false, explain why.

____ (5) Atoms that have four orbitals containing only one electron can form more bonds, and thus more molecules with different shapes, than can atoms with only two orbitals that each contain only one electron.

____ (6) Molecules are assembled into polymers by means of hydrolysis.

_____ (7) Carbon atoms are part of so many different substances because, in one of their electron configurations, there are four unpaired electrons in the outermost occupied energy level. Thus, each unpaired electron can form a covalent bond with a variety of other atoms, including other carbon atoms.

_____ (8) All organic molecules contain carbon.

_____ (9) When a ring like this ⬡ is shown, it is understood that a carbon atom occurs at every corner and that no other atoms are attached. (Think!)

_____ (10) In a branched chain, each carbon atom is never bonded to more than one other carbon atom. (Think!)

Short-Answer Essays

(11) Why does a carbon atom generally form four bonds? (Consult Figure 3.2, remembering what is required to form a bond, and page 40 of the main text.) For carbon to form four bonds consistently, what must happen to the distribution of electrons on the second energy level as carbon is about to bond with other atoms?

(12) What is a hydrocarbon? Give an example of a hydrocarbon.

(13) Interpret the meaning of ⬡ .

(14) (a) Define and contrast condensation and hydrolysis.
(b) Show how one process resembles the other in reverse.
(c) Name the class of proteins that speed both processes.
(d) State the role of water molecules in each case.

4-II
(pp. 51–53)

CARBOHYDRATES
    Monosaccharides
    Disaccharides
    Polysaccharides

Summary

Carbohydrates form some of the main structural materials (including cellulose and chitin) and foods (such as sucrose, glucose, starch, and glycogen) that serve as reservoirs of energy. *Monosaccharides* are simple sugars that are the basic subunits of the larger carbohydrates, the polysaccharides. *Polysaccharides* such as starch are digested by hydrolysis into simple sugars (such as glucose). Many simple sugars can be linked by condensation to make cellulose or starch.

Key Terms

carbohydrate	galactose	maltose
sugar	fructose	polysaccharide
monosaccharides	isomers	starch
ribose	disaccharide	cellulose
deoxyribose	sucrose	glycogen
glucose	lactose	chitin

Objectives

1. Explain how carbohydrates differ from hydrocarbons.
2. Define *carbohydrate*. State what the basic molecular subunits of carbohydrates are.

3. Distinguish among *monosaccharides*, *disaccharides*, and *polysaccharides* and name as many of each as you can.
4. Illustrate how a carbohydrate is assembled from its basic subunits.
5. Indicate how monosaccharides, disaccharides, and polysaccharides are used in living organisms. Explain how the structure of each allows organisms to use that form of carbohydrate for a specific purpose.

Self-Quiz Questions

True-False

If false, explain why.

___ (1) Glucose, a crystalline sugar, has the molecular formula $C_6H_{12}O_6$. The mass numbers of hydrogen, carbon, and oxygen are 1, 12, and 16, respectively. The molecular weight of glucose is 29.

___ (2) $(CH_2O)_6$ is just another way of writing $C_6H_{12}O_6$.

___ (3) Peptide bonds link monosaccharides into polysaccharides called *starches*.

___ (4) In living organisms, carbohydrates serve as structural supports and as food reserves.

___ (5) Polysaccharides can be broken down into simple sugars if the appropriate enzymes are present.

___ (6) Condensation is the process that assembles simple sugars into starches.

___ (7) Carbohydrates contain oxygen; hydrocarbons do not.

Matching

Match *all* applicable letters with the appropriate terms. A blank may contain more than one letter.

(8) ___ amylose
(9) ___ cellulose
(10) ___ chitin
(11) ___ fructose
(12) ___ galactose
(13) ___ glucose
(14) ___ glycogen
(15) ___ lactose
(16) ___ maltose
(17) ___ ribose
(18) ___ sucrose

A. Monosaccharide
B. Disaccharide
C. Polysaccharide
D. Used as a structural support
E. Used as a food reserve
F. Table sugar
G. Milk sugar
H. Used in brewing and found in germinating seeds
I. A five-carbon sugar
J. A six-carbon sugar

LIPIDS
   Lipids With Fatty Acid Components
   Lipids Without Fatty Acid Components

Summary

Lipids are oily or waxy substances such as true fats, waxes, steroids (such as cholesterol and estrogen), terpenes, phospholipids, and glycolipids. True fats are compact forms of stored energy. Waxes serve in waterproofing and protecting surfaces from potentially damaging agents. Phospholipids and glycolipids are important features of cell membranes.

Lipids with fatty acid (a long, unbranched hydrocarbon with a —COOH group at the end) components include *glycerides* (for example, fat), *phospholipids*, and *waxes* (for example, beeswax and cutin, a plant waterproofing material). Lipids without fatty acid components include *terpenes* (important parts of plant pigments) and *steroids* (for example, testosterone, estrogen, and cholesterol).

Key Terms

lipids	fats	waxes
fatty acid	oils	cutin
glyceride	saturated	terpenes
monoglyceride	unsaturated	steroids
diglyceride	phospholipid	cholesterol
triglyceride	glycerol	

Objectives

1. Explain how lipids differ from carbohydrates.
2. Define *lipid*, and identify the molecular subunits of lipids.
3. Distinguish among *triglycerides*, *waxes*, *steroids*, *terpenes*, and *phospholipids* in terms of structure and function.
4. Illustrate how the various classes of lipids function in living organisms. State the principal sources of each class of lipids.

Self-Quiz Questions

True-False

If false, explain why.

___ (1) Lipids contain oxygen, but carbohydrates lack oxygen.

___ (2) A true fat is broken down by hydrolysis into three fatty acid molecules and one molecule of glycerol.

___ (3) Saturated fats contain the maximum possible number of hydrogen atoms that can be covalently bonded to the carbon skeleton of their fatty acid tails.

___ (4) A polyunsaturated fat is usually an oil at room temperature and generally contains many double bonds in the fatty acid component.

___ (5) Waxes are long-chain alcohols combined with long-chain fatty acids. They help organisms prevent water loss and predation.

___ (6) Some of the male and female sex hormones belong to the group of lipids known as *terpenes*.

Matching   Choose *all* the appropriate answers for each.

(7) ___ cholesterol   A. Basic fabric for all membranes
                             B. Butter and bacon
(8) ___ cutin   C. Vegetable oil
(9) ___ phospholipid   D. Wax
                             E. Steroid
(10) ___ saturated fat

(11) ___ unsaturated fat

4-IV
(pp. 55–59)

PROTEINS
Primary Structure: A String of Amino Acids
Spatial Patterns of Protein Structure
Protein Denaturation

Summary

Some proteins, such as enzymes, speed chemical reactions. Other proteins transport substances or cause cells (or parts of cells) to move. Yet other proteins constitute structural elements in bone and cartilage, transmit chemical information, or protect vertebrates from disease agents. Each protein may be broken down into its own particular assemblage of some twenty different *amino acids*. The chemical nature, size, shape, and ordering of side groups projecting from amino acids dictate how a protein interacts chemically with other substances and, hence, what role the protein plays in a living system. The ultimate structure of a protein can be investigated on three (or four) levels: primary, secondary, tertiary, and quaternary. Disruption of the weak bonds on which the secondary, tertiary, and quaternary structures are based causes *denaturation*: a dramatic and irreversible loss of the three-dimensional structure of the protein.

Key Terms

protein	polypeptide chain	helical
amino acid	primary structure	hemoglobin
R group	secondary structure	collagen
peptide bond	tertiary structure	denaturation

Objectives

1. Define *protein* and identify the basic building blocks of proteins.
2. Describe the structure that all amino acids have in common and the structures that make each of the twenty amino acids different.
3. Explain how a peptide bond is formed.
4. Indicate the specific factors that determine (a) how a protein reacts with other substances and (b) what role the protein plays in a living organism.
5. List some specific proteins.
6. Illustrate how the primary, secondary, tertiary, and quaternary structures of a protein are formed.
7. Define *denaturation* and explain how it occurs.
8. Explain what enzymes do.

True-False

If false, explain why.

___ (1) Amino acids are linked by hydrolysis, a process that splits molecules of water as the amino acid subunits are linked together.

___ (2) Bone and cartilage are constructed, in part, of specific proteins.

___ (3) R groups projecting from the main carbon skeleton determine how a long-chain protein interacts chemically with other substances.

___ (4) The primary structure of a protein is formed principally by hydrogen bonds linking various amino acids.

___ (5) An amino group contains a nitrogen atom and two hydrogen atoms; a carboxyl group contains two oxygen atoms, a carbon atom, and a hydrogen atom.

4-V
(pp. 59–61)

NUCLEOTIDES AND NUCLEIC ACIDS
SUMMARY OF THE MAIN BIOLOGICAL MOLECULES

Summary

Each nucleotide contains at least a five-carbon sugar, a nitrogen-containing base, and a phosphate group. Nucleotide-based molecules serve as chemical messengers between cells (cAMP), as energy carriers (ATP), as transporters of hydrogen ions and electrons in metabolic reactions (NAD^+ and FAD), and as molecules that encode genetic instructions (DNA) or help translate these instructions (RNA) into the proteins upon which all forms of life are based.

Key Terms

nucleotide	deoxyribose	FAD
five-carbon sugar	adenosine phosphates	nucleic acids
nitrogen base	cAMP	DNA
pyrimidine	ATP	RNA
purine	nucleotide coenzymes	
ribose	NAD^+	

Objectives

1. Describe a nucleotide by discussing its components and the way the components are linked together.
2. Define *purine* and *pyrimidine* and distinguish them from each other.
3. List the three principal nucleotide-based molecules, give two examples of each, and tell how each functions in living organisms.
4. Define *nucleic acid* and describe its basic structure.

Self-Quiz Questions

True-False   If false, explain why.

__T__ (1) ATP is a temporary energy-storage molecule.

__F__ (2) A purine contains one carbon skeleton bent around to form a ring, whereas a pyrimidine contains two such rings.

__T__ (3) Adenosine phosphates act as chemical messengers between cells and as energy carriers.

__F__ (4) Nucleic acids are long chains of nucleotides strung together, with nitrogenous bases connecting the phosphates and with sugars sticking out to the side.

__T__ (5) DNA consists of two parallel nucleic acid strands twisted around each other and cross-linked by hydrogen bonds.

__T__ (6) DNA contains the code for constructing one or more proteins.

__T__ (7) RNA molecules translate the DNA code into a protein product.

Matching   Choose *all* the appropriate answers for each.

(8) __D__ adenosine phosphates
(9) __C__ nucleotide coenzymes
(10) __A__ nucleic acid

A. Long; single-stranded or double-stranded
B. ATP
C. Transport hydrogen ions and their associated electrons
D. cAMP, a chemical messenger
E. NAD$^+$ and FAD
F. RNA and DNA

CHAPTER TEST   **UNDERSTANDING AND INTERPRETING KEY CONCEPTS**

____ (1) Carbon is part of so many different substances because _____.
    (a) carbon generally forms two bonds with a variety of other atoms
    (b) carbon generally forms four bonds with a variety of atoms
    (c) carbon ionizes easily
    (d) carbon is a polar compound

____ (2) Proteins _____.
    (a) include all hormones
    (b) are composed of nucleotide subunits
    (c) are not very diverse in structure and function
    (d) include all enzymes

____ (3) Glucose dissolves in water because it _____.
    (a) ionizes
    (b) is a polysaccharide
    (c) forms many hydrogen bonds with the water molecules
    (d) has a very reactive primary structure

___ (4) Hydrolysis could be correctly described as the _____.
   (a) heating of a compound in order to drive off its excess water and concentrate its volume
   (b) breaking of a long-chain compound into its subunits by adding water molecules to its structure between the subunits
   (c) linking of two or more molecules by the removal of one or more water molecules
   (d) constant removal of hydrogen atoms from the surface of a carbohydrate

___ (5) DNA _____.
   (a) is one of the adenosine phosphates
   (b) is one of the nucleotide coenzymes
   (c) contains protein-building instructions
   (d) translates protein-building instructions into actual protein structures

___ (6) Amino acids are linked by _____ bonds to form the primary structure of a protein.
   (a) disulfide
   (b) hydrogen
   (c) ionic
   (d) peptide

___ (7) Lipids _____.
   (a) serve as food reserves in many organisms
   (b) include cartilage and chitin
   (c) include fats that are broken down into one fatty acid molecule and three glycerol molecules
   (d) are composed of monosaccharides

CHAPTER TEST

INTEGRATING AND APPLYING KEY CONCEPTS

(1) Humans can obtain energy from many different food sources. Do you think this ability is an advantage or a disadvantage in terms of long-term survival? Why?

(2) If the ways that atoms bond affect molecular shapes, do the ways that molecules behave toward one another influence the shapes of organelles? Do the ways that organelles behave toward one another influence the structure and function of the cells?

(3) If proteins have the most diverse shapes and the most complex structure of all molecules, why do you suppose that proteins are not the code molecules used to construct new proteins?

5

CELL STRUCTURE AND FUNCTION: AN OVERVIEW

General Objectives

1. Understand why cells generally fall into a predictable range of sizes.
2. Contrast the general features of prokaryotic and eukaryotic cells.
3. Describe the nucleus of eukaryotes with respect to structure and function.
4. Describe the organelles associated with the cytomembrane system, and tell the general function of each.
5. Contrast the structure and function of mitochondria and chloroplasts.
6. Describe the cytoskeleton of eukaryotes and distinguish it from the cytomembrane system.
7. List several surface structures of cells and tell how they help cells survive.

GENERALIZED PICTURE OF THE CELL
  Emergence of the Cell Theory
  Basic Aspects of Cell Structure and Function
  Cell Size and Cell Membranes

Summary

Crude systems of lenses developed early in the seventeenth century by Galileo, Hooke, and van Leeuwenhoek allowed observation of cells as small as bacteria and sperm. Since the 1820s, specific technological advances in lens design have enlarged our understanding of cellular structure and behavior. Various kinds of light and electron microscopes have their own particular uses. Low-resolution light microscopes are used to observe the cells of living organisms, whereas high-resolution electron microscopes form images of structures as small as 0.2 nanometers. Scanning electron microscopes have lower resolving power but remarkable depth of field; surface features of cells and organisms can be examined.

Scientists know that all organisms are composed of one or more cells, that the cell is the basic living unit of organization for all organisms, and that all cells arise from preexisting cells. Most cells are quite small but nevertheless highly organized. The smaller the cell, the more efficiently materials can enter the cell and be distributed through the cytoplasm. The surface-to-volume ratio determines the size and shape a cell reaches before it either stops growing or divides. Pathways of chemical communication link the plasma membrane, cytoplasm, and nuclear regions.

Key Terms

Galileo	nucleoid	phase contrast microscope
Hooke	cytoplasm	resolution
van Leeuwenhoek	micrometers	transmission electron
Schwann	nanometers	microscope
Schleiden	diffusion	high-voltage electron
cell theory	surface-to-volume ratio	microscope
Virchow	surface area	scanning electron
plasma membrane	compound light	microscope
nucleus	microscope	

Objectives

1. Briefly describe the contributions made by each of the following scientists to the modern understanding of cell biology: Galileo, Hooke, van Leeuwenhoek, Brown, Schwann, Schleiden, and Virchow.
2. a. Describe how each of these microscopes works: conventional compound light microscope, phase contrast light microscope, transmission electron microscope, and scanning electron microscope.
   b. Tell whether living cells can be viewed by each type of microscope and state the limitations of each type.
3. List the basic ideas of the cell theory stated by Schleiden, Schwann, and Virchow.
4. Consult Appendix II (main text) and determine how you would explain the range of cell sizes to a person unfamiliar with the metric system.
5. State what the plasma membrane, the nucleus, and the cytoplasm do in a cell.

Self-Quiz Questions

True-False

If false, explain why.

_T_ (1) The cell is the smallest independent living unit.

_F_ (2) The average cell is approximately as large as your thumbnail.

_T_ (3) A plasma membrane permits some substances to cross it and prevents other substances from crossing.

Matching

Choose the one best answer for each.

(4) ___ Galileo

(5) ___ Hooke

(6) ___ Schleiden and Schwann

(7) ___ van Leeuwenhoek

(8) ___ Virchow

A. Originated the term *cell*
B. Determined that all cells must come from preexisting cells
C. The first person to observe a great diversity of microscopic organisms
D. Said that all organisms are constructed of cells
E. The first person to record any biological observations under the microscope

Fill-in-the-Blanks

One problem with compound (9) _____ microscopes is spherical aberration: Tiny objects appear blurred when brought close to the objective lens. If you wish to observe a living cell, it must be small or thin enough for (10) _____ to pass through. (11) _____ is the property that dictates whether small objects close together can be seen as separate things. A(n) (12) _____ is one-billionth of a meter. With a (13) _____ electron microscope, a beam of electrons is transmitted through a prepared, thinly sliced section of a cell or organism. With a (14) _____ electron microscope, a narrow beam of electrons is played back and forth across a specimen's surface, which has been coated with a thin metal layer. The (15) _____ is a membrane-bound zone of hereditary control.

5-II
(pp. 67–70)

PROKARYOTIC CELLS—THE BACTERIA
EUKARYOTIC CELLS
    Function of Organelles
    Typical Components of Eukaryotic Cells

Summary

Every prokaryotic cell has a plasma membrane, cytoplasm, a nucleoid, and ribosomes; most also produce a cell wall and have a few internal membranes. Prokaryotic DNA in the nucleoid does not coil into chromosomes.

In eukaryotic cells, a true nucleus replaces the nucleoid, and additional membrane-bounded organelles are distributed throughout the cytoplasm. Unlike animals, many protistans, true fungi, and land plant cells have a cell wall surrounding the plasma membrane.

Key Terms

cell wall	Golgi bodies	central vacuole
ribosome	lysosomes	cytoskeleton
prokaryotic	microbodies	transport vesicles
eukaryotic	mitochondria	microtubules
organelles	plastids	microfilaments
endoplasmic reticulum	chloroplast	plasma membrane

Objectives

1. Look at Figures 5.8 and 5.9 (main text) and describe how plant cells differ from animal cells. Then look at Figure 5.7 (main text) and describe how a generalized moneran cell differs from plant and animal cells.
2. Explain how the term *prokaryote* is related to the moneran kingdom and how the term *eukaryote* is related to the other four kingdoms.
3. List the types of organisms that produce cell walls. State what cell walls are made of and explain what cell walls do for cells that have them.

**Self-Quiz
Questions**

Matching

Select the single best answer. A letter can be used more than once.

(1) _D_ endoplasmic reticulum
(2) _D_ Golgi complex
(3) _b_ lysosomes
(4) _F_ microbodies
(5) _C_ mitochondria
(6) _E_ nucleus

A. Photosynthesis occurs here
B. Digestion and disposal
C. Energy extraction
D. Material synthesis, modification, and distribution
E. Hereditary instructions for synthesis and cell operation
F. Material conversions and disposal

Fill-in-the-Blanks

(7) _____ have a membrane-bounded nucleus and other membrane-bounded organelles; (8) _____ lack such structures. (9) _____ cells do not produce walls, although some secrete products to the surface layer of tissues in which they are formed. The (10) _____ is the region of hereditary control in prokaryotic cells.

THE NUCLEUS
> Chromosomes
> Nucleolus
> Nuclear Envelope

CYTOMEMBRANE SYSTEM
> Endoplasmic Reticulum and Ribosomes
> Golgi Bodies
> Lysosomes
> Microbodies

Summary

All eukaryotic cells contain a nucleus that is the storehouse of information that promotes survival and reproduction. A *nuclear envelope* encloses the *nucleoplasm* and is regularly traversed by pores.

Chromatin is a mass of DNA and associated proteins that actively regulate the metabolic behavior of the cell; during cell division, chromatin becomes tightly coiled into wormlike structures called *chromosomes* for efficient distribution into the daughter cells. Prokaryotic DNA has few associated proteins and cannot condense into chromosomes.

The nucleolus is a dense region of nucleoplasm where the subunits that are later constructed into ribosomes are made.

Various organelles in eukaryotic cells affect intracellular substances in different ways. *Ribosomes* are the sites of protein synthesis. If ribosomes are on the *rough endoplasmic reticulum*, the proteins they synthesize will be exported from the cell. The smooth endoplasmic reticulum lacks ribosomes but contains enzymes that help assemble fats. The *smooth endoplasmic reticulum* also participates in isolating and transporting materials and in breaking down storage materials and potentially toxic materials. *Golgi bodies* make up a membrane system (generally located near the nucleus) that receives materials from the endoplasmic reticulum, packages them, and transports them. Most polysaccharides are synthesized in the Golgi bodies, and the finishing touches are put on glycoproteins there. *Lysosomes* are sacs of digestive enzymes in animal cells; they sometimes merge with membrane-enclosed materials and digest them. *Microbodies* convert excess amounts of some substances into different substances that are required but scarce.

Key Terms

nucleus	chromatin	ribosomal units
nuclear envelope	chromosomes	exocytic vesicle
nucleoplasm	smooth ER	endocytic vesicles
cytomembrane system	secretory	lysosome
ribosomes	sarcoplasmic reticulum	microbodies
endoplasmic reticulum	transport vesicles	perioxisomes
cisternal space	Golgi bodies	glyoxysomes
rough ER	nucleoprotein	
pores	nucleolus	

Objectives

1. Describe the basic organization of the nuclear envelope, the nucleolus, and the chromatin/chromosome system.
2. State the processes that occur in each of the nuclear components mentioned in Objective 1.
3. State the principal difference between prokaryotic and eukaryotic DNA.
4. Distinguish between *rough endoplasmic reticulum* and *smooth endoplasmic reticulum* in terms of structure and the substances each produces.

5. Draw a diagram of a Golgi body and explain how Golgi bodies accomplish their functions.
6. Trace the pathway traveled by a protein that has just been exported from a cell. Start at the point where its constituent amino acids are in the cytoplasm.
7. Explain how microbodies (peroxisomes) can be useful to the cells of your tissues.

Self-Quiz Questions

True-False

If false, explain why.

___ (1) The nucleolus is part of the nucleoplasm.

___ (2) If the chromatin network did not condense into chromosomes during cell division, many more tangles and breaks in the DNA material would occur.

___ (3) The nucleolus produces the materials from which the Golgi complex is later constructed in the cytoplasm.

___ (4) If you ate only fats and proteins and excluded all carbohydrates from your diet, microbodies in your cells would attempt to convert excess proteins and fats into carbohydrates in order to provide what your diet lacked.

___ (5) The rough endoplasmic reticulum is a membrane system that has many ribosomes associated with it.

___ (6) A protein due to be exported from the cell is made by ribosomes located in the rough endoplasmic reticulum and exported in vesicles pinched off from the endoplasmic reticulum.

___ (7) Hydrolytic enzymes that are usually stored in chromoplasts are able to break down virtually every large molecule used in cell architecture.

Fill-in-the-Blanks

The (8) _____ is the region where ribosomal subunits are synthesized. Masses of DNA and its associated proteins that extend throughout the nucleoplasm when it is not undergoing division are called (9) _____. During cell division, this material condenses into (10) _____.

Matching

Choose the best answer for each.

(11) ___ Golgi body

(12) ___ lysosome

(13) ___ microbody

(14) ___ sarcoplasmic reticulum

A. Stores and releases calcium ions in muscle cells
B. Synthesizes proteins to be exported from that cell
C. Converts excess amounts of some substances into different substances that are needed but not available

(15) —— rough endoplasmic reticulum

(16) —— smooth endoplasmic reticulum

D. Assembles fats and fat derivatives; breaks down glycogen

E. Assembles, packages, transports, exports substances

F. A small bag of hydrolytic enzymes found in animal cells

5-IV
(pp. 76–77)

MITOCHONDRIA
SPECIALIZED PLANT ORGANELLES
Chloroplasts and Other Plastids
Central Vacuoles

Summary

All eukaryotic cells contain *mitochondria*, which have membrane systems similar to those of chloroplasts except that the inner membrane forms shapes resembling projecting shelves (*cristae*) rather than stacks of coins. Mitochondria convert energy stored in carbon compounds into forms that the cell can use—principally ATP.

In plants and in some protistans, the reactions of photosynthesis occur in organelles called *chloroplasts*. Each chloroplast has an outer membrane and a complex, much-folded inner membrane thrown into configurations that resemble stacks of coins (*grana*). Apparently the grana are the sites where some of the energy of sunlight is absorbed by chlorophyll and passed on to be stored in the bonds of ATP and $NADPH_2$. The energy released by ATP and $NADPH_2$ can subsequently be used in the *stroma* to build carbon compounds such as sugar molecules from carbon dioxide and water.

In photosynthetic prokaryotes, light-trapping reactions and ATP formation occur on a portion of the plasma membrane that has been folded back into the cytoplasm, and the remainder of the energy-transformation reactions occur in the cytoplasm.

Storage organelles concentrate materials not intended for export in places that are out of the way of metabolic activity. Plastids are plant cell storerooms that may contain starch grains, plant pigments, oil, and other high-energy food reserves. The storage organelles in animal and protistan cells are generally known as *vacuoles*, as are the large fluid-filled sacs that promote environmental contact in many plant cells.

Key Terms

aerobic	chloroplast	chlorophyll
mitochondrion, -dria	stroma	chromoplasts
matrix	granum, grana	amyloplasts
plastids	starch grains	central vacuole

Objectives

1. Describe the basic structure of the mitochondrion. Indicate the processes that occur in mitochondria and the approximate size of mitochondria.
2. State which types of cells contain (a) chloroplasts and (b) mitochondria.
3. Describe the basic structure of the chloroplast, indicate how the grana differ from the stroma and what happens in each, and tell where the chlorophyll is located.
4. Explain how cells store materials.
5. Describe how cells increase their size and interior surface area through the use of vacuoles.

Self-Quiz
Questions

True-False

If false, explain why.

___ (1) Muscle cells and other cells that demand high-energy output generally have many more chloroplasts than less active cells have.

___ (2) Mitochondria and chloroplasts are organelles that have both inner and outer membranes.

___ (3) Enzymes and molecules involved in ATP formation are located on the outer membranes.

___ (4) Chloroplasts always appear green because of the chlorophyll they contain.

___ (5) Chlorophyll is found in all chloroplasts.

Matching

Choose the one best answer for each.

(6) ___ chromoplast
(7) ___ amyloplast
(8) ___ vacuole
(9) ___ vesicle
(10) ___ mitochondrion

A. Colorless; accumulates starch grains
B. A tiny bag that transports substances through the cytoplasm
C. Stores plant pigments
D. A large storage compartment that generally stores fluids
E. A narrow, hollow cylinder
F. Tubulins
G. Provides the ATP necessary for rapid movements

Fill-in-the-Blanks

In photosynthetic prokaryotes such as bacteria and blue-green algae, light-trapping reactions and ATP formation occur on the (11) _____ _____ _____. In photosynthetic eukaryotes these reactions occur in an organelle called the (12) _____ _____ _____. Two energy-rich molecules produced in the grana are (13) _____ and (14) _____. An energy-rich molecule produced by a mitochondrion is (15) _____.

_____

5-V
(pp. 78–82)

THE CYTOSKELETON
   Structure and Function of the Cytoskeleton
   Flagella and Cilia
   What Organizes the Cytoskeleton?
CELL SURFACE SPECIALIZATIONS
   Cell Walls
   The Extracellular Matrix and Cell Junctions

Summary

Cellular contents stream about in living eukaryotic cells, and proper cell functioning requires organization based on a three-dimensional lattice that pervades the cytoplasm.

Microfilaments are extremely long, thin structural elements composed of contractile proteins. The activities of microfilaments can cause cells to pinch in two, crawl across a culture dish, or change shape, or they can make organelles move around within the cytoplasm. *Microtubules* are tiny hollow cylinders that are thicker and more rigid than microfilaments. Microtubules help dictate the shape of many cells and cell extensions; they are also involved in the movement of chromosomes during nuclear division and in the beating of *cilia* and *flagella* that move cells through their environment. *Bacterial flagella* are threadlike extensions of the cell surface that act as propellers. *Eukaryotic flagella* and *cilia* are assembled from microtubules that are anchored in basal bodies; they help small organisms move through liquid environments. Cilia and flagella also help attract food particles to nonmotile animals such as some mollusks and all sponges; they also move fluids and suspended particles along the surfaces of epithelium that line the digestive, respiratory, and reproductive tracts of many types of animals.

Cell walls provide support, resist mechanical pressure, and confer tensile strength when cells absorb water and expand. Porous cell walls occur in all kingdoms except Animalia and are quite diverse structurally. Cell-to-cell junctions enable cells to interact with their cellular neighbors.

Key Terms

myosin	transient	cell walls
cytoskeleton	flagella, flagellum	primary cell wall
microtubules	cilia, cilium	secondary cell wall
tubulins	bacterial flagellum	extracellular matrix
microfilament	flagellin	tight junctions
actin	dynein	adhering junctions
intermediate filaments	microtubule organizing	gap junctions
keratin	centers (MTOCs)	plasmodesmata,
fluorescence microscopy	centrioles	plasmodesma
permanent	basal bodies	

Objectives

1. Distinguish *microfilaments* from *microtubules* in terms of structure and function. Mention actin and tubulins in your explanations.
2. Explain how microfilaments and microtubules influence cell shape, motion, and growth.
3. Describe the cytoskeleton and explain why you think that it cannot be seen very well using ordinary transmission electron microscopy.
4. State the benefits and limitations imposed on a cell by the presence of cell coats, capsules, sheaths, and walls.
5. List the major groups of organisms that have walls surrounding their cells.
6. Describe the structure of cell walls.
7. Explain how prokaryotic flagella differ from eukaryotic flagella.
8. Describe the relationship between centrioles and basal bodies.
9. List the major groups of organisms that have eukaryotic cilia and flagella and state some of the major functions provided by those organelles.
10. Compare the three main types of cell junctions.

Fill-in-the-Blanks

The principal subcellular structures involved in establishing and maintaining the shapes of cells and extensions from cells are (1) _____. (2) _____ are long, thin structural elements composed of contractile proteins. The (3) _____ is a three-dimensional network that pervades the cytoplasm. (4) _____ are assembled from protein subunits called (5) _____.

Matching

Link each letter with its appropriate blank(s). Each blank should have only one letter. The same letter may be used in more than one blank.

(6) ___ adhering junction

(7) ___ basal body

(8) ___ bacterial flagellum

(9) ___ cell wall

(10) ___ centriole

(11) ___ cilium

(12) ___ eukaryotic flagellum

(13) ___ gap junction

(14) ___ plasmodesmata

(15) ___ tight junction

A. Short, barrel-shaped organelle that organizes the interior of a cilium or flagellum
B. In plants, a channel that extends across adjacent cell walls
C. A whiplike extension that acts as a propeller; flagellin present
D. Lignin and cellulose; cutin
E. Assembled from microtubules in a 9 + 2 array within a sheath that is continuous with a plasma membrane
F. Interconnect all living cells in multicelled plants
G. Rows of membrane proteins match up and form sealing strands between adjacent epithelial cells
H. Desmosome; abundant in epithelial cells of animals; resemble welded spots
I. Exists in a pair that is pushed apart by spindle formation

5-VI
(pp. 82–83)

SUMMARY

Summary

As time passed, cells became more complex. There were selective advantages for cells that increased their internal surface area with infoldings and outfoldings of membrane on which metabolic reactions could occur. Such developments have enabled cells to acquire energy and materials in highly controlled and specialized ways.

Objectives

1. List the basic minimal parts a cell must have in order to carry on metabolism and live.
2. Explain the advantages that accrue for cells that develop many internal compartments.

3. Consult and memorize Table 5.2 in your main text. Decide which kingdom has the least complex cells and which has the most complex cells. Indicate which other kingdom shares the most features with plants and which other kingdom most strongly resembles the animals.

Self-Quiz Questions

Fill-in-the-Blanks   If the cell structure is present in all or most members of that group, put a check (√) on its line; if the cell structure is present in some of that group, put a cross (+) on its line. If it is not present, leave the line blank.

	Monera	Protista	Fungi	Plantae	Animalia
Cell wall	(1) __	(2) __	(3) __	(4) __	(5) __
Plasma membrane	(6) __	(7) __	(8) __	(9) __	(10) __
Photosynthetic pigments	(11) __	(12) __	(13) __	(14) __	(15) __
Chloroplasts	(16) __	(17) __	(18) __	(19) __	(20) __
Mitochondria	(21) __	(22) __	(23) __	(24) __	(25) __
Ribosomes	(26) __	(27) __	(28) __	(29) __	(30) __
Endoplasmic reticulum	(31) __	(32) __	(33) __	(34) __	(35) __
Cytoskeleton	(36) __	(37) __	(38) __	(39) __	(40) __
Complex cilia or flagella	(41) __	(42) __	(43) __	(44) __	(45) __
DNA molecules	(46) __	(47) __	(48) __	(49) __	(50) __
Chromatin condensed into chromosomes	(51) __	(52) __	(53) __	(54) __	(55) __

True-False   If false, explain why.

___ (56) Some cells can live if they contain only a plasma membrane, DNA molecules, and cytoplasm that lacks other cell structures.

___ (57) All living organisms are composed of one or more cells.

___ (58) New cells can arise only from cells that already exist.

___ (59) There are no essential differences in the ways that prokaryotes and eukaryotes acquire and process energy and materials.

Labeling   Identify each indicated part of the accompanying illustration.

(60) _____

(61) _____ _____

(62) _____ _____ _____

(63) _____

(64) _____ _____

(65) _____

(66) _____

(67) _____ _____

(68) _____ _____

(69) _____ _____

(70) _____

(71) _____

(72) _____ _____

(73) _____

(74) _____ _____

(75) _____ _____

(76) _____

(77) _____

(78) _____

(79) _____

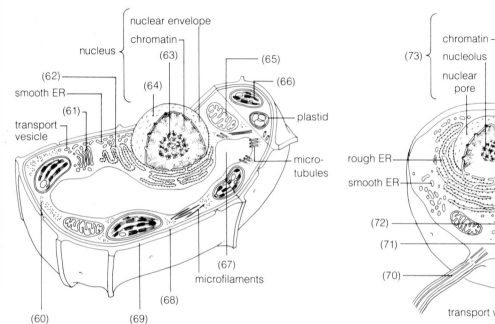

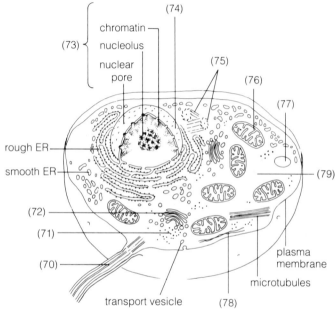

CHAPTER TEST

UNDERSTANDING AND INTERPRETING KEY CONCEPTS

____ (1) Animal cells dismantle and dispose of waste materials by _____.
   (a) using centrally located vacuoles
   (b) several lysosomes fusing with a sac that encloses the wastes
   (c) microvilli packaging and exporting the wastes
   (d) mitochondrial breakdown of the wastes

____ (2) Root hairs lengthen through _____ enlargement.
   (a) chromoplast
   (b) chromosome
   (c) vesicle
   (d) vacuole

___ (3) The nucleolus is a dense region of _____ where the subunits that later will be constructed into _____ are made.
   (a) nucleoplasm, ribosomes
   (b) cytoplasm, vesicles
   (c) cytoplasm, chromosomes
   (d) nucleoplasm, chromatin

___ (4) Which of the following is *not* present in all cells?
   (a) Cell wall
   (b) Plasma membrane
   (c) Ribosomes
   (d) DNA molecules

___ (5) A nanometer is _____ of a meter.
   (a) one-ninth
   (b) one-tenth
   (c) one one-hundredth
   (d) one one-billionth

___ (6) Mitochondria convert energy stored in _____ to forms that the cell can use, principally ATP.
   (a) water
   (b) carbon compound
   (c) $NADPH_2$
   (d) carbon dioxide

CHAPTER TEST

INTEGRATING AND APPLYING KEY CONCEPTS

(1) Which parts of a cell constitute the minimum necessary for keeping the simplest of living cells alive?

(2) How did the existence of a nucleus, compartments, and extensive internal membranes confer selective advantages on cells that developed these features?

6

MEMBRANE STRUCTURE
AND FUNCTION

**General
Objectives**

1. Understand the essential structure and function of the cell membrane.
2. Know the forces that cause water and solutes to move across membranes passively (that is, without expending energy).
3. Understand which types of substances move by simple diffusion and which by bulk flow. Understand the importance of osmosis to all cells.
4. Know the mechanisms by which substances are moved across membranes against a concentration gradient.
5. Understand how material can be imported into or exported from a cell by being wrapped in membranes.

6-I
(pp. 85–88)

FLUID MEMBRANES IN A LARGELY FLUID WORLD
    The Lipid Bilayer
    Membrane Proteins
    Summary of Membrane Structure and Function

Summary

Cell membranes are composed of lipids (predominantly phospholipids) and proteins. Lipid molecules surrounded by water show self-assembling and self-sealing behavior. This behavior maintains the integrity of the plasma membrane and organelles. Membrane lipids have hydrophilic heads and hydrophobic tails; when surrounded by water they assemble spontaneously into a bilayer. All heads are at the two outer faces of a lipid bilayer, and all tails are sandwiched between them. Lipid molecules move about within a bilayer, thereby contributing to membrane fluidity. The lipid bilayer provides the basic *structure* of all cell membranes and acts as a hydrophobic barrier between the fluid regions. Membrane functions are carried out largely by proteins associated with the bilayer. Some proteins transport water-soluble substances across the membrane. Some are enzymes. Others are receptors for chemical signals or for specific substances.

Key Terms

plasma membrane	fluid mosaic model of	freeze fracture
internal cell membranes	membrane structure	freeze etching
phospholipid	channel protein	metal shadowing
hydrophilic head	gate, gated channel	glycolipids
hydrophobic tail	transport protein	surface receptors
lipid bilayer	electron-transfer protein	hydrophobic barrier
glycoproteins	impermeability	

Objectives

1. Describe the structure of a plasma membrane as portrayed by the fluid mosaic model.
2. Explain how the formation of lipid bilayers contributes to shapes commonly seen in cells, such as tiny bubbles and membrane sheets.
3. Does a punctured plasma membrane release its cytoplasm or seal itself? Explain why.
4. List three functions carried out by a plasma membrane.
5. Understand how freeze-fracturing and freeze-etching techniques contribute to our understanding of membrane structure.
6. Explain what would occur if receptor proteins on membrane surfaces suddenly were to vanish from all of your cells.

Self-Quiz Questions

Fill-in-the-Blanks

(1) _____-_____ and (2) _____-_____ are two relatively new methods of preparing membranes for study with the electron microscope. A (3) _____ _____ serves as a sort of fluid sea matrix in which diverse (4) _____ are suspended like icebergs. Together, the two components form the (5) "_____." (6) _____ have fatty acid tails (which repel water) and heads with (7) _____ and alcohol groups (which dissolve in water). Membrane surface receptors are often (8) _____ groups attached to regions of membrane (9) _____ and membrane lipids. When activated, some receptors bring about changes in cell (10) _____ or behavior. In multicelled organisms, surface receptors also function in identifying cells of like type during the formation of (11) _____.

True-False   If false, explain why.

   F (12) In a plasma membrane, the hydrophilic tails point inward, tail to tail, and form a region that excludes water.

   T (13) The tails of phospholipids are hydrophobic.

6-II
(pp. 89–92)

DIFFUSION
  Gradients Defined
  Simple Diffusion
  Bulk Flow
OSMOSIS
  Osmosis Defined
  Tonicity
  Water Potential

Summary

Maintaining internal concentrations of water and solutes depends on *passive transport mechanisms* such as diffusion, osmosis, and bulk flow, which do not alter in any way the direction in which a substance is moving on its own. No direct energy outlay by the cell is required for these processes.

Diffusion is a random movement of like molecules along a concentration gradient from their region of greater concentration to a region of lesser concentration. It accounts for the greatest volume of substances that are moved into and out of cells, and it is also an important transport process within cells. In *bulk flow*, different ions and molecules present in a fluid move together in the same direction—often in response to a pressure gradient. *Osmosis* is the passive movement of *water* across a differentially permeable membrane in response to solute concentration gradients and/or pressure gradients. Osmotic movements across cell membranes are influenced by the concentrations of dissolved substances inside and outside of the membrane being considered. Red blood cells shrivel when placed in a *hypertonic* solution but swell and burst when immersed in a *hypotonic* solution.

Key Terms

concentration gradient	pressure gradient	hypertonic
simple diffusion	solute	hemolyze
bulk flow	tonicity	turgor pressure
differentially permeable	isotonic	water potential
osmosis	hypotonic	

Objectives

1. Define *diffusion* and explain what causes diffusion of any sort.
2. List three factors that influence the rate of diffusion and state what sorts of substances diffuse readily across plasma membranes.
3. Distinguish *bulk flow* from *osmosis*.
4. Understand the effects of osmosis on living cells.
5. Differentiate between active and passive forms of transport and give two examples of each. Describe how cells expend energy.
6. Explain what is meant by *concentration gradient* and explain how such gradients are formed.
7. Distinguish between *permeable* and *differentially permeable*.

Self-Quiz Questions

Fill-in-the-Blanks

(1) _____ refers to the number of molecules (or ions) of a substance in a given volume of space. A (2) _____ can exist between two regions that differ in (3) _____, pressure, temperature, or net electric charge. Diffusion is driven by the (4) _____ _____ _____ inherent in all individual molecules as they move from a region of (5) _____ concentration to a region of (6) _____ concentration. In (7) _____ _____, different ions and molecules present in a fluid move together in the same direction, often in response to a pressure gradient.

The plasma membrane is (8) _____ _____; some molecules travel rapidly across the membrane, others cross it more slowly, and some are kept from crossing it at all. (9) _____ is one of the few molecules that can move freely into and out of the cell.

Red blood cells shrivel and shrink when placed in a(n) (10) [choose one] hypotonic ☐ isotonic ☐ hypertonic ☐ solution.

True-False

If false, explain why.

____ (11) Diffusion accounts for the greatest volume of substances that are moved into and out of cells.

____ (12) Osmosis occurs in response to a concentration gradient that involves unequal concentrations of water molecules.

6-III
(pp. 92–96)

MOVEMENT OF WATER AND SOLUTES ACROSS CELL MEMBRANES
    The Available Routes
    Facilitated Diffusion
    Active Transport
    Exocytosis and Endocytosis
SUMMARY
    Membrane Structure
    Membrane Functions
    Movement of Water and Solutes Across Membranes

Summary

In *facilitated diffusion*, proteins embedded in the plasma membrane assist passively in the passage of small molecules across the membrane, but maintaining internal concentrations of water and solutes depends on *active transport mechanisms* such as active transport, exocytosis, and endocytosis, all of which work to move a substance in a direction contrary to its spontaneous direction of movement. These mechanisms cannot operate without direct energy outlays by the cell.

When ions and molecules required by a cell are scarce, the cell must expend energy in order to stockpile these nutrients by getting them to move against their concentration gradient. In *active transport*, proteins serve as carriers or as fixed channels for moving ions or molecules across the plasma membrane. As much as 70 percent of the energy readily available in a cell may be devoted to active transport. Cells that must move relatively large amounts of solids or fluids across the plasma membrane enclose the substances to be transported in membrane-bounded compartments and move them through the membrane to the opposite side. *Endocytosis* and *exocytosis* move substances into and out of the cell, respectively.

Key Terms

membrane transport proteins
carrier proteins
active transport
passive transport
facilitated diffusion

sodium-potassium pump
calcium pump
exocytosis
endocytosis
phagocytic
endocytic vesicle

lysosomes
pinocytosis
receptor-mediated endocytosis
lipoproteins

Objectives

1. Explain how a cell gets the ions and molecules it requires if they are not abundant in the environment.
2. Explain how a cell excretes ions and molecules it does not need when there are many more of them outside the cell than inside.
3. Distinguish *active transport* from *bulk transport*, *endocytosis* from *exocytosis*, and *phagocytosis* from *pinocytosis*.
4. State which cells depend on the sodium-potassium pump and which depend on the calcium pump.
5. List six functions of the plasma membrane.

Self-Quiz Questions

True-False

If false, explain why.

___ (1) Active transport depends on proteins that serve either as carriers or as fixed channels across the plasma membrane.

___ (2) Bulk flow requires an expenditure of ATP to move large amounts of materials across membranes and may occur in response to a pressure gradient.

___ (3) To receive messages from other cells, nerve cells depend on the sodium-potassium pump mechanism.

___ (4) The secretion of mucus is achieved by exocytosis.

___ (5) Facilitated diffusion is a type of diffusion that requires the cell to expend ATP molecules.

Fill-in-the-Blanks

(6) _____, (7) _____, (8) _____, a few other simple molecules, and some (9) _____ diffuse readily across plasma membranes.

(10) _____ _____ _____ cannot operate without direct outlays by the cell.

UNDERSTANDING AND INTERPRETING KEY CONCEPTS

___ (1) White blood cells use _____ to devour disease agents invading your body.
  (a) diffusion
  (b) bulk flow
  (c) osmosis
  (d) phagocytosis

___ (2) _____ cells depend on the calcium-pump mechanism.
  (a) Intestinal mucous-secreting
  (b) Nerve
  (c) Muscle
  (d) *Amoeba*

___ (3) Water is such an excellent solvent primarily because _____.
  (a) it forms spheres of hydration around substances and can form hydrogen bonds with many nonpolar substances
  (b) it has a high heat of fusion
  (c) of its cohesive properties
  (d) it is a liquid at room temperature

___ (4) In a lipid bilayer, _____ tails point inward and form a region that excludes water.
  (a) acidic
  (b) basic
  (c) hydrophilic
  (d) hydrophobic

___ (5) A protistan adapted to life in a freshwater pond is collected in a bottle and transferred to a saltwater bay. Which of the following is likely to happen?
  (a) The cell bursts.
  (b) Salts flow out of the protistan cell.
  (c) The cell shrinks.
  (d) Enzymes flow out of the protistan cell.

___ (6) Which of the following is *not* a form of active transport?
  (a) Sodium–potassium pump
  (b) Endocytosis
  (c) Exocytosis
  (d) Bulk flow

___ (7) Which of the following is *not* a form of passive transport?
  (a) Osmosis
  (b) Facilitated diffusion
  (c) Bulk flow
  (d) Exocytosis

INTEGRATING AND APPLYING KEY CONCEPTS

(1) Name three aspects of the biological world that would be changed if water were not a polar molecule.

(2) If there were no such thing as active transport, how would the lives of organisms be affected?

7

GROUND RULES OF METABOLISM

**General
Objectives**

1. Know two laws that govern the way energy is transferred from one substance to another.
2. Provide an example of a metabolic pathway and explain what kinds of substances regulate activity of the pathway.
3. Tell exactly what enzymes do and how they do it.
4. Explain how a molecule can "carry" energy.

7-I
(pp. 97–99)

THE NATURE OF ENERGY
    Two Laws Governing Energy Transformations
    Living Systems and the Second Law

Summary

All events in our universe are governed by two laws of energy. The first law of thermodynamics states that the total amount of energy in the universe never changes. Energy may be converted from one form into another, but it can be neither created nor destroyed. Living organisms can channel the ways in which energy changes from one form into another, first hoarding it temporarily and then letting go of what is already there.

The second law of thermodynamics says that, left to itself, any system and its surroundings spontaneously undergo conversions to a less-organized form. Each time this happens, some energy is randomly dispersed in a form that is not readily available to do work: *entropy*. Thus, although the total amount of energy in the universe stays the same, the amount available in useful forms is dwindling. The entropy of any local region can be lowered as long as that region is resupplied with usable energy that is being lost from some other place. Through energy transfusions from the sun, the universal trend toward increased entropy can be postponed here on Earth.

Key Terms

metabolism	low-quality energy	entropy
first law of thermodynamics	second law of thermodynamics	system
high-quality energy	spontaneous	surroundings
		randomly dispersed

Objectives

1. State the first and second laws of thermodynamics.
2. Explain energy conversion by giving an example.
3. Explain what *spontaneous* means in relation to energy.
4. Trace the path of energy flow from the sun to your muscle movement.
5. Explain how, if the universe is becoming progressively disordered, a human embryo can grow into an infant and an infant can grow into an adult.

Self-Quiz Questions

True-False

If false, explain why.

___ (1) The first law of thermodynamics states that entropy is constantly increasing in the universe.

___ (2) Your body steadily gives off heat equal to that from a 100-watt light bulb.

___ (3) When you eat a potato, some of the stored chemical energy of the food is converted into mechanical energy that moves your muscles.

7-II
(pp. 99–101)

METABOLIC REACTIONS: THEIR NATURE AND DIRECTION
Energy Changes in Metabolic Reactions
Reversible Reactions
Metabolic Pathways

Summary

A *metabolic reaction* is a form of internal energy change in the cell. How efficiently energy is used in any system depends on the precise route by which one energy form is transferred or converted to another form. If the route consists of a big explosive reaction that occurs in a single step, much of the potential energy is converted to entropy rather than energy that can be used to do work. Cells generally convert the potential chemical energy stored in nutrient molecules in a *series* of steps, each of which releases tiny amounts of kinetic energy. The energy from such *exergonic reactions* is used to drive energy-requiring (*endergonic*) reactions in the cell. By relying on such metabolic pathways, a cell temporarily conserves some energy that might otherwise be lost as entropy during its continual chemical conversions. Products from one or more reactions can serve as intermediates for subsequent reactions in the pathway.

Chemical reactions are generally reversible under certain conditions. The greater the concentration of reactants, the faster the forward reaction. The greater the concentration of products, the faster the reverse reaction. An increase in concentration, temperature, or pressure may cause some "product" molecules to revert to reactant molecules. *Dynamic equilibrium* may be reached in which the reaction is proceeding as quickly in reverse as it is forward, and there is no further change in the net concentration of reactants or products. At *equilibrium*, product concentrations may be greater or lower than reactant concentrations, depending on how much energy is fed into or released from the reaction. In cells the availability of raw materials (reactants) shifts, and requirements for different products may vary from minute to minute. In a changing environment, cellular homeostasis is maintained only through constant adjustments in the metabolic pathways that sustain life. Degradative pathways break apart larger molecules and release larger molecules, which may be stored or used to build parts of the cell.

There are limits to the number of compounds that can be formed during metabolic reactions. Compounds must be produced in concentrations high enough to let a reaction run to completion but low enough to prevent their use in unnecessary side reactions.

Key Terms

exergonic reaction	rate of reaction	metabolites
endergonic reaction	equilibrium constant	enzymes
dissociates	metabolic pathways	cofactors
reversible reaction	degradative	energy carriers
dynamic equilibrium	biosynthetic	end product
net change	reactants	

Objectives

1. Explain what might happen to cells if they tried to use the most direct route for releasing the potential chemical energy stored in their nutrient molecules.
2. Explain what causes a chemical reaction to occur.
3. Distinguish between *exergonic* and *endergonic* reactions by defining and giving an example of each.
4. State what the requirements are for a system to achieve equilibrium. Then decide whether you think a living system can ever be in a true state of equilibrium.
5. List some ways by which cells can manage their metabolic pathways when the availability of raw materials shifts.

Self-Quiz Questions

Matching

Match the most appropriate letter to its number.

(1) _B_ dynamic equilibrium

(2) _E_ endergonic

(3) ___ energy carriers

(4) ___ equilibrium constant

A. Ratio of product concentration to reactant molecule concentration
B. Reaction showing a net loss in energy
C. ATP and NADP$^+$ mainly
D. Rate of forward reaction = rate of reverse reaction

(5) __β__ exergonic

(6) ___ metabolite

E. Compound traveling through a series of chemical reactions

F. Reactants show a net gain in energy

7-III
(pp. 101–104)

ENZYMES
   Enzymes Defined
   Enzyme Structure
   Enzyme Function
   Regulation of Enzyme Activity

Summary

An enzyme is a protein catalyst that speeds the rate of a chemical reaction by lowering the activation energy that must be reached before the reaction will occur. In living systems, metabolic reactions must occur within a certain range of low temperatures so that substances with shapes that depend on hydrogen bonding are not altered beyond redemption. An enzyme holds the reacting molecules in an orientation so favorable that collisions between the molecules are more likely to break certain bonds and cause atoms to become rearranged into the product molecules. In addition, an enzyme bonding to a reactant molecule can sometimes strain certain bonds so that they are forced to break apart. Enzymes accelerate the rate at which a reaction approaches equilibrium, but they do not alter the proportions of reactants and products that will be present once equilibrium is reached. Enzymes can attach only to specific molecules; thereby they can cause high rates of required reactions even though overall concentrations of the reactants are low. The induced-fit model states that the active site of an enzyme and its *substrate* become fully complementary to each other. Some enzymes require assistance from nonprotein components known as *coenzymes*.

Key Terms

enzymes	enzyme-substrate	transition state
catalyst, catalytic	complex	regulatory enzymes
substrate	induced-fit model	allosteric
controls	activation energy	feedback inhibition
active site	energy hill diagram	inhibitor

Objectives

1. Describe what must happen for reactant molecules to become product molecules.
2. Explain what is meant by *spontaneous reaction*.
3. Describe the precise role that enzymes play in speeding chemical reactions and explain why enzymes make particularly effective catalysts.
4. Name two factors that influence rates of enzyme activity.
5. Describe what happens when an enzyme is denatured.
6. Explain why parents become very concerned when small children develop very high body temperatures as a result of severe infections.
7. Suppose that you switch to a diet that contains only fats, proteins, and nucleic acids and omits carbohydrates. Suppose also that your body can make the several different kinds of carbohydrates it needs from fats and proteins. Explain how allosteric enzymes might be of use to you in making the required carbohydrates.
8. List three ways that enzymes encourage reactions to occur more quickly.

Fill-in-the-Blanks

(1) _Enzyme_ _____ are proteins that act as (2) _catalyst_ _____, which means they greatly enhance the rate at which specific reactions approach (3) _equilibrium_ _____. The substance upon which an enzyme acts is called its (4) _substrate_ _____; the substance fits into the enzyme's crevice, which is called its (5) _____.

(6) _____ and (7) _____ are two factors that influence the rates of enzyme activity. During severe viral infections, extremely high fevers can cause (8) _____ of proteins, which lead to cell death. When denaturation occurs, (9) _____ holding the protein in its coiled secondary structure break, and the protein chains unwind. When carbohydrates are in your diet, there is no need to construct them; thus the (10) _____ enzyme should be rendered inactive by the (11) _____ binding to the regulatory site, changing shape of the (12) _____ site so that no substrate molecules can bind to it to be processed into carbohydrate. When the carbohydrate concentration falls, what is bound to the enzyme lifts off, the shape of the (13) _____ site is again restored, and substrate can again bond to the enzyme and be converted into carbohydrate.

True-False

If false, explain why.

_T_ (14) Enzyme shape may change during catalysis.

_F_ (15) The active site is a groove on the reactant molecule.

___ (16) For two reactant molecules to become product molecules, the reactant molecules must first collide with a certain minimum energy.

_T_ (17) The striking of a match to start its wood burning is an example of a spontaneous reaction.

_T_ (18) The appropriate enzyme lowers the amount of energy that must be supplied to enable the reactants to be converted to products, and it also lowers the amount of energy required by the reverse reaction.

7-IV
(pp. 104–107)

COFACTORS
ATP: THE MAIN ENERGY CARRIER
    Structure and Function of ATP
    The ATP/ADP Cycle
ELECTRON TRANSPORT SYSTEMS
SUMMARY

Summary

Cofactors are nonprotein substances that either help enzymes to catalyze specific reactions or serve fleetingly as transfer agents. Hydrolysis of the high-energy phosphate bonds of ATP releases larger amounts of useful energy than does hydrolysis of other kinds of covalent bonds. Reactions that do not proceed on their own can be driven indirectly by energy-releasing reactions such as the splitting of ATP. ATP molecules are used over and over again in such reactions. ATP is sometimes called the universal energy currency of cells because it transfers usable energy to reactions concerned with energy metabolism, biosynthesis, active transport, and cellular movement.

Energy is embodied not only in phosphate bonds but also in some of the electrons associated with carrier molecules such as NADH, NADPH, $FADH_2$, and the cytochromes. An *oxidation* reaction strips from an atom or molecule one or more electrons, which are simultaneously gained by other atoms or molecules in a *reduction* reaction. Sometimes hydrogen ions or atoms are transferred as well as the electron. Because the loss and gain happen simultaneously, these two events are regarded as being coupled in an oxidation-reduction reaction, which generates usable forms of energy. Free-moving electron carriers such as NAD^+ and $NADP^+$ accept electrons and hydrogen at one reaction site in the cell and transfer them to different reaction sites concerned with ATP production or biosynthesis. Sometimes electron carriers such as the cytochromes are arranged in a series known as an electron transport chain, and oxidation-reduction reactions can occur one after another in organized sequence. These chains of membrane-bound electron carriers are the basis of transport systems that establish pH and electric gradients, which help drive ATP formation.

Key Terms

cofactors	ATP	oxidized
coenzymes	cytochromes	reduced
FAD	ATP/ADP cycle	oxidation-reduction
NAD^+	phosphorylation	reaction
$NADP^+$	electron transport	
metal ions	systems	

Objectives

1. Explain how the hydrolysis of ATP coupled with a reaction in which two simple molecules are synthesized into one larger molecule resembles the connection of an electric motor to a sewing machine.
2. Describe the process of phosphorylation and show how it is important to cells.
3. Name three energy-requiring activities of cells.
4. Describe the structure of ATP and name its five components.
5. Distinguish an *oxidation* reaction from a *reduction* reaction; show how the two must be coupled into an oxidation-reduction reaction.
6. Explain how phosphorylation reactions and oxidation-reduction reactions are related to the transfer of energy.
7. Describe an electron transport chain. Tell which types of molecules participate in such chains and how they participate.
8. State what electron transport chains accomplish for the cells that contain them.
9. Describe how NAD^+, $NADP^+$, FAD, and cytochromes take part in energy transfers. Write an equation for a generalized reaction that shows these energy-poor molecules becoming energy-rich molecules. Do this by translating each of the preceding molecules into a capital letter. Then show what the letter combines with to yield a product molecule.

Self-Quiz
Questions

Fill-in-the-Blanks

Nonprotein substances that aid enzymes in their catalytic task are called
(1) _____. (2) _____ is a coenzyme. ATP is constructed of
(3) _____, (4) _____, and three (5) _____ groups. When ATP is
hydrolyzed, a molecule of (6) _____ in the presence of an appropriate
(7) _____ is used to split ATP into (8) _____, a (9) _____ group,
and (most important) usable (10) _____, which is easily transferred to
other molecules in the cell.

An (11) _____ _____ chain is a series of (12) _____ carrier
molecules that carries out an organized sequence of (13) _____ - _____
reactions. The first molecule in line accepts an excited (14) _____ from a
donor molecule outside the chain. It gets transferred from the first molecule to
a series of (15) _____ molecules, releasing usable (16) _____ at each
step. At certain transfer points, the amount of energy given off is sufficient to
do useful work, such as attaching a (17) _____ group to ADP. The last
molecule in the chain gives up the electron to an (18) _____ acceptor
molecule. Removing the electron in this way keeps the transport chain clear
for operation. (19) _____, (20) _____, and (21) _____ can accept
two electrons and then later release those electrons as part of a step-by-step
reaction series. There are also some iron-containing protein compounds that
are collectively referred to as (22) _____ and are located largely in
chloroplasts and mitochondria.

True-False

If false, explain why.

__F__ (23) A substance that has been oxidized has gained one or more electrons.

__T__ (24) Phosphorylation is a chemical reaction in which a phosphate group is
attached to another molecule, thus increasing the potential chemical
energy of that molecule.

__T__ (25) Reduction is a chemical reaction in which one or more electrons or
hydrogen atoms is attached to a molecule, thus increasing its potential
chemical energy.

CHAPTER TEST

UNDERSTANDING AND INTERPRETING KEY CONCEPTS

____ (1) The laws of thermodynamics state that _____.
(a) energy can be transformed into matter, and because of this, we *can*
get something for nothing

(b) energy can only be destroyed during nuclear reactions, such as those that occur inside the sun

(c) if energy is gained by one region of the universe, another place in the universe also must gain energy in order to maintain the balance of nature

(d) matter tends to become increasingly more disorganized

___ (2) Essentially, the first law of thermodynamics states that _____.
(a) one form of energy cannot be converted into another
(b) entropy is increasing in the universe
(c) energy cannot be created or destroyed
(d) energy cannot be converted into matter or matter into energy

___ (3) Which is not true of enzyme behavior?
(a) Enzyme shape may change during catalysis.
(b) The active site of an enzyme orients its substrate molecules, thereby promoting interaction of their reactive parts.
(c) All enzymes have an active site where substrates are temporarily bound.
(d) An enzyme can catalyze a wide variety of different reactions.

___ (4) When NAD^+ combines with hydrogen, the NAD^+ is _____.
(a) reduced
(b) oxidized
(c) phosphorylated
(d) denatured

___ (5) A substance that gains electrons is _____.
(a) oxidized
(b) a catalyst
(c) reduced
(d) a substrate

___ (6) A pyrophosphate bond _____.
(a) absorbs a large amount of free energy when the phosphate group is attached during hydrolysis
(b) is formed when ATP is hydrolyzed to ADP and one phosphate group
(c) is usually found in each glucose molecule; that is why glucose is chosen as the starting point for glycolysis
(d) releases a large amount of usable energy when the phosphate group is split off during hydrolysis

___ (7) An allosteric enzyme _____.
(a) has an active site where substrate molecules bind and another site that binds with intermediate or end-product molecules
(b) is an important energy-carrying nucleotide
(c) carries out either oxidation reactions or reduction reactions but not both
(d) raises the activation energy of the chemical reaction it catalyzes

---

CHAPTER TEST

INTEGRATING AND APPLYING KEY CONCEPTS

A piece of dry ice left sitting on a table at room temperature vaporizes. As the dry ice vaporizes into CO_2 gas, does its entropy increase or decrease? Tell why you answered as you did.

8

ENERGY-ACQUIRING PATHWAYS

General Objectives

1. Understand the main pathways by which energy from the sun or from specific chemical reactions enters organisms and passes from organism to organism and/or back into the environment.
2. Know the steps of the light-dependent and light-independent reactions. Know the raw materials needed to start each phase and know the products made by each phase.
3. Explain how autotrophs use the intermediates as well as the products of photosynthesis in their own metabolism.

8-I
(p. 108–110)

FROM SUNLIGHT TO CELLULAR WORK: PREVIEW OF THE MAIN PATHWAYS

PHOTOSYNTHESIS
    Simplified Picture of Photosynthesis

Summary

The three major pathways (photosynthesis, glycolysis, and cellular respiration) are linked by energy flowing through them. In *photosynthesis*, sunlight energy is trapped by pigment molecules (such as chlorophyll) and is used to form ATP and NADPH, which are intermediate energy carriers. These carriers transfer some of their energy to reactions in which carbon dioxide and water from the environment are converted into food molecules. Glycolysis and aerobic respiration are two interconnected ways of releasing for use in cellular work energy stored in food molecules.

    Photosynthesis consists of two sets of chemical reactions. In the *light-dependent reactions*, energy from sunlight is absorbed by molecules such as chlorophyll and converted into potential chemical energy stored briefly in chemical bonds of ATP and NADPH. In the *light-independent reactions*, ATP and NADPH are used to assemble sugars and other organic compounds. From the light-dependent reactions to the light-independent reactions, ATP carries energy and NADPH carries electrons and hydrogen ions.

Key Terms

autotrophic organism
photosynthetic autotroph
chemosynthetic autotroph
heterotrophic organism

photosynthesis
glycolysis
respiration
light-dependent reactions

light-independent
   reactions
intermediates

Objectives

1. Study Figure 8.1 until you can remember how the reactants and products of each of the three major energy-trapping and energy-releasing pathways are interrelated. Then reproduce the diagram from memory on another piece of paper.
2. Name the principal raw materials needed to begin each of the three processes and identify the products formed at the end of each process.

Self-Quiz Questions

Fill-in-the-Blanks

(1) _Photosynthetic autotrophs_ obtain energy from sunlight. (2) _Chemosynthetic autotrophs_ oxidize inorganic substances such as ammonium ions or iron or sulfur compounds to obtain energy. Photosynthetic autotrophs include all plants, some protistans, and some (3) _bacteria_. Chemosynthetic autotrophs are limited to a few kinds of (4) _bacteria_. (5) _Heterotrophic_ organisms feed on autotrophs, each other, or organic wastes. Energy stored in organic compounds such as glucose may be released by the two interconnected pathways (6) _glycolysis_ and (7) _aerobic respiration_, which also produce carbon dioxide and water.

8-II
(pp. 110–113)

PHOTOSYNTHESIS (cont.)
   Chloroplast Structure and Function
LIGHT-DEPENDENT REACTIONS
   Light Absorption in Photosystems

Summary

Photosynthesis occurs on thylakoid membranes, which are enclosed in the chloroplasts of eukaryote cells or located near the cell membrane of prokaryotic cells. In eukaryotes, the thylakoid membrane is folded into a system of stacked disks and flattened channels, the interior spaces of which are open to each other and store hydrogen ions. Embedded in the thylakoid membranes may be various pigments that absorb wavelengths of light (colors) ranging from about 400 to 750 nanometers. Chlorophylls absorb blue and red wavelengths but transmit green; carotenoids absorb violet and blue but transmit yellow.

A photosystem is a cluster of pigment molecules. The first event of photosynthesis is the transfer of an electron from a photosystem to an acceptor molecule embedded in the thylakoid membrane. The transfer occurs when the absorption of light energy excites the electron.

Key Terms

thylakoid membrane	chlorophyll *a*	color
stroma	chlorophyll *b*	absorption spectrum
granum, grana	carotenoids	pigments
photon	absorb	T. Englemann
nanometer	transmit	photosystems

Objectives

1. Describe the internal structure of a chloroplast. Contrast the grana and stroma in terms of structure and function.
2. Describe the role chlorophyll and the accessory pigments play in the light-dependent reactions. After consulting Figure 8.3, state which colors of the visible spectrum are absorbed by (a) chlorophyll *a*, (b) chlorophyll *b*, (c) carotenoids, and (d) phycocyanin.
3. State what T. Englemann's 1882 experiment with *Cladophora* revealed.

Self-Quiz Questions

Fill-in-the-Blanks

Photosynthesis takes place on (1) _Thylakoid membrane_, which in (2) _eukaryotes_ cells are folded into a system of stacked disks and flattened channels; this system forms a single compartment separate from the (3) _stroma_, where the actual assembly of sugars and other organic compounds occurs.

Light-trapping proteins are called (4) _pigment_; one of these is (5) _chlorophyll_, which absorbs blue and red wavelengths but transmits green. (6) _carotenoid_ absorb violet and blue but transmit yellow. A cluster of 200 to 300 of these proteins is a (7) _photosystem_. When pigments absorb (8) _light_ energy, an (9) _electron_ is transferred from a photosystem to an (10) _acceptor molecule_.

= a cluster of

8-III
(pp. 111–116)

LIGHT-DEPENDENT REACTIONS (cont.)
Two Pathways of Electron Transfer
A Closer Look at ATP Formation
Summary of the Light-Dependent Reactions

Summary

Most autotrophs carry out photosynthesis of one sort or another. Light absorbed by two types of photosystems activates the transfer of electrons from one or the other photosystem to a nearby acceptor molecule, which passes them on to the nearest electron transport system. As electrons are transferred down through the chain, some of the energy released is used to form ATP from ADP and inorganic phosphate. Eventually, the electrons lose their excess energy and return to the photosystem.

During *cyclic photophosphorylation*, excited electrons flow from photosystem I through a transport system and then reenter photosystem I. Every two electrons that enter the cyclic photophosphorylation pathway can cause one ATP molecule to be formed. The production of ATP is powered by the flow of hydrogen ions down concentration and electrical gradients established by hydrogen ions accumulating in the thylakoid disks during electron transfers through the transport system. The plant can then use this ATP to do various forms of work. Primitive organisms that use only the cyclic pathway *can* synthesize organic compounds, but their synthetic pathways are inefficient; hence, these organisms never evolved into large, complex, energy-demanding beings.

Most modern autotrophs carry out *noncyclic photophosphorylation*, which includes two photosystems (photosystem II and photosystem I) and two transport chains. When the two photosystems function together, electrons do not flow in a cycle. ADP is still phosphorylated to form ATP as electrons are transported from II to I. When the electrons are reexcited in photosystem I, they are sent to the second transport system, at the end of which they help to reduce $NADP^+$ to NADPH. NADPH carries hydrogen ions and electrons (energy) to the light-independent reactions, where they help convert PGA to PGAL.

Light absorption at the start of the pathway indirectly drives the splitting of water molecules. Electrons released from water replace electrons being expelled from photosystem II. Every two new electrons from water molecules are used to make two ATP molecules via the first transport chain. Then they are reexcited in the second photosystem, and, as they pass down the second transport chain, the two electrons end up in a molecule of NADPH and are exported from the light reactions and used to drive one of the light-independent reactions. Because they are exported, the electrons do not follow a cyclic path. Some of the ATP is also used to drive one of the light-independent reactions.

Key Terms

electron transport system	P700	photolysis
photophosphorylation	noncyclic	electric gradient
cyclic	photophosphorylation	concentration gradient
photophosphorylation	photosystem II	channel proteins
photosystem I	P680	chemiosmotic theory

Objectives

1. Contrast cyclic and noncyclic photophosphorylation in terms of the substances produced, the number of photosystems involved, and the number of transport chains used.
2. Explain what the water split during photolysis contributes to both cyclic photophosphorylation and noncyclic photophosphorylation.
3. Name the two energy-carrier molecules produced during noncyclic photophosphorylation and indicate how they will be used later.
4. Explain how the chemiosmotic theory is related to thylakoid compartments and the production of ATP.

Self-Quiz
Questions

Fill-in-the-Blanks

Today, land plants rely mostly on (1) _____, which creates ATP and NADPH as energy carriers. In photosynthesis, (2) _____ _____ and (3) _____ are the raw materials that are converted into (4) _____ and energy-rich (5) _____ _____; energy from (6) _____ is trapped by (7) _____ _____ and is used to form the intermediate energy carriers (8) _____ and (9) _____.

8-IV
(pp. 116–120)

LIGHT-INDEPENDENT REACTIONS
    Carbon Dioxide Fixation and the Calvin-Benson Cycle
    Summary of the Light-Independent Reactions
    How Autotrophs Use Intermediates and Products of Photosynthesis
    C4 Plants
CHEMOSYNTHESIS
SUMMARY

Summary

During the light-independent reactions, a photosynthetic cell can use energy carriers such as ATP and NADPH to produce organic food molecules. In the first stage of these reactions, carbon dioxide from the air is combined with (that is, fixed to) ribulose biphosphate and incorporated into stable intermediate compounds (PGA). In the second stage, ATP and NADPH supply chemical energy to convert PGA to PGAL; some of this PGAL is then used to form glucose, and some is used to make more ribulose biphosphate.

Tropical grasses that are adapted to high light levels, high temperatures, and limited water have complex leaf structures and a carbon fixation system that precedes the Calvin-Benson cycle. In such plants, called *C4 plants*, one leaf cell type specializes in preliminary CO_2 uptake and rapid, active transport of the fixed carbon dioxide to another leaf cell type where the Calvin-Benson cycle operates. C4 plants can continue to photosynthesize even on hot, dry days, when sunlight is intense and when water must be conserved.

Some bacteria can harness energy released during the oxidation of such inorganic substances as ammonium ions (NH_4^+) and sulfur compounds. Such chemosynthesizers are found in the soils of terrestrial ecosystems and in the sediments of aquatic ecosystems, where their activities enhance the cycling of nitrogen and sulfur through natural communities.

Key Terms

Calvin-Benson cycle	inorganic	mesophyll cells
carbon dioxide fixation	ammonium ions, NH_4^+	bundle-sheath cells
PGAL	nitrifying bacteria	C3 plants
RuBP	photorespiration	ammonia molecules, NH_3
chemosynthesis	C4 plants	nitrite, NO_2^-

Objectives

1. Explain why the light-independent reactions are called by that name.
2. Describe the process of carbon dioxide fixation by stating which reactants are necessary to get the process going and what stable products result from this process alone.
3. Describe the Calvin-Benson reaction series in terms of its reactants and products.
4. Explain what happens to each of the products of photosynthesis.
5. Describe the mechanism by which C4 plants thrive under hot, dry conditions.
6. Distinguish chemosynthesis from photosynthesis and state which groups of organisms carry on each process.

Self-Quiz Questions

Fill-in-the-Blanks

The light-independent reactions can proceed without sunlight as long as (1) _____ and (2) _____ are available. The reactions begin when an enzyme links (3) _____ _____ to (4) _____, a five-carbon compound. The resulting six-carbon compound is highly unstable and breaks apart at once into two molecules of a three-carbon compound, (5) _____. This entire reaction sequence is called (6) _____ _____ _____ . For every (7) _____ carbon dioxide molecules fixed, twelve molecules of the three-carbon compound are produced. The twelve (8) _____ molecules now enter a reaction series that uses twelve (9) _____ and twelve (10) _____ molecules from the (11) _____-_____ _____ to convert PGA into (12) _____. Two of these molecules are rearranged and linked together to form one (13) _____ molecule, which is regarded as the principal end product of the light-independent reactions. Six more ATP molecules are used to regenerate (14) _____, the five-carbon molecule to which CO_2 was initially linked.

Chemosynthetic bacteria are (15) _____, for they can live completely on an (16) _____ diet. Unlike photosynthesizers, they harness energy released during the (17) _____ of such inorganic substances as (18) _____ ions and (19) _____ compounds.

Some (20) _____ bacteria use ammonia molecules as their energy source, stripping them of hydrogen ions and their associated electrons.

UNDERSTANDING AND INTERPRETING KEY CONCEPTS

___ (1) The electrons that are passed to NADPH during noncyclic photo-phosphorylation were obtained from _____.
- (a) water
- (b) CO_2
- (c) glucose
- (d) sunlight

___ (2) Cyclic photophosphorylation functions mainly to _____.
- (a) fix CO_2
- (b) make ATP
- (c) produce PGAL
- (d) regenerate ribulose biphosphate

___ (3) Chemosynthesis involves the oxidation of such inorganic substances as _____.
- (a) PGA
- (b) PGAL
- (c) ammonium ions
- (d) water

___ (4) The ultimate electron and hydrogen acceptor in noncyclic photophosphorylation is _____.
- (a) $NADP^+$
- (b) ADP
- (c) O_2
- (d) H_2O

___ (5) PEP-carboxylase _____.
- (a) is important in the CO_2 fixation phase of C3 plants
- (b) catalyzes the addition of CO_2 to ribulose biphosphate
- (c) is an enzyme that catalyzes the splitting of water during photolysis
- (d) catalyzes part of the carbon fixation system that precedes the Calvin-Benson cycle in the tropical grasses

___ (6) Chlorophyll is _____.
- (a) on the outer chloroplast membrane
- (b) inside the mitochondria
- (c) in the stroma
- (d) in the thylakoids

___ (7) Thylakoid disks are stacked in groups called _____.
- (a) grana
- (b) stroma
- (c) lamellae
- (d) cristae

___ (8) Plant cells produce O_2 during photosynthesis by _____.
  (a) splitting CO_2
  (b) splitting water
  (c) degradation of the stroma
  (d) breaking up sugar molecules

___ (9) Plants need _____ and _____ to carry on photosynthesis.
  (a) oxygen, water
  (b) oxygen, CO_2
  (c) CO_2, H_2O
  (d) sugar, water

CHAPTER TEST

INTEGRATING AND APPLYING KEY CONCEPTS

Suppose that humans acquired all the enzymes needed to carry out photosynthesis. Speculate about the attendant changes in human anatomy, physiology, and behavior that would be necessary for those enzymes actually to carry out photosynthetic reactions.

9

ENERGY-RELEASING PATHWAYS

General Objectives	1. Understand what kinds of molecules can serve as food molecules.
	2. Know the relationship of food molecules to glucose and thus to glycolysis.
	3. Understand the fundamental differences between glycolysis + fermentation and glycolysis + aerobic respiration. Know the factors that determine whether an organism will carry on fermentation or aerobic respiration.
	4. Know the raw materials and products of each of these processes: glycolysis, fermentation, the Krebs cycle, and electron transport phosphorylation.

9-I
(pp. 121–126)

OVERVIEW OF THE MAIN ENERGY-RELEASING PATHWAYS
    Glycolysis: First Stage of the Energy-Releasing Pathways
    Aerobic Respiration
    Anaerobic Electron Transport
    Fermentation Pathways

Summary

Energy-rich carbohydrates and other food molecules are stockpiled in autotrophic organisms. Eventually, autotrophic cells and the cells of the heterotrophs that dine on them will break down these molecules and use the stored energy to drive life processes. Although many different types of molecules are stored and eventually used, glucose breakdown and the associated energy transfers leading to ATP formation are central to the functioning of almost all prokaryotic and eukaryotic cells. Glucose is partially dismantled by the glycolytic pathway; during this process, some of its stored energy ends up in two ATP molecules. At the end of glycolysis, glucose has been converted to two molecules of pyruvic acid (pyruvate); also, substrates give up electrons and hydrogen ions to two

NAD^+ coenzymes, which yields two molecules of the electron carrier NADH. If oxygen—O_2—is not present in sufficient amounts, pyruvate enters fermentative pathways in which pyruvate is converted into *fermentation* products such as lactate or ethanol and carbon dioxide. Fermentative processes ensure that the essential carrier molecule NAD^+ is regenerated and glycolysis can continue. Some prokaryotes use another kind of pathway, *anaerobic electron transport*, in which an inorganic compound other than oxygen is the final acceptor of electrons from NADH. If sufficient oxygen and the appropriate enzymes are present, pyruvate enters the *aerobic pathways* (the Krebs cycle + electron transport phosphorylation), and thirty-six additional ATP molecules may be generated. The entire aerobic pathway can generate nineteen times as many ATP molecules as does glycolysis plus any of the fermentative processes, so cells and organisms that carry out aerobic glucose metabolism have the means to accomplish much more work than can anaerobic organisms.

Key Terms

phosphorylation	net yield	fermentation
oxidation-reduction	facultative aerobes	alcoholic fermentation
glycolysis	anaerobes	lactate fermentation
pyruvate	strict anaerobes	ADP/ATP
aerobic respiration	anaerobic electron	NAD^+/NADH
aerobe	transport	$FAD/FADH_2$

Objectives

1. Contrast the anaerobic and aerobic pathways of glucose metabolism in terms of (a) the organisms that use the pathways, (b) the number of ATP molecules generated by each, (c) the names of the components of each type of pathway, and (d) the products of each type of pathway.
2. Explain why glucose breakdown and ATP formation are so important to the functioning of almost all prokaryotic and eukaryotic cells.
3. List some places where there is very little oxygen present and where anaerobic organisms might be found.
4. Describe what happens to pyruvate in anaerobic organisms. Then explain the necessity for pyruvate to be changed to any of a variety of fermentative products.
5. State which factors determine whether the pyruvate (pyruvic acid) produced at the end of glycolysis will enter into the alcoholic fermentation pathway, the lactate fermentation pathway, or the acetyl-coenzyme A formation pathway.
6. Calculate how many net ATP molecules are produced during glycolysis + fermentation for each glucose molecule that began the trip.
7. Calculate how many net NADH molecules are produced during glycolysis + fermentation for every glucose molecule that enters the pathway.

Self-Quiz Questions

Fill-in-the-Blanks

(1) _____ organisms can synthesize and stockpile energy-rich carbohydrates and other food molecules from inorganic raw materials.

(2) _____ is partially dismantled by the glycolytic pathway, during which process some of its stored energy ends up in two (3) _____ molecules. Some of the energy of glucose is released during the breakdown reactions and is used in forming the energy carriers (4) _____ and (5) _____. These

reactions take place in the cytoplasm. At the end of glycolysis, the starting molecule has been converted into two molecules of (6) _____. If (7) _____ is not present in sufficient amounts, the end product of glycolysis enters (8) _____ pathways, in which it is converted into such products as (9) _____ or (10) _____ and (11) _____ _____.
If sufficient oxygen is present, the end product of glycolysis enters (12) _____ pathways (the (13) _____ cycle + (14) _____ _____ _____), during which process (15) _____ additional (16) _____ molecules are generated. In the (17) _____ _____, the food molecule fragments are further broken down into (18) _____ _____ bits. During the reactions, hydrogen atoms (with their (19) _____) are stripped from the fragments and are transferred to the energy carriers (20) _____ and (21) _____. The electrons are then sent down a (22) _____ _____, and this energy that they give off is used in forming (23) _____. Electrons leaving the chain combine with hydrogen ions and (24) _____ to form water. These reactions occur only in (25) _____.

9-II
(pp. 126–132)

A CLOSER LOOK AT AEROBIC RESPIRATION
  First Stage: Glycolysis
  Second Stage: The Krebs Cycle
  Third Stage: Electron Transport Phosphorylation
  Glucose Energy Yield

Summary

During glycolysis, which occurs in the cytoplasm, one molecule of glucose is broken down into two molecules of pyruvic acid. In the process, four ATP molecules are made for every two ATPs used, and two NADH molecules bear away from glycolysis the hydrogen ions and electrons recently acquired from an intermediate of glucose breakdown, PGAL. Glycolysis would quickly halt if the process ran out of NAD^+, which serves as the hydrogen and electron acceptor. When sufficient oxygen is present in eukaryotic cells, pyruvate from glycolysis is completely oxidized into carbon dioxide during two phases called *acetyl-coenzyme A* formation and the *Krebs cycle* (Figure 9.7). Enzymes catalyze each step, and the intermediate produced serves as a substrate for the next enzyme in the series. In these reactions, which occur in the matrix within the inner membrane of the mitochondrion, NAD^+ and FAD serve as temporary acceptors for hydrogen ions and electrons released during the oxidations and are reduced to NADH and $FADH_2$. NADH and $FADH_2$, in turn, feed their hydrogen ions and electrons into the process known as electron transport phosphorylation, which occurs on the surface of the inner mitochondrial membrane, where NADH and $FADH_2$ are stripped of hydrogen ions and the high-energy electrons associated with those ions. The hydrogen ions are pumped into a reservoir that includes all of the space between the outer and inner mitochondrial membranes; thus, a proton gradient is established, and the energy from this drives the phosphorylation of ADP, producing ATP. Energy released from the passage of the electrons down an electron transport chain is used to form ATP molecules.

Most NADH molecules that enter the oxidative-phosphorylation sequence yield three ATP molecules; every $FADH_2$ molecule yields two. In heart and liver cells, each incoming NADH generates three ATP molecules. In other eukaryotic cells, NADH molecules made during glycolysis must be transported into the mitochondrion in order to enter the electron transport chain at the same point as $FADH_2$, and so each NADH generates only two ATP molecules (like $FADH_2$).

Key Terms

Krebs cycle	substrate-level	GDP/GTP
electron transport	phosphorylation	electric gradient
phosphorylation	PEP	ATP synthetase system
energy-requiring	acetyl CoA	electron carriers
energy-releasing	matrix	cytochromes
PGAL	oxaloacetate	chemiosmotic theory of
DHAP	citrate	ATP formation

Objectives

1. Explain the purpose served by molecules of ATP reacting first with glucose and then with fructose-6-phosphate in the early part of glycolysis.
2. Explain why your text says that four ATP molecules are made for every two used during glycolysis. Consult Figure 9.5 in your main text.
3. Consult Figure 9.7 in your main text. State the events that happen during acetyl-coenzyme A formation and explain how the process of acetyl-CoA formation relates glycolysis to the Krebs cycle.
4. State the factors that cause pyruvic acid to enter the acetyl-CoA formation pathway.
5. State what happens to the CO_2 produced during acetyl-CoA formation and the Krebs cycle.
6. Consult Figure 9.7 in your main text and predict what will happen to the NADH produced during acetyl-CoA formation and the Krebs cycle.
7. Calculate the number of ATP molecules made during acetyl-CoA formation and during the Krebs cycle for each glucose molecule that enters glycolysis.
8. Summarize the basic ideas of the chemiosmotic theory and tell what the theory helps to explain.
9. Briefly describe the process of oxidative phosphorylation by stating what reactants are needed and what the products are. State the exchange value of NADH and $FADH_2$.

Self-Quiz Questions

Short-Answer Essay

(1) What is achieved by molecules of ATP reacting first with glucose and then with fructose-6-phosphate in the early part of glycolysis?
(2) What happens to dihydroxyacetone phosphate midway through glycolysis?
(3) What happens to the ADP produced during the first part of glycolysis?
(4) What happens to the ATP produced during the last part of glycolysis?
(5) What are the potential uses of NADH produced during glycolysis?
(6) What role do NAD^+ and FAD serve in acetyl-coenzyme A formation and the Krebs cycle?
(7) Write the reduced forms of NAD^+ and FAD.
(8) Into which process are the reduced forms of NAD^+ and FAD fed?
(9) How many ATP molecules are made during acetyl-coenzyme A formation for every glucose molecule that enters glycolysis? During the Krebs cycle? During glycolysis (net)?

(10) How many NADH molecules are made during acetyl-coenzyme A formation for every glucose molecule that enters glycolysis? During the Krebs cycle? During glycolysis?

(11) How many $FADH_2$ molecules are made during acetyl-coenzyme A formation for every glucose molecule that enters glycolysis? During the Krebs cycle?

(12) In heart and liver cells, how many ATP molecules can each NADH molecule generate during electron transport phosphorylation? How many ATP molecules can each $FADH_2$ molecule generate during electron transport phosphorylation?

(13) How many ATP molecules are generated directly during glycolysis and the Krebs cycle?

(14) How many ATP molecules are generated in heart and liver cells as a result of NADH molecules acting as energy carriers?

(15) How many ATP molecules are generated as a result of $FADH_2$ molecules acting as an energy-carrier molecule?

(16) What happens to aerobic respiration when molecular oxygen becomes scarce?

(17) How many more ATP molecules can be made as a result of glycolysis + aerobic respiration in ordinary eukaryotic cells than can be made as a result of glycolysis + fermentation?

Labeling

In problems 18–23, identify the structure or location; in problems 24–27, identify the chemical substance involved.

(18) _____ _____ _____ _____

(19) _____ _____ _____ _____

(20) _____ _____ _____ _____

(21) _____ _____ _____ _____

(22) _____

(23) _____

(24) _____

(25) _____

(26) _____

(27) _____ _____ _____

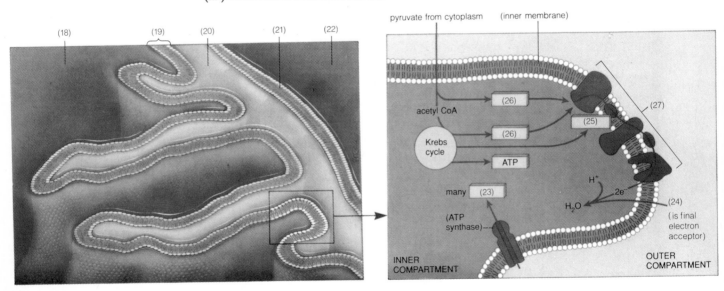

FUELS OR BUILDING BLOCKS? CONTROLS OVER CARBOHYDRATE METABOLISM

Summary

Many energy sources other than glucose can be fed into the glycolytic, CoA formation, or Krebs cycle pathways. Controls over enzymes that catalyze key steps in glycolysis and the Krebs cycle govern whether molecules are degraded as an ATP-generating energy source or converted to intermediate forms for use in biosynthesis. Carbohydrates can be converted into glucose or some other compound and thus enter glycolysis. Fats are first digested into glycerol and fatty acids, which enter glycolysis and CoA formation. Proteins are digested into a variety of amino acids; some enter glycolysis, but many enter the Krebs cycle at various points by being converted into Krebs cycle intermediates. Metabolizing a gram of fat generally yields about twice as much ATP as metabolizing either a gram of carbohydrate or a gram of protein, so fats are efficient long-term energy-storage molecules. Cells generally use simple sugars, amino acids, nucleotides, fatty acids, and glycerol to synthesize as many of the more complex molecules as they need; the remainder of the nutrients taken in are sent down respiratory pathways to make ATP and waste products or are converted into lipids or starches for longer-term energy storage.

Objectives

1. List some sources of energy (other than glucose) that can be fed into the respiratory pathways.
2. Explain what cells do with simple sugars, amino acids, fatty acids, and glycerol that exceed what the cells need for synthesizing their own assortments of more complex molecules.
3. Predict what your body would do to synthesize its needed carbohydrates and fats if you switched to a diet of 100 percent protein.

Self-Quiz Questions

True-False

If false, explain why.

____ (1) Glucose is the only carbon-containing molecule that can be fed into the glycolytic pathway.

____ (2) Fats are efficient long-term energy-storage molecules.

____ (3) Simple sugars, fatty acids, and glycerol that remain after a cell's biosynthetic needs have been met are generally sent to the cell's respiratory pathways for energy extraction.

____ (4) Carbon dioxide and water, the products of aerobic respiration, generally get into the blood and are carried to gills or lungs, kidneys, and skin, where they are expelled from the animal's body.

Identify the process or substance indicated in the accompanying illustration.

(5) _____ _____

(6) _____

(7) _____

(8) _____ _____

(9) _____ _____

(10) _____ _____

(11) _____

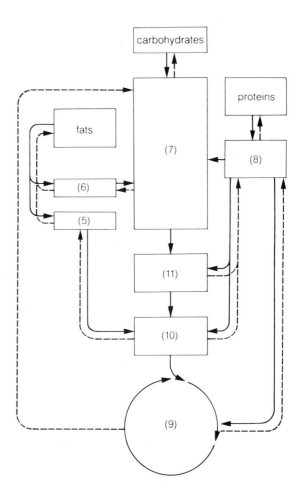

9-IV
(pp. 133–136)

PERSPECTIVE

SUMMARY

Summary

Life apparently originated when the environment was rich in molecules containing carbon, hydrogen, oxygen, and nitrogen. Presumably, the first forms of life were heterotrophic prokaryotes that utilized glycolysis to extract energy from the early carbon compounds. As increasing demands were made on these carbon resources, alternative forms of harnessing energy, such as photosynthesis and chemosynthesis, evolved among the prokaryotes. Glycolysis followed by fermentation is an inefficient method of energy extraction that produces lactic acid or alcohol as accumulating by-products. Photosynthesis produces oxygen as a waste product.

Advances in the complexity of cells caused cells to become compartmentalized, and a variety of complex processes could occur in these well-organized regions. Cellular respiration was one such complex process that used the waste products of glycolysis and photosynthesis to extract much more energy and produce carbon dioxide and water, the raw materials of photosynthesis. Thus, the cycling of carbon, hydrogen, and oxygen came full circle, and life became more self-sustaining and balanced with the *carbon cycle*. Similar cycles have come to exist for other essential elements such as nitrogen and phosphorus.

Once molecules are assembled into the cells of organisms, their organization must be sustained by outside energy derived from food, water, and air. All living forms are part of an interconnected web of energy use and materials cycling that permeates all levels of biological organization. Should energy fail to reach any of these levels, life will become increasingly disordered. Energy flows only from forms rich in potential energy to forms having fewer and fewer usable stores of it. Life is no more and no less than a wonderfully complex system of prolonging order. It can do so because hereditary instructions are passed from generation to generation; thus, even though individual organisms die, life and order continue.

Objectives

1. Outline the supposed evolutionary sequence of energy-extraction processes.
2. Scrutinize Figure 9.11 closely; then reproduce the carbon cycle from memory.

Self-Quiz Questions

True-False

If false, explain why.

____ (1) Energy is recycled along with materials.

____ (2) The first forms of life on Earth were most probably photosynthetic eukaryotes.

____ (3) Photosynthesis produces molecular oxygen as a by-product.

____ (4) Energy flows only from forms rich in potential energy to forms with fewer usable stores of energy.

CHAPTER TEST

UNDERSTANDING AND INTERPRETING KEY CONCEPTS

____ (1) Glycolysis would quickly halt if the process ran out of _____, which serves as the hydrogen and electron acceptor.
(a) $NADP^+$
(b) ADP
(c) NAD^+
(d) H_2O

___ (2) The ultimate electronic acceptor in aerobic respiration is _____.
  (a) NADH
  (b) CO_2
  (c) $1/2\ O_2$
  (d) ATP

___ (3) When glucose is used as an energy source, the largest amount of ATP is used by the _____ portion of the entire respiratory process.
  (a) glycolytic
  (b) acetyl-CoA formation
  (c) Krebs cycle
  (d) oxidative phosphorylation

___ (4) The process by which about ten percent of the energy stored in a sugar molecule is released as it is converted into two small organic-acid molecules is _____.
  (a) photolysis
  (b) glycolysis
  (c) oxidative phosphorylation
  (d) the dark reactions

___ (5) During which of the following phases of aerobic respiration is ATP produced directly?
  (a) Glucose formation
  (b) Ethyl-alcohol production
  (c) Acetyl-CoA formation
  (d) The Krebs cycle

___ (6) What is the name of the process by which reduced NADH transfers electrons along a chain of acceptors to oxygen so as to form water and in which the energy released along the way is used to generate ATP?
  (a) Glycolysis
  (b) Acetyl-CoA formation
  (c) The Krebs cycle
  (d) Oxidative phosphorylation

___ (7) Pyruvic acid can be regarded as the end product of _____.
  (a) glycolysis
  (b) acetyl-CoA formation
  (c) fermentation
  (d) the Krebs cycle

___ (8) Which of the following is *not* ordinarily capable of being reduced at any time?
  (a) NAD
  (b) FAD
  (c) Oxygen, O_2
  (d) Water

CHAPTER TEST

INTEGRATING AND APPLYING KEY CONCEPTS

How is the "oxygen debt" experienced by runners and sprinters related to aerobic and anaerobic respiration in humans?

Crossword
Number Two

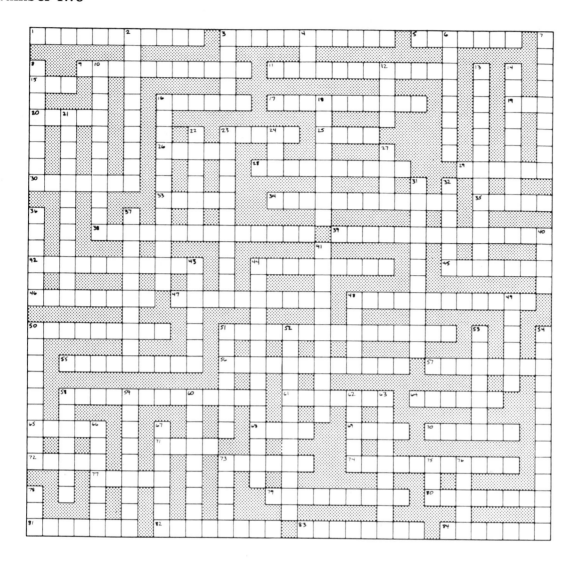

The terms in this puzzle are not necessarily biology terms.

Across

1. organelle in which photosynthetic reactions occur
3. substances composed of fat and carbohydrate that are important constituents of cell membranes
5. lipid assembled from five-carbon subunits
9. the principal light-trapping pigment
11. organelles that release ninety percent of the energy stored in glucose
15. a liquid dispersion of ions and water molecules around cell components, particularly proteins
16. a polysaccharide that forms the exoskeleton of insects and their relatives
17. the enzyme-controlled process that breaks down large molecules into their subunits by reaction with water molecules
19. one of a series of teeth on a gear
20. surname of English royal family from Henry VII (1485) through Elizabeth I (1603)

23. both _____ and flagella have the 9 + 2 arrangement of microtubules
25. _____ flow; different kinds of molecules present in a fluid move together in the same direction in response to a gradient
26. a length of time established by the moon's revolution around the earth
28. an enzyme that works as a control agent by distorting the active site
29. fluid _____ model of membrane structure
30. glucose + fructose = _____
33. gradient
34. an anaerobic pathway that generally converts pyruvic acid to either lactic acid or ethyl alcohol and carbon dioxide
35. a kingdom is divided into _____
38. harnessing energy released during the oxidation of such inorganic substances as ammonium ions and sulfur compounds
39. the study of heat and other forms of energy
42. net protein _____, a means of comparing proteins from different sources
44. able to let some substances pass through
45. a pure substance composed of only one kind of atom
46. negatively charged particle located outside the nucleus
47. consists of nitrogenous base + a phosphate group + a five-carbon sugar
48. sugar or starch is a _____
50. a gain of electrons
51. attaching a phosphate group to a substance
55. chromosomes disperse into the _____ network
56. a substance composed of simple sugars and amino acids; important in cell recognition
57. a six-carbon sugar found in fruits and elsewhere
58. an extremely fine threadlike structural element involved in cell movement
61. a _____ bond covalently links two amino acids
64. a system that prevents large changes in pH
65. *Syringa vulgaris*, widely cultivated for its clusters of fragrant purplish or white flowers
68. a town in Buckinghamshire, England, on the Thames opposite Windsor; the site of public school for boys
69. Greek goddess of victory
70. this is one of two substances that compose microfilaments
71. Fe; an important constituent of hemoglobin
72. aboriginal people of New Zealand, of Polynesian-Melanesian descent
73. not basic; caustic
74. accessory light-trapping pigment molecules
77. French composer, 1866–1925
79. the passive movement of particles of a substance from a region of high concentration to a region of low concentration
80. a membranous bag of hydrolytic enzymes
81. a complex substance composed of amino acids
82. sticking together
83. a _____ bond is formed when two atoms share a pair of electrons
84. passive transport of water across membranes in response to concentration gradients and/or pressure gradients

Down

2. all monerans are organisms of this type
3. _____ complex; a membrane system that assembles, packages, and transports substances
4. permit
6. having the capacity to combine chemically with some other atoms

7. releasing energy
8. atoms that have the same atomic number but different atomic weights
10. substances that do not mix easily with water
12. contains organ of smell and the beginning of the respiratory tract
13. P—O—P linkages; _____ bonds
14. the _____ structure of a protein is established by repeated hydrogen bonding
16. the _____ theory explains how H^+ ions are involved in oxidative phosphorylation
18. an organelle involved in protein synthesis
21. a substance that contains two simple sugars
22. a measure of the randomness, disorder, or chaos in a system
23. a steroid component of most cell membranes
24. sick
27. Bhagavad _____
31. vesicles of extracellular fluid are brought into a cell by bulk transport
32. a protein catalyst
36. that which is dissolved
37. all of the physical and chemical processes involved in the maintenance of life
40. place
41. an organism that cannot synthesize its own food
43. electrically neutral particle in the nucleus of an atom
44. the principal energy-trapping pathway used by autotrophs
48. an underground vault or chamber
49. osmotically induced internal pressure
50. endoplasmic _____, a network of membranes
51. a substance that reflects certain colors of light
52. assembled from glycerol, fatty acids, a phosphate group and usually a nitrogen-containing alcohol
53. the control center of a cell
54. forms of the same kind of molecule (e.g., glucose) that differ in the way certain functional groups are covalently bonded to a carbon atom
58. a double sugar composed of two glucose subunits
59. the loss of electrons
60. an organism that can synthesize its own food
62. of, or relating to, atoms that have gained or lost one or more electrons
63. the cells of such an organism have membrane-bound nuclei and other organelles
66. projecting shelves formed by the inner membrane of the mitochondrion
67. of, or related to, motion
73. the smallest units of an element that still have the physical and chemical properties of that element
75. a fish of the genus *Anguilla*
76. a fertile or green area with water in a desert or wasteland
78. a derivative of a nucleotide used in energy storage

10

CELL REPRODUCTION

OVERVIEW OF DIVISION MECHANISMS
    Prokaryotic Fission
    The Eukaryotic Chromosome
    Mitosis, Meiosis, and the Chromosome
    Number
WHERE MITOSIS OCCURS IN THE CELL
CYCLE

STAGES OF MITOSIS
    Prophase: Mitosis Begins
    Metaphase
    Anaphase
    Telophase
CYTOKINESIS: DIVIDING UP THE
CYTOPLASM
SUMMARY

General Objectives

1. Understand the factors that cause cells to reproduce.
2. Know the differences that occur in prokaryotic and eukaryotic cell division.
3. Understand what is meant by *cell cycle* and be able to visualize where mitosis fits into the cell cycle.
4. Be able to describe each phase of mitosis.
5. Explain how the apportioning of cytoplasm to the daughter cells follows mitosis, which is a nuclear event.

10-I
(pp. 138–142)

OVERVIEW OF DIVISION MECHANISMS
    Prokaryotic Fission
    The Eukaryotic Chromosome
    Mitosis, Meiosis, and the Chromosome Number
WHERE MITOSIS OCCURS IN THE CELL CYCLE

Summary

Whether single-celled or multicellular, each living organism follows a life cycle characteristic of its species: It undergoes sequential changes in form and function from the moment it emerges as a distinct entity until its death. The point on which all life cycles pivot is *reproduction*: making a copy of itself. Instructions for producing each new cell reside in the DNA, in which the sequence of nucleotide bases represents instructions for assembling certain RNA molecules, which in turn contain instructions for building proteins. Enzymes are proteins that give the cell a means to control the assembly of organic molecules from simple building blocks present in its surroundings. Thus, directly or indirectly, DNA governs which organic materials will occur in a cell and, in many cases, how those materials will become arranged as the cell grows and develops.

The continuity of life depends on the replication of a cell's set of DNA molecules before that cell reproduces. When the cell does divide, one DNA set ends up in each daughter cell, along with a portion of the parental cell cytoplasm that contains enzymes involved in reading the instructions contained in DNA and RNA.

Bacteria carry out *prokaryotic fission,* in which an original DNA molecule and its replica are both attached to the plasma membrane; *membrane growth* between their attachment sites separates them and allows distribution into daughter cells. In eukaryotes, hereditary instructions are divided up among a set of two or more different DNA molecules, which vary in length and shape. There can be one, two, or more of these sets in a cell. With *mitosis* the number of DNA sets in the nucleus does not change. With *meiosis* the number of DNA sets in each daughter nucleus is exactly half of what it was in the parent nucleus.

Parceling out DNA without errors to a new generation is a remarkable feat, even for prokaryotes that have a single circular DNA molecule. Unwinding, unfolding, untwisting, assembling a duplicate strand, rewinding, refolding, and separating must occur in long, easily tangled DNA molecules. More than 1 billion years ago, some fundamental changes in structure arose that improved a cell's ability to manage increased amounts of DNA. These changes occurred in the ancestors of modern eukaryotes. The physical isolation of DNA in the nucleus, chromosome packaging, and a microtubular system for moving many chromosomes at a time form the basis of reproduction in single-celled eukaryotes, as well as of physical growth in multicelled eukaryotes. Each chromosome contains a DNA molecule and different proteins grouped together as a single, intact fiber— a *nucleoprotein.* During cell construction and maintenance activities, the nucleoprotein is unfolded and uncoiled so that replication of DNA and synthesis of proteins can occur in accordance with DNA's instructions; in this extended state, the nucleoprotein is referred to as *chromatin.* Before and during nuclear division, the nucleoprotein is folded and coiled into the condensed form called a *chromosome.* The number of chromosomes differs from one species to the next, but within one multicellular organism, each cell (except the sex cells) contains the same number of chromosomes as the other. Mitosis is only a small slice of a larger series of events known as the *cell cycle.* Interphase, when two phases of protein synthesis and one phase of DNA replication occur, occupies most of the cell cycle.

Key Terms

reproduction	chromosome	homologous
life cycle	sister chromatids	chromosomes
daughter cell	centromere	haploid
prokaryotic fission	kinetochores	diploid
mitosis	somatic cells	interphase
meiosis	germ cells	gap phases, G_1 and G_2
binary fission	gametes	S phase
membrane growth	zygote	

Objectives

1. Define the term *life cycle* and use it in an example.
2. Explain how DNA is related to reproduction in cells.
3. Explain how DNA is related to protein synthesis.
4. Describe how cells will use the instructions in DNA to synthesize molecules such as carbohydrates and lipids.
5. Define *replication* and tell why it must occur before a cell reproduces.
6. Explain why it is important that daughter cells receive a significant portion of the parental cytoplasm during reproduction.
7. List in order the events that occur during prokaryotic fission.

8. List two of the evolutionary changes that occurred in cellular strucure that allowed increased amounts of hereditary material to be managed more efficiently.
9. Scrutinize Figure 10.4 (main text); then reproduce the entire figure of the cell cycle from memory.
10. List the events of interphase.

Self-Quiz Questions

Fill-in-the-Blanks

Each living organism undergoes sequential changes in form and function characteristic of its species, called a (1) __life__ __cycle__. Making a copy of oneself is called (2) __reproduction,__ and the process begins at the molecular level. Instructions for producing each new cell reside in (3) __DNA__. (4) _____, which are specific DNA regions that tell how to build different kinds of (5) _____ _____ (hence different kinds of proteins), are activated from the whole collection. The ones activated function in particular cells at particular times. (6) _____ instructs only how to build proteins directly; some of the proteins produced in this way are the cellular catalysts called (7) _____. These proteins may assist reactions that lead to other, nonprotein molecules needed by the cell—for instance, (8) _____ and (9) _____. The continuity of life depends on the (10) _____ of a cell's set of DNA molecules before that cell reproduces so that the cell acquires two complete sets.

Two evolutionary changes in cellular construction that promoted more efficient management of hereditary material were the packaging of (11) _____ and the development of a system of (12) _____ for moving many (13) _____ at a time. Prior to nuclear division, each chromosome is (14) _____. The original chromosome and the copy remain attached at the (15) _____. As long as they are attached, they are called (16) _____ _____ .

True-False

If false, explain why.

____ (17) Prokaryotes reproduce by prokaryotic fission.

____ (18) Asexual processes such as prokaryotic fission do not promote diversification in prokaryotes.

____ (19) Within a single organism, each cell except the sex cells contains the same number of chromosomes as the others.

____ (20) Immediately prior to cell division, the hereditary material becomes extended and less condensed.

____ (21) Each prokaryote generally has a single, circular chromosome.

____ (22) Replication is the process that builds cellular proteins.

Labeling

Identify the stage in the cell cycle indicated by each number.

(23) _____

(24) _____

(25) _____

(26) _____

(27) _____

(28) _____

(29) _____

(30) _____

(31) _____

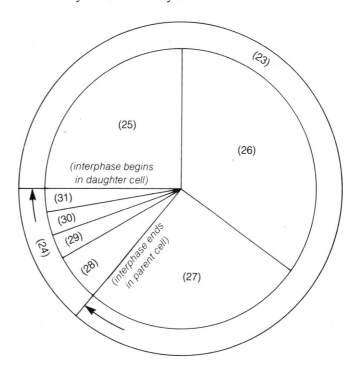

Matching

Link each time span identified below with the most appropriate number in the preceding labeling section.

____ (A) Period after replication of DNA during which the cell prepares for division by further growth and protein synthesis

____ (B) Period of asexual cell division

____ (C) DNA replication occurs now

____ (D) Period of cell growth before DNA replication

____ (E) Period when chromosomes are not visible

10-II
(pp. 143–149)

STAGES OF MITOSIS
    Prophase: Mitosis Begins
    Metaphase
    Anaphase
    Telophase
CYTOKINESIS: DIVIDING UP THE CYTOPLASM
SUMMARY

Summary

DNA replication precedes mitosis and is followed at some point by condensing of the duplicated material—the sister chromatids—into chromosomes. The stages of mitosis (prophase, metaphase, anaphase, and telophase) do not have distinct boundaries; nevertheless, the chromosomes ultimately are guided into separate daughter cells in two equivalent parcels. The microtubules forming the mitotic spindle are responsible for separating sister chromatids from each other. Once the two parcels of separated chromatids arrive at opposite poles, newly forming nuclear membranes surround each parcel, and two nuclei eventually form. The chromosome content of each daughter nucleus is identical with that of the original nucleus. As soon as nuclear division (mitosis) is complete, cytokinesis (apportioning of cytoplasm to each of the daughter nuclei) generally occurs. Cytokinesis is the basis for reproduction in single-celled eukaryotes and is the basis for physical growth in multicellular eukaryotes.

Key Terms

mitosis	anaphase	spindle equator
prophase	telophase	microfilaments
centriole	cytokinesis	cell plate formation
metaphase	cleavage furrow	vesicles
spindle apparatus		

Objectives

1. Describe the process of mitosis and what it accomplishes.
2. Describe the appearance of hereditary material (a) immediately before and during nuclear division and (b) when the cell is carrying out construction and maintenance activities.
3. List the events of each phase of mitosis.
4. Describe microtubule behavior during metaphase and anaphase as chromatids are lined up and then pulled apart.
5. Contrast the way cytokinesis occurs in animals with the way it occurs in plants.

Self-Quiz Questions

True-False

If false, explain why.

___ (1) In plant cells, a cleavage furrow forms around each cell's midsection.

___ (2) Mitosis *must* be followed by cytokinesis so that the cytoplasm is apportioned equally between two daughter nuclei.

___ (3) During early metaphase, two sets of microtubules become attached to each chromosome at the centromere; one set extends from one pole to a chromatid, and one set extends from the opposite pole to the other chromatid.

___ (4) The centrioles divide during anaphase.

___ (5) During anaphase, the microtubules that connect the centromere to its pole shorten.

Matching

Choose the one most appropriate answer for each.

(6) ___ anaphase   A. First phase of protein synthesis

(7) ___ centromere   B. Cytoplasm allotted to and divided between two nuclei

(8) ___ chromatid
(9) ___ cytokinesis
(10) ___ G₁ phase
(11) ___ G₂ phase
(12) ___ metaphase
(13) ___ prophase
(14) ___ S phase
(15) ___ telophase

C. Chromosomes condense; mitotic spindle starts to form
D. Cytokinesis occurs now
E. Final phase of mitosis
F. Connects two sister chromatids
G. Second phase of protein building
H. Elongate half of a chromosome
I. Sister chromatids separate and move to opposite spindle poles now
J. Chromosomes lined up along equator of cell

CHAPTER TEST

UNDERSTANDING AND INTERPRETING KEY CONCEPTS

___ (1) The replication of DNA occurs_____.
   (a) between the growth phases of interphase
   (b) immediately before prophase of mitosis
   (c) during prophase of mitosis
   (d) during prophase of meiosis

___ (2) Cytokinesis _____.
   (a) in animal cells begins with various deposits of material associated with groups of microtubules at each pole of the nucleus
   (b) in animal cells occurs by the plasma membrane being pulled inward by a ring of microtubules that has become attached to the cell plate
   (c) usually accompanies nuclear division
   (d) in plant cells begins with the deposition of a very rigid lipid bilayer, which is the major constituent of the cell wall

___ (3) In eukaryotic cells, which of the following can occur during mitosis?
   (a) The duplication of centrioles
   (b) The replication of DNA
   (c) Synapsis and crossing over
   (d) The nuclear envelope and nucleolus disappear

___ (4) A gene is _____.
   (a) a sequence of nucleotides that instructs the cell how to make a particular protein
   (b) a piece of RNA that indicates how to synthesize lipids of carbohydrates
   (c) a special kind of enzyme that catalyzes the assembly of organic molecules into new cellular components
   (d) the RNA molecule connected with many proteins that forms a worm-like structure called a chromosome

___ (5) If a parent cell has sixteen chromosomes and undergoes meiosis, the resulting cells will have _____ chromosomes.
   (a) sixty-four
   (b) thirty-two
   (c) sixteen
   (d) eight
   (e) four

_C_ (6) Which of the following processes is absolutely necessary for sexual reproduction to occur in a life cycle, but is not necessarily required for organisms that reproduce asexually?
  (a) Prokaryotic fission
  (b) Mitosis
  (c) Meiosis
  (d) Cytokinesis
  (e) Karyokinesis

_e_ (7) Chromosomes are duplicated during the _____ phase.
  (a) M
  (b) D
  (c) G_1
  (d) G_2
  (e) S

CHAPTER TEST

INTEGRATING AND APPLYING KEY CONCEPTS

Runaway cell division is characteristic of cancer. Imagine the various points of the mitotic process that might be sabotaged in cancerous cells in order to halt their multiplication. Then try to imagine how one might discriminate between cancerous and normal cells in order to guide the methods of sabotage most effective in combating cancer.

11

A CLOSER LOOK AT MEIOSIS

General Objectives

1. Contrast asexual and sexual types of reproduction that occur on the cellular and multicellular organism levels.
2. Understand the effect that meiosis has on chromosome number.
3. Describe the events that occur in each meiotic phase.
4. Compare mitosis and meiosis; cite similarities and differences.
5. Show where meiosis generally occurs in plant life cycles and contrast this with where it generally occurs in animal life cycles.

11-I
(pp. 150–156)

ON ASEXUAL AND SEXUAL REPRODUCTION
OVERVIEW OF MEIOSIS
    Think "Homologues"
    Overview of the Two Divisions
STAGES OF MEIOSIS
    Prophase I Activities
    Separating the Homologues
    Separating the Sister Chromatids
    More Gene Shufflings at Fertilization

Summary

Prokaryotic fission and eukaryotic mitosis are examples of *asexual reproduction* in which offspring cells are virtually identical to each other and to the parent cell. Most forms of *sexual reproduction* distribute hereditary instructions into offspring by means of meiosis, gamete formation, and fertilization.

    In organisms that reproduce sexually, the diploid parent organisms must, at some point, produce haploid reproductive cells (gametes), which find each other and fuse (an event known as *fertilization*), thus forming a new diploid cell, the *zygote. Mitosis* occurs in diploid as well as haploid organisms and maintains whatever number of chromosomes the parent cell had (either 2N or 1N), but *meiosis* reduces the parental diploid number (2N) to the haploid state (1N). If there were no such reduction, fertilization would produce a zygote with twice the amount of hereditary information required to build a new individual.

Complete meiosis is composed of two sequences of events. The first sequence halves the number of chromosomes, most of which have undergone some changes in their assortments of genes through processes known as crossing over and genetic recombination (to be discussed later). The second sequence then separates the chromatids by means of a process that resembles mitosis mechanically—although four haploid daughter cells form, rather than the two diploid cells that result from mitosis.

Eukaryotes rely on both sexual and asexual reproductive processes at various stages of the life cycle. The sexual process requires both gamete formation and fertilization, but the timing of these separate events varies considerably during the life cycles of different species. Most animals form haploid gametes; soon thereafter, one male and one female gamete generally fuse during fertilization.

Key Terms

asexual reproduction	testis, testes	genetic recombination
sexual reproduction	sister chromatids	zygote
meiosis	prophase I	chiasma, chiasmata
genes	synapsis	metaphase I
gamete	crossing over	maternal
homologous	nonsister chromatids	paternal
chromosomes	alleles	anaphase I
anther	interkinesis	metaphase II
ovary, ovaries	meiosis I	fertilization

Objectives

1. Contrast the life cycle of a haploid prokaryote with the life cycle of a diploid animal or plant.
2. List the differences between prophase I of meiosis and prophase of mitosis. State how the events of meiotic prophase I increase diversity within a species.
3. List the events that occur in the eight stages of meiosis.
4. Scrutinize Figure 11.4 (main text) and note the stage in which crossing over and genetic recombination occur.
5. Contrast the overall achievements of mitosis with those of meiosis.

Self-Quiz Questions

Matching

Assume for the following linkages that the cell starts with two pairs of already-duplicated chromosomes and that each chromosome consists of two chromatids. Choose the one most appropriate answer for each term.

(1) __B__ anaphase I

(2) __C__ anaphase II

(3) __C__ metaphase I

(4) __H__ metaphase II

A. Meiosis I is completed by this stage.
B. One member of each of the homologous pairs is separated from its mate; both are guided to opposite poles.
C. Two pairs of chromosomes are lined up at the spindle apparatus equator.

(5) _D_ prophase I
(6) _A_ prophase II
(7) _G_ telophase I
(8) _E_ telophase II

D. Chromosomes become clearly visible; nuclear region has two pairs of already-duplicated chromosomes.
E. Each centromere splits; sister chromatids are separated and moved to opposite poles.
F. Each daughter cell ends up with one chromosome of each type and may function as a gamete.
G. Spindle apparatus reforms; each cell has two chromosomes, each of which consists of two chromatids.
H. Two chromosomes are lined up at spindle apparatus equator.

True-False

If false, explain why

_T_ (9) The first diploid cell formed after fertilization is called a *gamete*.

_T_ (10) Most eukaryotic cells are diploid.

_T_ (11) Prokaryotic fission is a form of sexual reproduction.

_F_ (12) Haploid organisms have only a single copy of each of the genes needed for cell construction, growth, and maintenance.

11-II
(pp. 156–159)

MEIOSIS AND THE LIFE CYCLES
    Animal Life Cycles
    Plant Life Cycles
SUMMARY
    Key Features of Sexual Reproduction
    Stages of Meiosis
    Meiosis Compared With Mitosis

Summary

Gametogenesis occurs only in special sex cells in male and female reproductive organs. Spermatogenesis produces four tiny, equal-sized haploid spermatids; oogenesis produces one larger functional egg and three tiny nonfunctional polar bodies.

Plants differ from animals in certain features of their life cycles. The body of the more highly evolved plants is diploid, with multicellular organs in which meiosis occurs. But in plants, meiosis does not result directly in the formation of gametes. Instead, haploid spores are produced, each of which, if it encounters suitable germinating conditions, grows into a multicellular body composed exclusively of haploid cells. On one or more of these multicellular gametophytes, multicellular sex organs form, in which eggs or sperm are produced by *mitosis*. The gametes are haploid, like the gametophyte. When two unlike gametes fuse, a diploid zygote is formed that will grow into the diploid spore-producing sporophyte generation. Plant life cycles, then, contain alternating prolonged haploid and diploid stages.

Key Terms

gametogenesis	spermatogonium	spermatids
zygote	primary spermatocyte	sperm
spermatogenesis	secondary spermatocyte	oogenesis

oogonium   egg   meiospores
primary oocyte   ovum   sporophyte
secondary oocyte   polar bodies   gametophyte

Objectives

1. Describe and distinguish between the two forms of gametogenesis. Explain why daughter cells of unequal size are produced during oogenesis.
2. Contrast the generalized life cycle for plants with that for animals. Make sure that terms such as *sporophyte* and *gametophyte* are used and defined in your explanation.

Self-Quiz Questions

Sequence

Arrange the following entities in correct order of development, putting a 1 by the stage that appears first and a 5 by the stage that finishes the process of spermatogenesis.

_(2)_ (1) primary spermatocyte

_(5)_ (2) sperm

_(4)_ (3) spermatid

_(1)_ (4) spermatogonium

_(3)_ (5) secondary spermatocyte

True-False

If false, explain why.

_T_ (6) The stage that ends meiosis II in male animals is the spermatid stage.

_T_ (7) The spore in a plant life cycle grows into the sporophyte generation.

_T_ (8) The gametophyte generation of more highly evolved plants produces haploid gametes as a result of mitosis.

_T_ (9) All cells in you are diploid except for any gametes you may produce.

_F_ ___ (10) In plants the haploid sporophyte generation alternates with the diploid gametophyte generation.

should be the diploid zygote

CHAPTER TEST

UNDERSTANDING AND INTERPRETING KEY CONCEPTS

___ (1) Which of the following does *not* occur in prophase I of meiosis?
(a) Interkinesis
(b) Tetrad formation
(c) Synapsis
(d) Crossing over

___ (2) Crossing over is one of the most important events in meiosis because _____.

(a) it produces new arrays of alleles on chromosomes

(b) homologous chromosomes must be separated into different daughter cells

(c) the number of chromosomes allotted to each daughter cell must be halved

(d) homologous chromatids must be separated into different daughter cells

___ (3) Crossing over _____.
(a) generally results in synapsis and binary fission
(b) is never accompanied by gene transcription events
(c) involves breakages and exchanges being made between sister chromatids
(d) alters the composition of chromosomes and results in new combinations of alleles being channeled into the daughter cells

___ (4) Chiasmata provide evidence of _____.
(a) meiosis
(b) crossing over
(c) chromosomal aberration
(d) fertilization
(e) spindle fiber formation

___ (5) The mature ovum is produced by maturation of the _____.
(a) oogonium
(b) primary oocyte
(c) secondary polar body
(d) polar body I
(e) none of the above

___ (6) A pine tree is called a sporophyte because it _____.
(a) develops from a germinated spore
(b) produces spores by meiosis
(c) is haploid
(d) undergoes fertilization
(e) reproduces by both sexual and asexual means

___ (7) Fertilization of plant gametes produces a _____.
(a) zygote
(b) gametophyte
(c) spore
(d) meiospore
(e) multicellular haploid plant

CHAPTER TEST

INTEGRATING AND APPLYING KEY CONCEPTS

What do you think is the environment to which humans are best adapted? Did you base your answer on biological features or cultural features?

12

OBSERVABLE PATTERNS OF INHERITANCE

General Objectives

1. Know Mendel's principles of dominance, segregation, and independent assortment.
2. Understand how to solve genetics problems that involve monohybrid and dihybrid crosses, as well as sex-linked inheritance.
3. Understand the variations that can occur in observable patterns of inheritance.

12-I
(pp. 162–165)

MENDEL'S INSIGHTS INTO THE PATTERNS OF INHERITANCE
    Mendel's Experimental Approach
    Some Terms Currently Used in Genetics
    The Concept of Segregation

Summary

In 1865 Gregor Mendel presented the results of his now-famous experiments establishing the physical and mathematical rules that govern inheritance. Through his combined talents in plant breeding and mathematics, he perceived patterns in the expression of traits from one generation to the next. Mendel examined seven different traits (flower color, for instance), each of which had two different strains (white-flowered and purple-flowered), in sexually reproducing garden pea plants. Not only can garden pea flowers fertilize themselves, but the gametes from one individual can fertilize the gametes of another individual (in a process known as cross-fertilization). Mendel obtained seeds from strains in which all self-fertilized offspring displayed the same form of a trait, generation after generation; these are known as true-breeding strains. When two true-breeding organisms displaying different forms of a trait are crossed, the offspring are called *hybrids* because the two genetic units inherited, one from each parent, are unlike each other. Because Mendel first concentrated on studying a *single* trait, his initial experiments were called *monohybrid* crosses. The results of monohybrid cross experiments demonstrated that, in sexually reproducing organisms, one member from each pair is separated (segregated) from the other, and each ends up in a different gamete. This statement is the Mendelian *principle of segregation*.

Biochemical experiments and improvements in microscopy have advanced our understanding of the segregation process since Mendel's work in the mid-1800s, and the terms used to describe genetic events have changed accordingly. *Genes* are distinct units of heredity that dictate the traits of a new individual. There is one gene for each trait in a gamete (which is haploid). Following the fusion of the gametes (egg and sperm), the new individual has two genes for each trait; that is, the individual is diploid.

Alternative forms of genes for a given trait are known as *alleles*. In a diploid individual, two alleles for a given trait may be identical (in which case the individual is said to be homozygous for that trait) or two alleles may be different (in which case the individual is heterozygous). In heterozygous individuals, expression of one allele may mask expression of the other; the masking allele is *dominant* and the masked allele is *recessive*. Even so, both alleles retain their physical identity, and all the pairs of alleles constitute the individual's *genotype* (genetic makeup) for a particular assemblage of traits. The term *phenotype* refers to the collection of structural and behavioral features that are expressed, not masked.

Key Terms

trait	alleles	homologous
the "blending hypothesis"	homozygous	chromosomes
Mendel	heterozygous	hybrids
true-breeding strains	homozygous dominant	homozygote
stamen	dominant	heterozygote
carpel	recessive	monohybrid cross
cross-fertilization	dominant allele	homozygous genes
genes	recessive allele	dihybrid cross
locus	genotype	Mendelian principle of
gene pair	phenotype	segregation

Objectives

1. Explain why the blending hypothesis is not regarded as an acceptable explanation of genetic behavior today.
2. Distinguish self-fertilizing organisms from cross-fertilizing organisms. Explain the relation of true-breeding strains to self-fertilization.
3. Describe Mendel's monohybrid cross experiments with different true-breeding strains of garden peas (Figure 12.4 in main text). Give the numbers and ratios he obtained in his F_1 and F_2 generations when he crossed a round-seeded parent with a wrinkled-seeded parent.
4. Distinguish between the terms *dominant* and *recessive* and state what happens to the recessive gene in a hybrid.
5. Define and distinguish between *genes* and *alleles*.
6. Contrast the number of genes in a gamete with the number of genes in a zygote.
7. Define and distinguish between *homozygous* and *heterozygous*.
8. State the Mendelian principle of segregation and explain what would happen if segregation did not occur.

Self-Quiz Questions

Fill-in-the-Blanks   Mendel perceived patterns in the expression of (1) _____ from one generation to the next. He examined seven different traits, each with two different (2) _____ in sexually reproducing garden (3) _____ plants, which can carry out both self-fertilization and (4) _____-_____. Self-fertilized plants that all display the same form of a trait generation after generation are (5) _____-_____ strains. When two such strains that display different forms of a trait are crossed, the offspring are called

their offspring are (7) _____; that is, they again show the trait that seemed to have disappeared in their parents' generation. Through such studies, Mendel demonstrated that the widely believed (8) _____ theory was untrue. Mendel's initial experiments were called (9) _____ crosses because he concentrated his examinations on a single trait. With such experiments, he was able to demonstrate that, in sexually reproducing organisms, two units of heredity control each trait. During gamete formation, the two units of each pair are separated from each other and end up in different gametes; the statement of this fact is the Mendelian principle of (10) _____.

12-II
(pp. 166–169)

MENDEL'S INSIGHTS INTO THE PATTERNS OF INHERITANCE (cont.)
    Probability: Predicting the Outcome of Crosses
    Testcrosses
    The Concept of Independent Assortment

Summary   To determine whether an individual that shows the dominant form of a trait is heterozygous or homozygous, a *testcross* is performed: the individual is mated with an individual that possesses the homozygous *recessive* genotype. If the unknown individual is homozygous, all offspring will show the dominant trait; if it is heterozygous, half the offspring will show the recessive trait. The *Punnett-square method* is the easiest way to predict the probable ratio of traits in offspring of monohybrid crosses.

    Through experiments with dihybrid crosses, Mendel began to perceive how sexual reproduction might foster genetic diversity. The Mendelian principle of independent assortment states that, when gametes are formed, distribution of alleles for a given trait that have been segregated into one gamete or another is independent of, and does not interfere with, distribution of alleles for other traits that have been segregated.

Key Terms

P, parental generation

F₁, first generation

F₂, second generation

Punnett-square method

probability

testcross

dihybrid cross

Mendel's principle of

independent

assortment

Objectives

1. Explain why an individual's phenotype does not always express its genotype.
2. State the rules for symbolizing the genotype of a particular pair of alleles in which one allele is dominant and the other is recessive. Then use the stated rules to write the genotype (the pair of letter symbols) for each of the following traits:
   a. black fur (dominant) vs. white fur (recessive) in rabbits
   b. freckled (dominant) vs. unfreckled (recessive) in humans
   c. black (dominant) vs. white (recessive) feathers in Andalusian chickens; the heterozygote appears bluish-gray.
3. Distinguish the P, F_1, and F_2 generations from each other.
4. Explain why, as a result of monohybrid crosses, Mendel always obtained an approximate 3:1 phenotypic ratio of dominant to recessive traits in the F_2 generation.
5. Define *probability* and explain what it has to do with genetics.
6. In one of Mendel's studies, the gene for green pod color was found to be dominant over its allele for yellow pods. Cross a homozygous green-podded plant with a yellow-podded plant. Show by a diagram (with the letter symbols) both the genotype and the phenotype for the P, F_1, and F_2 generations. Use the Punnett-square method.
7. If Mendel had crossed one of the F_1 hybrids of the cross in Objective 6 with a yellow-podded plant, what would have resulted? State the general name for such a cross and state the question answered by such a cross. Show in a diagram the genotype and phenotype of one generation.
8. Define *dihybrid cross* and distinguish it from *monohybrid cross*.
9. Explain why Mendel obtained a 9:3:3:1 phenotypic ratio of characteristics in the F_2 generation as a result of a dihybrid cross.
10. State the principle of independent assortment as formulated by Mendel.
11. Solve the following problems—all of which deal with two different traits on different chromosomes—by using the Punnett-square method or by multiplying the separate probabilities.
    a. The abilities to bark and to erect their ears are dominant traits in some dogs; keeping silent while trailing prey and droopy ears are recessive traits. If a barking dog with erect ears (heterozygous for both traits) is mated with a droopy-eared dog that keeps silent while trailing, what kinds of puppies can result?
    b. A spinach plant with straight leaves and white flowers was crossed with another spinach plant with curly leaves and yellow flowers. All of the F_1 generation have curly leaves and white flowers. If any two individuals of the F_1 are crossed, what phenotypic ratio will result? Show all genotypes.

Self-Quiz Questions

True-False

If false, explain why.

___ (1) There are never more than two alleles for a particular trait in any given population of organisms.

___ (2) The phenotypic ratio of a testcross that involves monohybrids is 1:1 if the dominant-appearing parent is heterozygous.

___ (3) Albinos cannot form the pigments that normally produce skin, hair, and eye color, so albinos have white hair and pink eyes and skin (because the blood shows through). To be an albino, one must be homozygous recessive for the pair of genes that codes for the key enzyme in pigment production. Suppose that a woman of normal pigmentation with an albino mother marries an albino man. State the possible kinds of pigmentation possible for this couple's children and specify the ratio of each kind of children the couple is likely to have. Show the genotype(s) and state the phenotype(s).

___ (4) In heterozygous individuals, _____.
- (a) the expression of one allele may mask the expression of the other for any given trait
- (b) both alleles for a given trait retain their identity throughout the individual's life cycle
- (c) if one allele masks the expression of the other, the former allele is dominant and the latter is recessive
- (d) All of the above are correct.

___ (5) When independent assortment of alleles does occur, _____.
- (a) the alleles that code for the differing traits of an individual are parceled out independently of one another into separate gametes
- (b) the process of meiosis retains the diploid number of chromosomes
- (c) the alleles that code for a single trait are parceled out into the same gamete
- (d) Both (a) and (c) are correct.

___ (6) A true-breeding red-flowered, tall plant is crossed with a true-breeding dwarfed, white-flowered plant. When two of the resulting F_1 plants are crossed, the phenotypic ratio in the F_2 offspring is approximately a _____ ratio. Red flowers and tall plants are dominant.
- (a) 1:1:1:1
- (b) 1:2:1
- (c) 1:3:1
- (d) 9:3:3:1

___ (7) In genetics, the symbol for the dominant allele is usually the first letter of the trait, capitalized. The symbol for the recessive allele is _____.
- (a) the first letter of the recessive trait, capitalized
- (b) the first letter of the recessive trait, in lowercase
- (c) the first letter of the dominant trait, in lowercase
- (d) none of the above

12-III
(pp. 170–172)

VARIATIONS ON MENDEL'S THEMES
Dominance Relations
Interactions Between Different Gene Pairs
Multiple Effects of Single Genes

Summary

Since Mendel's time, studies have shown that interactions between genes can lead to many phenotypic possibilities in addition to the clear-cut fully dominant or fully recessive ("all-or-nothing") examples investigated by Mendel. Cases of *incomplete dominance* and *codominance* tell us that there can be a range of dominance when only two alleles are under consideration. Moreover, for one trait, dominance can also be expressed in a variety of forms if the trait is governed by a *multiple allele system*. Fruit fly eye color and the blood types that result from the ABO blood group are examples of multiple allele traits. For the sake of simplicity, most examples of heritable traits mentioned in texts are controlled by a single gene locus, but there are many traits (human skin color, for instance) that are governed by several active alleles at different loci; such traits are examples of *polygenic inheritance*.

Epistasis is a situation in which one gene pair masks the effect of one or more other pairs; calico cats, in which the spotting allele masks either black or yellow, produces white patches in whatever region of the cat's coat the spotting allele is behaving epistatically to the alleles for coat color. Single genes also may have major or minor influences on the expression of more than one trait—a situation known as *pleiotropism*. Sickle-cell anemia and mutations in genes that code for proteins necessary in a variety of structures and/or processes are examples of pleiotropy.

Key Terms

incomplete dominance	cleft lip	albino
codominance	variable expressivity	pleiotropy
multiple allele system	polydactyly	sickle-cell anemia
ABO blood group	epistasis	hemoglobin A, HbA
incomplete penetrance	melanin	

Objectives

1. Explain why inheritance of many traits is no longer viewed as being under the control of purely dominant or purely recessive agents.
2. List and explain several examples of inheritance in which expression of a given trait occurs as a spectrum of intermediate aspects of the trait, rather than just as either of the two extremes (dominant or recessive).
3. State whether each of the examples you gave in Objective 2 is controlled by one pair of genes or by more than one pair.
4. Define *epistasis* and cite an example.
5. Explain the meaning of *pleiotropy* and discuss how it is related to sickle-cell anemia. Name some of the phenotypic characteristics that are part of the sickle-cell syndrome.

True-False

If false, explain why.

___ (1) If a true-breeding, red-flowered snapdragon is crossed with a white-flowered snapdragon, all of the F_1 generation will be red-flowered.

___ (2) If two of the F_1 red-flowered snapdragons are crossed, three-fourths of their offspring (the F_2) will be red-flowered, and one-fourth will be white-flowered.

___ (3) Hemophilia is an example of a disorder caused by incomplete dominance.

___ (4) The expression of flower color in snapdragons is an example of codominance.

___ (5) A child of blood type AB has a mother with type A blood. The child's father could not have type O.

___ (6) Human skin color is an example of incomplete dominance that demonstrates the validity of the "blending" hypothesis.

12-IV
(pp. 173–175)

VARIATIONS ON MENDEL'S THEMES (cont.)
  Environmental Effects on Phenotype
  Continuous and Discontinuous Variation
SUMMARY

Summary

Throughout any individual's life cycle, genes must interact with the environment in the expression of phenotype. Genes provide the chemical messages for growth and development, but neither growth nor development can proceed without environmental contributions to the living form. Gradations of dominance exist between two alleles for a given gene locus, so the expression of one or both may be fully dominant, or one may be completely dominant over the other. A hierarchy of dominance can exist among all the alternative forms that exist for any given type of gene. How a gene behaves in space or time may be influenced by other genes, with all of their positive and negative actions producing some effect on the phenotype. The activity of a single gene may have major or minor effects on more than one trait (pleiotropy). The environment profoundly influences the expression of all genes and their contributions to phenotype.

Key Terms

discontinuous variation   continuous variation   quantitative inheritance

Objectives

1. Cite several examples of environmental alteration of an organism's basic phenotype.
2. Contrast discontinuous variation and continuous variation. Discuss the relationship between quantitative inheritance and polygenes.
3. Explain why frequency distribution diagrams are used to portray continuous variation in populations.

Self-Quiz Questions

True-False

If false, explain why.

___ (1) A person's becoming thirty pounds overweight is one way that the external environment can influence the phenotypic expression of human genotype.

___ (2) The fact that it is possible to change the extent to which genotypic potential is expressed in the phenotype demonstrates that acquired phenotypic characteristics can be inherited.

CHAPTER TEST

UNDERSTANDING AND INTERPRETING KEY CONCEPTS

___ (1) Mendel's principle of independent assortment states that _____.
  (a) one allele is always dominant to another
  (b) hereditary units from the male and female parents are blended in the offspring
  (c) the two hereditary units that influence a certain trait separate during gamete formation
  (d) each hereditary unit is inherited separately from other hereditary units

___ (2) One of two or more alternative forms of a gene for a single trait is a(n) _____.
  (a) chiasma
  (b) allele
  (c) autosome
  (d) locus

___ (3) In the F_2 generation of a monohybrid cross involving complete dominance, the expected *phenotypic* ratio is _____.
  (a) 3:1
  (b) 1:1:1:1
  (c) 1:2:1
  (d) 1:1

___ (4) In the F_2 generation of a cross between a red snapdragon (homozygous) and a white snapdragon, the expected phenotypic ratio of the offspring is _____.
  (a) three-fourths red, one-fourth white
  (b) 100 percent red
  (c) one-fourth red, one-half pink, one-fourth white
  (d) 100 percent pink

___ (5) The results of a testcross reveal that all of the offspring resemble the parent being tested. That parent necessarily is _____.
  (a) heterozygous
  (b) polygenic
  (c) homozygous
  (d) recessive

___ (6) When gametes from one individual undergo sexual fusion with gametes from another, this is called _____.
  (a) blending
  (b) cross-fertilization
  (c) true-breeding
  (d) independent assortment

___ (7) A single gene that affects several seemingly unrelated aspects of an individual's phenotype is said to be _____.
  (a) pleiotropic
  (b) epistatic
  (c) mosaic
  (d) continuous

___ (8) Suppose two individuals, each heterozygous for the same characteristic, are crossed. The characteristic involves complete dominance. The expected genotypic ratio of their progeny is _____.
  (a) 1:2:1
  (b) 1:1
  (c) 100 percent of one genotype
  (d) 3:1

___ (9) If the two homozygous classes in the F_1 generation of the preceding cross are allowed to mate, the observed genotypic ratio of the offspring will be _____.
  (a) 1:1
  (b) 1:2:1
  (c) 100 percent of one genotype
  (d) 3:1

___ (10) Assume that genes A, B, and C are located on different chromosomes. How many different types of gametes could an individual of genotype AaBbCc produce during meiosis if crossing over did not occur?
  (a) Two
  (b) Eight
  (c) Six
  (d) Twelve

CHAPTER TEST

INTEGRATING AND APPLYING KEY CONCEPTS

Solve the following genetics problems. Show all setups, genotypes, and phenotypes.

(1) A husband sues his wife for divorce, arguing that she has been unfaithful. His wife gave birth to a girl with a fissure in the iris of her eye, a sex-linked recessive trait. Both parents have normal eye structure. Can the genetic facts be used to argue for the husband's suit? Explain your answer.

(2) Holstein cattle have spotted coats due to a recessive gene; a solid-colored coat is caused by a dominant gene. If two spotted Holsteins are mated, what types of offspring would be likely? Explain your statement with a diagram.

(3) Give the genotypes of the parents in this problem, labeling appropriately. Cocker spaniels are black due to a dominant gene and red due to its recessive allele. Solid color depends on a dominant gene, and the development of white spots depends on its recessive allele. A solid red female was mated with a black-and-white male, and five puppies resulted: one black, one red, two black-and-white, and one red-and-white.

13

CHROMOSOMAL THEORY OF INHERITANCE

General Objectives

1. Describe the chromosomal theory of inheritance and explain how the theory helps to account for events that compose mitosis and meiosis.
2. Name some ordinary and extraordinary chromosomal events that can create new phenotypes (outward appearances).
3. Know how fruit fly experiments have helped us understand chromosomal behavior.
4. Understand how changes in chromosome structure and number can affect the outward appearance of organisms.

13-I
(pp. 177–178)

RETURN OF THE PEA PLANT
THE CHROMOSOMAL THEORY

Summary

During the late 1800s, several cytologists described the events that compose mitosis (Flemming) and meiosis (Weismann, Flemming). Mitotic and meiotic phenomena were linked with Mendel's concepts of segregation and independent assortment, and it was suggested that Flemming's "threads" were likely the carriers of Mendel's "hereditary units."

The chromosomal theory of inheritance states that each chromosome contains a linear sequence of genes that govern the development of phenotype. Diploid organisms inherit one set of chromosomes from each parent; these double sets are mingled as homologous pairs during merging of gametes, as a new diploid nucleus is formed. Later, during meiosis, when gametes are formed, the diploid number of chromosomes is halved. Each member of a homologous pair is segregated into different gametes, and the chromosomes of each pair assort independently of the chromosomes of other parts. New genotypic combinations are produced by crossing over, by independent assortment, and by chromosomal aberrations; these chromosomal events give rise to new phenotypes upon which selective agents can act.

Key Terms

cytology	genes	*n*
Flemming	homologues	assort independently
chromosome	alleles	crossing over
mitosis	meiosis	chromosomal aberrations
Weismann	diploid	genotype
chromosomal theory of	2*n*	phenotype
inheritance	haploid	

Objectives

1. Identify the cytologists who described mitosis and meiosis. State what was so useful about Flemming's contribution.
2. List the major tenets of the chromosomal theory of inheritance.

Self-Quiz Questions

Fill-in-the-Blanks

(1) _____ described the behavior of threadlike bodies (chromosomes) during mitosis in salamander cells. (2) _____ first proposed that chromosomal number must be reduced by half during gamete formation; the process became known as (3) _____. (4) _____ chromosomes resemble each other in length, shape, and gene sequence.

Matching

Choose the single most appropriate letter.

(5) ___ crossing over

(6) ___ deletion

(7) ___ duplication

(8) ___ independent assortment

(9) ___ inversion

(10) ___ segregation

(11) ___ translocation

A. Chromosomal mutation; segment of chromosome lost
B. Homologous chromosomes are channeled into different gametes.
C. Chromosomal aberration; chromosomal segment transferred to a nonhomologous chromosome
D. Chromosomal mutation; repeat of DNA sequence in same chromosome or in nonhomologous chromosome
E. Exchange of parts of chromatids between nonsister chromatids during prophase I of meiosis
F. Chromosomal mutation; chromosomal segment excised and reattached backward.
G. Chromosomes from one pair have no effect on the channeling into gametes of chromosomes from another pair.

CLUES FROM THE INHERITANCE OF SEX
    Sex Determination
    Sex-Linked Genes
    Morgan's Studies of Inheritance

Summary

Thomas Hunt Morgan's research group kept under culture fruit flies with a variety of normal and mutant traits. They discovered that the X chromosome (the one chromosome in male grasshoppers that does not have a partner) carries many genes besides the ones involved in sex determination; any gene on an X chromosome is known as a *sex-linked gene*. In organisms that have only one sex chromosome, all genes on the X chromosome are expressed regardless of whether they are dominant or recessive because there are no homologous genes on any companion chromosome that could mask the recessive instructions. In later studies, they discovered *sex-determining genes* that influenced the expressions of maleness and femaleness on autosomes (chromosomes other than X or Y chromosomes), as well. Developmental events ensure that genes are expressed largely through the stimulation or inhibition produced by different sex homones.

Key Terms

autosomes	TDF	sex-linked
sex chromosomes	committed	Thomas Hunt Morgan
X chromosome	sex-determining genes	white-eyed male
Y chromosome	X-linked	reciprocal cross
fruit fly	Y-linked	wild-type
gender		

Objectives

1. Describe the basic types of experiments that Morgan's research group carried out and state two major discoveries the group made that dramatically advanced our understanding of genetics.
2. Define *sex chromosomes* and *autosomes*; then distinguish the types of alleles found on each, if possible.
3. Distinguish between *sex-linked genes* and *sex-determining genes*.
4. Contrast the X and Y chromosomes in terms of length; then explain what, if anything, matches up with or masks the "extra" part of the longer chromosome. Finally, explain the consequences of this lone chromosomal segment.

**Self-Quiz
Questions**

Multiple Choice

____ (1) Most animal and some plant species have sex chromosomes that _____.
    (a) are called autosomes
    (b) always consists of one X and one Y
    (c) differ in number or in kind between males and females
    (d) All of the above are correct.

____ (2) Normally, all eggs from a human female contain _____.
    (a) one X chromosome
    (b) one Y chromosome
    (c) one X chromosome and one Y chromosome
    (d) only autosomes

If false, explain why.

—— (3) Individual genes always assort independently of one another during meiosis I.

—— (4) In most species, males transmit the Y chromosome to their sons but not to their daughters; males receive their X chromosome only from their mothers.

13-III
(pp. 181–183)

LINKAGE
CROSSING OVER AND LINKAGE MAPPING OF CHROMOSOMES

Summary

Morgan's research group also discovered that some groups of genes do not assort independently—as Mendel had said—because they are physically linked as part of the same chromosome. These and later studies with other organisms have made it clear that the number of *linkage groups* corresponds to the number of chromosomes characteristic of the species, which further confirms Sutton's hypothesis that chromosomes are the vehicles that transport the genes through generations. Different genes physically located on the same chromosome tend to end up together in the *same* gamete; they do *not* assort independently. Different genes located on *nonhomologous* chromosomes can assort independently of each other into gametes.

Crossing over happens during the first division of meiosis, in which two nonsister chromatids of homologous pairs of chromosomes exchange corresponding segments of breakage points. Crossing over results in genetic recombination— the formation of new combinations of alleles in a chromosome, hence in gametes, and finally in zygotes. Janssens's description of the formation of chiasmata (crossovers) gave Sturtevant the idea that locations of genes on their chromosomes could be mapped because the farther apart on the chromosome two linked genes may be, the more likely chiasma formation and crossing over will be to disrupt the original combination of linked alleles. The probability that crossing over and recombination will occur at a point somewhere between two genes located on the same chromatid is directly proportional to the distance that separates them. Genes are carried in linear array on chromosomes.

Key Terms

marker	linkage, linked	linkage mapping
nonhomologous	crossing over	Sturtevant
chromosomes	nonsister chromatids	map units

Objectives

1. Define *crossing over*, state when it occurs, and explain its significance for the chromosomal theory of inheritance.
2. Explain why it is important that crossover involves nonsister chromatids rather than sister chromatids. State why crossover between sister chromatids would not likely cause evolution in populations.
3. Explain how Janssens's description of chiasmata formation during meiotic prophase I gave Sturtevant an idea that has since contributed greatly to our understanding of genetic behavior.
4. State the relationship between the probability of crossing over (and subsequent recombination) and the distances between two genes located on the same chromatid.

5. Define *linkage* and explain how the concept became related to the chromosomal theory of inheritance.

Self-Quiz Questions

Multiple Choice

____ (1) In the first stage of meiosis, two sister chromatids of one chromosome are drawn into a locus-by-locus alignment with the two sister chromatids of its homologue. This process of alignment and attachment is called _____.
(a) crossing over
(b) linkage
(c) synapsis
(d) polygenic inheritance

____ (2) _____ is a consequence of crossing over.
(a) A dihybrid cross
(b) Linkage
(c) Dependent assortment
(d) Genetic recombination

____ (3) The closer together two genes are on one chromosome, _____.
(a) the fewer times there will be crossing over and genetic recombination occurring between them
(b) the more times there will be crossing over and genetic recombination occurring between them
(c) the farther apart they will appear in a linkage group
(d) the greater the chance will be that point mutations will happen to either of them

____ (4) Genetic recombination occurs as a result of crossing over between nonsister chromatids of homologous chromosomes during _____.
(a) metaphase of mitosis
(b) prophase I of meiosis
(c) syngamy (fusion of gametes)
(d) prophase II of meiosis

____ (5) Plotting the positions of genes on chromosomes according to their segregation patterns during meiosis is called _____.
(a) linkage mapping
(b) karyotyping
(c) G-banding
(d) transposing

____ (6) A standard map unit is _____.
(a) ten crossovers per 100 bands
(b) one crossover per 100 gametes
(c) 1 mm between the closest G-bands
(d) a 1 percent frequency of recombination between the genes involved

____ (7) Which of the following did Morgan and his students *not* do?
(a) They isolated and kept under culture fruit flies with the sex-linked recessive white-eyed trait.

(b) They developed the technique of amniocentesis.
(c) They developed the technique of linkage mapping.
(d) They discovered and described crossing over.

13-IV
(pp. 183–187)

CHANGES IN CHROMOSOME STRUCTURE
Chromosome Banding and Karyotypes
Deletions, Duplications, and Other Structural Rearrangements
CHANGES IN CHROMOSOME NUMBER
Missing or Extra Chromosomes
Polyploidy
SUMMARY

Summary

Rapid advances in microscopy enabled cytogeneticists to visualize the structure of chromosomes. Observations of *polytene chromosomes* connected specific chromosome regions with specific phenotypic traits. *G-banding* and the preparation of *karyotypes* permitted ordinary chromosomes to be more clearly identified.

Crossing over normally occurs during gamete formation. On very rare occasions, chromosomal rearrangements known as *chromosomal aberrations*—deletions, duplications, translocations, and inversions—also occur. They invariably affect fairly large chromosomal segments of the total chromosome number characteristic of the species. Besides these rare internal arrangements, from time to time whole chromosomes or chromosome sets go astray.

Aneuploidy (in which a particular type of chromosome is either absent entirely or present three or more times in the diploid chromosome set) arises when homologous chromosomes fail to segregate properly during prophase I of meiosis. *Polyploidy* (in which there are three or more whole sets of chromosomes either in certain tissues or in all somatic cells of an individual) commonly occurs in flowering plants and other eukaryotes; it can be induced by exposing plants to colchicine.

Domestic hybrids of wheat have been developed by manipulating hybridization, polyploidy, and exposure to colchicine.

Key Terms

polytene chromosome	inversion	monosomy
G-banding	translocation	polyploidy
karyotype	nondisjunction	triploid
duplication	aneuploidy	autopolyploidy
deletion	trisomy	allopolyploidy

Objectives

1. Define *chromosomal aberration* and list four types of such alterations. Explain how such changes occur.
2. Define *polyploidy*. Tell what agents cause this situation and how it differs from the four types of chromosomal changes.
3. Explain how the wheat from which bread is made is thought to have evolved.
4. Tell what role all of these chromosomal rearrangements play in variation within populations and, thus, in natural selection.
5. Describe how G-banding and karyotyping are done and tell what the two techniques have contributed to cytogenetics.

Fill-in-the-Blanks

A (1) _____ is the loss of a piece of a chromosome that happens during gamete formation. Segments of chromosomes occasionally break off and move to a nonhomologous chromosome; such an event is known as a(n) (2) _____. A(n) (3) _____ occurs when a deleted piece of chromosome gets turned around and rejoins the chromosome at the same place, but backward. Hybridization in wheat causes multiple sets of chromosomes in gametes, a condition known as (4) _____. (5) _____ is a chemical that induces polyploidy in many plants. (6) _____ causes (7) _____, in which three chromosomes of the same kind are present in the chromosome set. Homologous chromosomes can exchange parts as a result of (8) _____-_____. A cell with four sets of one kind of chromosome is (9) _____.

Multiple Choice

___ (10)   Plotting the positions of genes on chromosomes according to their segregation patterns during meiosis is called _____.
   (a) linkage mapping
   (b) karyotyping
   (c) G-banding
   (d) transposing

CHAPTER TEST

UNDERSTANDING AND INTERPRETING KEY CONCEPTS

___ (1) All the genes located on a given chromosome compose a(n) _____.
   (a) karyotype
   (b) bridging cross
   (c) wild-type allele
   (d) linkage groups

___ (2) Chromosomes other than those involved in sex determination are known as _____.
   (a) nucleosomes
   (b) heterosomes
   (c) alleles
   (d) autosomes

___ (3) If two genes are very close to each other on the same chromatid, _____.
   (a) crossing over and recombination occurs between them quite often
   (b) they act as if they are assorted independently into gametes
   (c) they will most probably be in the same linkage group
   (d) they will be segregated into different gametes when meiosis occurs

___ (4) Linkage mapping is a technique that determines _____.
- (a) genetic linkages between sister chromatids that are undergoing synapsis
- (b) the positions of genes on chromosomes relative to one another
- (c) the positions of the chiasmata between homologous chromosomes during the last stage of prophase I
- (d) the sex of the offspring

___ (5) A color-blind man and a woman with normal vision whose father was color blind have a son. Color blindness, in this case, is caused by a sex-linked recessive gene. If only the male offspring are considered, the probability that their son is color blind is _____.
- (a) .25 (or 25 percent)
- (b) .50 (or 50 percent)
- (c) .75 (or 75 percent)
- (d) 1.00 (or 100 percent)
- (e) none of the above

___ (6) Susan, a mother with type B blood, has a child with type O blood. She claims that Craig, who has type A blood, is the father. He claims that he cannot possibly be the father. Further blood tests ordered by the judge revealed that the father is homozygous. The judge rules that _____.
- (a) Susan is right and Craig must pay child support
- (b) Craig is right and doesn't have to pay child support
- (c) Susan cannot be the real mother of the child; there must have been an error made at the hospital
- (d) it is impossible to reach a decision based on the limited data available

___ (7) Red-green color blindness is a sex-linked recessive trait in humans. A color-blind woman and a man with normal vision have a son. (1) What are the chances that the son is color blind? (2) If the parents ever have a daughter, what is the chance for each birth that the daughter will be color blind? (Consider only the female offspring.)
- (a) (1) 100 percent, (2) 0 percent
- (b) (1) 50 percent, (2) 0 percent
- (c) (1) 100 percent, (2) 100 percent
- (d) (1) 50 percent, (2) 100 percent
- (e) none of the above

___ (8) If you want to produce a plant that has larger than normal fruits, exploitation of which of the following features is most likely to be an effective research strategy to pursue?
- (a) Preparing karyotypes
- (b) Administering colchicine
- (c) Translocation and crossing over
- (d) Chromosome fusion

___ (9) In his experiments with *Drosophila melanogaster*, Morgan demonstrated that _____.
  (a) fertilized eggs have two sets of chromosomes, but eggs and sperm only have one set in each gamete
  (b) aneuploidy exists in karyotypes that have undergone deletions and inversions in specific chromosomes
  (c) colchicine is effective in producing polyploidy in F_2 generations
  (d) certain genes are located only on an X chromosome and have no corresponding alleles on the Y chromosome

---

CHAPTER TEST

INTEGRATING AND APPLYING KEY CONCEPTS

The parents of a young boy bring him to their doctor. They explain that the boy does not seem to be going through the same vocal developmental stages as his older brother. The doctor orders a common cytogenetics test to be done, and it reveals that the young boy's cells contain two X chromosomes and one Y chromosome. Describe the test that the doctor ordered and explain how and when such a genetic result—XXY—most logically occurred.

14

HUMAN GENETICS

General Objectives

1. Distinguish *autosomal recessive inheritance* from *sex-linked recessive inheritance*.
2. Give two examples of each of the above types of inheritance.
3. Explain how changes in chromosomal number can occur and present an example of such a change.
4. List three examples of phenotypic defects and describe how each can be treated.
5. Explain how knowing about modern methods of genetic screening can minimize potentially tragic events.

14-I
(pp. 190–194)

DISORDERS ARISING FROM GENE MUTATIONS
    Autosomal Recessive Inheritance
    Autosomal Dominant Inheritance
    X-Linked Recessive Inheritance
DISORDERS ARISING FROM CHANGES IN CHROMOSOME NUMBER
    Down Syndrome
    Turner Syndrome
    Klinefelter Syndrome
    XYY Condition

Summary

For victims of some human genetic diseases—such as phenylketonuria, galactosemia, and some forms of diabetes—stopgap measures called *phenotypic cures* can often preserve their lives if the disorder is detected early enough. Galactosemia is a disorder caused by a recessive allele carried on an autosome (a nonsex chromosome). The disorder appears to be transmitted according to the rules of simple Mendelian inheritance, because it is expressed in homozygotes of both sexes.

Hemophilia and green color blindness (green weakness) are examples of *sex-linked recessive* inheritance: the female almost always carries the recessive allele, and the male almost always is the one afflicted with the disorder.

Chromosomal mutations occur in all organisms, even in humans. Improper separation of chromosomes during gamete formation is the cause of Down syndrome, trisomic XXY, trisomic XYY, and Turner syndrome.

Key Terms

pedigree
phenotypic treatment
genotypic cure
abnormal
disorder
disease
galactosemia
autosomal recessive
   inheritance

autosomal dominant
   inheritance
achondroplasia
Huntington's disorder
hemophilia A
X-linked recessive
   inheritance
nondisjunction

trisomy 21
Down syndrome
Turner syndrome
sex chromosome
   abnormality
Klinefelter syndrome
XYY condition

Objectives

1. Distinguish *phenotypic treatments* from *genotypic cures* and state which kind of therapy is used most at this time.
2. Explain how a genotypic cure is achieved.
3. List and describe the causes and phenotypic features of two human disorders that result from improper separation of chromosomes during gamete formation.
4. List and describe the causes and phenotypic features of two human disorders that are due to mutations of a single gene in each case.
5. List three features of a genetic disorder that would suggest it is transmitted as a form of autosomal recessive inheritance.
6. State the feature(s) of a genetic disorder that would suggest it is a form of *sex-linked* inheritance.

Self-Quiz Questions

Matching

Choose the single most appropriate letter. You can use the same letter more than once.

(1) ___ Down syndrome

(2) ___ galactosemia

(3) ___ hemophilia A

(4) ___ Huntington's disorder

(5) ___ XYY condition

(6) ___ Turner syndrome

A. Autosomal recessive inheritance
B. X-linked recessive inheritance
C. Nondisjunctive inheritance involving autosomes
D. Autosomal dominant inheritance
E. Nondisjunctive inheritance involving sex chromosomes

PROSPECTS AND PROBLEMS IN HUMAN GENETICS
    Treatments for Phenotypic Defects
    Genetic Screening
    Genetic Counseling and Prenatal Diagnosis
RFLPs
SUMMARY

Summary

A variety of *therapeutic measures* (diet modification, environmental adjustment, surgical correction, and chemotherapy) are used in treating genetically based disorders. *Preventive* measures include mutagen reduction, genetic screening, genetic counseling, and prenatal diagnosis; gene replacement therapy also shows promise but is still in the research stage. Normal alleles would then be substituted for mutant alleles in sperm or eggs. Through *amniocentesis* (the sampling of the fluid in the mother's uterus), it is possible to determine whether the developing fetus has a severe genetic disorder. If a severe disorder is discovered, the parents may elect to induce abortion of the fetus, if specific requirements are met.

Key Terms

diet modification	genetic screening	chorion
phenylketonuria	amniocentesis	RFLPs
cleft lip	amnion	polymorphism
Wilson's disorder	chorionic villi sampling	genetic fingerprints

Objectives

1. Discuss some of the ethical predicaments that have arisen with our increased knowledge of human genetics and our increased technical abilities.
2. Describe the phenotypic cures in current use for any of the human disorders known; tell how the "cure" is administered.
3. Define *amniocentesis* and state one potential benefit and one potential drawback of the technique.
4. Describe the use of RFLPs to detect specific genes on specific chromosomes.

Self-Quiz Questions

Matching

Choose the one most appropriate answer for each. You can use the same letter more than once.

(1) _E_ albinism

(2) _E_ amniocentesis

(3) _E_ chorionic villi sampling

(4) _C_ cleft lip

(5) _A_ galactosemia

(6) _A_ phenylketonuria

A. A phenotypic defect that can be helped by diet modification
B. A phenotypic defect that can be helped by environmental adjustments
C. A phenotypic defect that can be helped by surgical correction
D. A phenotypic defect that can be helped by chemotherapy
E. Used to arrive at a diagnosis

(7) ___ radioactive probe

(8) ___ RFLPs

(9) ___ sickle-cell anemia

(10) ___ Wilson's disorder

F. Used to determine the location of a specific gene on a specific chromosome

---

CHAPTER TEST

UNDERSTANDING AND INTERPRETING KEY CONCEPTS

___ (1) Gene-replacement therapy _____.
   (a) has not yet been used successfully with mammals
   (b) is a surgical technique that separates chromosomes that have failed to segregate properly during meiosis II
   (c) has been used successfully to treat victims of Huntington's chorea by removing the dominant damaging autosomal allele and replacing it with a harmless one
   (d) substitutes defective alleles with normal ones

___ (2) Karyotype analysis is _____.
   (a) a means of detecting and reducing mutagenic agents
   (b) is a surgical technique that separates chromosomes that have failed to segregate properly during meiosis II
   (c) used in prenatal diagnosis to detect chromosomal mutations and metabolic disorders in embryos
   (d) substitutes defective alleles with normal ones

___ (3) Amniocentesis is _____.
   (a) a surgical means of repairing deformities
   (b) a form of chemotherapy that modifies or inhibits gene expression or the function of gene products
   (c) used in prenatal diagnosis to detect chromosomal mutations and metabolic disorders in embryos
   (d) a form of gene-replacement therapy

___ (4) Of all phenotypically normal males in prisons, 2 percent have _____.
   (a) trisomic XXY
   (b) trisomic XYY
   (c) Turner syndrome
   (d) Down syndrome

___ (5) A woman heterozygous for color blindness (a sex-linked recessive allele) marries a man with normal color vision. What is the probability that their first child will be color blind?
   (a) 25 percent
   (b) 50 percent
   (c) 75 percent
   (d) 100 percent

___ (6) Nondisjunction involving the X chromosome occurs during oogenesis and produces two kinds of eggs. If normal sperm fertilize two types, which genoptypes are possible?
  (a) XX and XY
  (b) XXY and XO
  (c) XYY and XO
  (d) XYY and YO

___ (7) Suppose that a hemophilic male (sex-linked recessive allele) and a female carrier for the hemophilic trait have a nonhemophilic daughter with Turner syndrome. Nondisjunction could have occurred in _____.
  (a) both parents
  (b) neither parent
  (c) the father only
  (d) the mother only

___ (8) Suppose that the parents described in Question 7 had produced a nonhemophilic male child apparently normal in all respects. Nondisjunction could have occurred in _____.
  (a) both parents
  (b) neither parent
  (c) the father only
  (d) the mother only

___ (9) Suppose the parents in Question 7 had produced a nonhemophilic male child with trisomic $X^H X^h Y$. Nondisjunction could have occurred in _____.
  (a) both parents
  (b) neither parent
  (c) the father only
  (d) the mother only

CHAPTER TEST

INTEGRATING AND APPLYING KEY CONCEPTS

Suppose that you are a genetics researcher who discovers in humans a genetic sequence that codes for the production of several proteins that bring about aging and cell death. How might you use this information in writing a proposal to the National Science Foundation for a grant that would finance research investigating ways to prevent cells, tissues, and organisms from aging and dying? What specific research program would you propose? What equipment and personnel do you think you would need? How long would you be doing this research, where would you propose to conduct this study, and how much money would you ask for? You may wish to consult appropriate previous chapters and their bibliographic references in answering this question. Do not assume that famous researchers would not want to be included in your research team.

15

THE RISE OF MOLECULAR GENETICS

DISCOVERY OF DNA FUNCTION
  A Puzzling Transformation
  Bacteriophage Studies
DNA STRUCTURE
  Components of DNA
  Patterns of Base Pairing

DNA REPLICATION
  How DNA Is Duplicated
  A Closer Look at Replication
SUMMARY

**General
Objectives**

1. Understand how experiments using bacteria and viruses demonstrated that instructions for producing heritable traits are encoded in DNA.
2. Know the parts of a nucleotide and know how they are linked together to make DNA.
3. Understand how DNA is replicated and what materials are needed for replication.

15-I
(pp. 199–201)

DISCOVERY OF DNA FUNCTION
  A Puzzling Transformation
  Bacteriophage Studies

Summary

During its life span, each cell deploys thousands of different enzymes in translating its hereditary instructions (its genotype) into specific physical traits (its phenotype). Each cell comes into the world with a limited set of specific instructions that promote survival in a particular kind of habitat. Hereditary instructions must be of a sufficiently stable chemical nature to ensure constancy in structure and function, and yet must allow room for subtle change. In the search for the hereditary molecule, two principal candidates were considered as likely: *proteins* because they are diverse and complex and *nucleic acids* because—although they have only a simple repeating-unit structure—they are abundant in the nuclear regions of cells where the hereditary instructions were thought most likely to reside. After a variety of experiments, in almost all organisms, DNA (deoxyribonucleic acid) was demonstrated to be the molecule that contains the genetic code.

Key Terms

deoxyribonucleic acid, DNA
Griffith
Diplococcus pneumoniae

Avery
Delbruck
Hershey and Chase
bacteriophages

^{35}S
^{32}P
lytic pathway

Objectives

1. State which two classes of molecules were (before 1952) suspected of housing the genetic code.
2. Identify the types of research carried out by Miescher, Griffith, Avery and colleagues, and Hershey and Chase, and state the specific advances in our understanding of genetic behavior made by each.
3. Explain how bacteriophages were used to study genetics.
4. State which piece of research demonstrated that DNA, not protein, governed inheritance.

Self-Quiz Questions

True-False

If false, explain why.

___ (1) Miescher first isolated and described nucleic acids obtained from pus and fish sperm during the 1920s.

___ (2) ^{35}S can be used to label DNA, and ^{32}P can be used to label almost any protein.

___ (3) DNA is the only hereditary molecule found in viruses.

___ (4) Griffith demonstrated that DNA was the substance that had been permanently transformed when the rough strain of bacteria was transformed to the disease-causing smooth strain of bacteria.

___ (5) Bacteriophages are bacteria that eat viruses.

___ (6) DNA contains sulfur but not phosphorus.

___ (7) Hershey and Chase were the first to demonstrate that DNA was unquestionably the molecule that contained the genetic code.

15-II
(pp. 202–204)

DNA STRUCTURE
   Components of DNA
   Patterns of Base Pairing

Summary

Only four different kinds of nucleotides are found in a DNA molecule, and each has the same phosphate group and the same five-carbon sugar (deoxyribose). Each nucleotide has one of four different nitrogen-containing bases. Cytosine and thymine are single-ring pyrimidines; adenine and guanine are double-ring purines. In a DNA molecule, nucleotides are strung together into a long chain, with each phosphate group of one nucleotide linking with the sugar of the next nucleotide; when nucleotides are thus assembled, the purines and pyrimidines project to one side. Chargaff and Franklin, by using a variety of physical and chemical experimental techniques, provided Watson and Crick with enough clues for them to build a scale model, revealing that DNA must be double-stranded, with two nucleotide strands wound helically about each other like a spiral staircase. Watson and Crick stated that the two strands of nucleotides were held together by hydrogen bonds linking purines (A and G) from one strand, with pyrimidines (C and T) on the other strand and vice versa; this statement has become known as the *principle of base pairing*. Any purine-pyrimidine pair can follow any other in the DNA chain, so an extremely large number of code sequences can result from different arrangements of these two pairs of nitrogenous bases. In all species, there is a constancy in base pairing (adenine to thymine, guanine to cytosine) between the two strands of the DNA double helix. The DNA of different species shows variation in which base follows which in a strand.

Key Terms			
	nucleotide	pyrimidines	X-ray diffraction
	deoxyribose	adenine	Franklin
	nitrogen-containing base	guanine	Watson and Crick
	cytosine	purines	principle of base pairing
	thymine	Chargaff	

Objectives

1. Draw the basic shape of a deoxyribose molecule and show how a phosphate group is joined to it when forming a nucleotide.
2. Define, diagram, and distinguish *purine* and *pyrimidine*.
3. Describe how cytosine differs from thymine and how adenine differs from guanine. Then show how each would be joined to the sugar–phosphate combination drawn in Objective 2.
4. List both of the clues that Chargaff's research demonstrated.
5. Describe the technique of X-ray diffraction and state the kinds of information about DNA structure that Rosalind Franklin discovered by using X-ray diffraction.
6. List the particular contributions of Watson and Crick to the modern understanding of DNA structure and behavior.
7. Explain what is meant by the pairing of nitrogen-containing bases (base pairing), and state the feature that causes adenine to pair with thymine and guanine to pair with cytosine.
8. Explain how the Watson-Crick model obeyed Chargaff's rules and fulfilled Franklin's structural predictions.
9. Describe how the Watson-Crick model reflects the constancy in DNA observed from species to species, yet allows for variations from species to species.

Self-Quiz Questions

True-False

If false, explain why.

___ (1) Each nucleotide consists of a six-carbon sugar, a phosphate group, and a nitrogen-containing base.

___ (2) DNA is composed of only four different types of nucleotides, each of which contains adenine, thymine, cytosine, or guanine.

___ (3) In every species, the amount of adenine present always equals the amount of thymine, and the amount of cytosine always equals the amounts of guanine (A = T, and C = G).

___ (4) In a nucleotide, the phosphate group is attached to the nitrogen-containing base, which is attached to the five-carbon sugar.

___ (5) Watson and Crick built their paper model of DNA in the early 1950s.

___ (6) Rosalind Franklin discovered that DNA was double-stranded.

___ (7) Guanine pairs with cytosine by forming three hydrogen bonds.

15-III
(pp. 204–207)

DNA REPLICATION
    How DNA Is Duplicated
    A Closer Look at Replication
SUMMARY

Summary

Once DNA's double-stranded nature had been deduced, Watson and Crick immediately saw how such a molecule could be duplicated prior to mitosis or meiosis. If the hydrogen bond(s) that connect a purine with a pyrimidine in a base pair are broken by the appropriate enzyme and separated, each base can attract and form hydrogen bonds with its complementary nucleotide already synthesized and present in the cell. Each parent DNA strand remains intact, and a new companion strand is assembled on each one. The only sequence that could be attached to one parent strand (say, ATTCGC) would be an exact complement of the base sequence (TAAGCG) that occurs on the other parent strand. Each single parent strand ends up forming a new complementary strand, with which it remains linked by means of hydrogen bonds. In each of the two resulting DNA molecules, one old DNA strand has been conserved as a partner for the newly formed strand; this semiconservative replication was demonstrated by the experiments of Meselson and Stahl.

Replication begins at an origin and proceeds simultaneously in two directions from the origin. DNA synthesis occurs at replication forks. Discontinuous DNA synthesis occurs on both parent strands; new nucleotides are added wherever an −OH group projects from the 3' carbon atom of a nucleotide already on track.

A variety of enzymes participate in DNA replication; some serve to detach the purine-pyrimidine pairs, and others act to link new nucleotides with the parent strands. Still other enzymes form a sort of quality-control inspection team and act to correct errors that may have occurred during replication.

Key Terms

free nucleotides	DNA polymerases	continuous DNA assembly
template	initiation site	discontinuous DNA assembly
parent strand	origin	"origin"
double helix	bidirectional	
semiconservative replication	replication fork	
	Okasaki	

Objectives

1. Assume that the two parent strands of DNA have been separated and that the base sequence on one parent strand is ATTCGC. State the sequence of new bases that will complement the parent strand.
2. Describe how double-stranded DNA replicates itself by bidirectional base pairing.
3. Describe Okasaki's description of DNA assembly on two different templates.
4. List four jobs that enzymes perform during the process of replication.

Self-Quiz Questions

True-False

If false, explain why.

_F_ (1) The hydrogen bonding of adenine to guanine is an example of complementary base pairing.

_F_ (2) The replication of DNA is considered semiconservative because the same four nucleotides are used again and again during replication.

_T_ (3) Each parent strand remains intact during replication, and a new companion strand is assembled on each one.

_T_ (4) Some of the enzymes associated with DNA assembly repair errors during the replication process.

Multiple Choice

___ (5) A substance acts as a mold for the formation of a substance complementary in form; the mold substance is known as a _____.
(a) hybrid molecule
(b) plasmid
(c) template
(d) clone

Labeling

Identify each indicated part of the accompanying illustration.

(6) _A_   (9) _C_   (11) _Sugar_

(7) _T_   (10) _phosphate_ (12) _nucleotide_

(8) _G_

CHAPTER TEST

UNDERSTANDING AND INTERPRETING KEY CONCEPTS

_D_ (1) Each DNA strand has a backbone that consists of alternating _____.
(a) purines and pyrimidines
(b) nitrogen-containing bases
(c) hydrogen bonds
(d) sugar and phosphate molecules

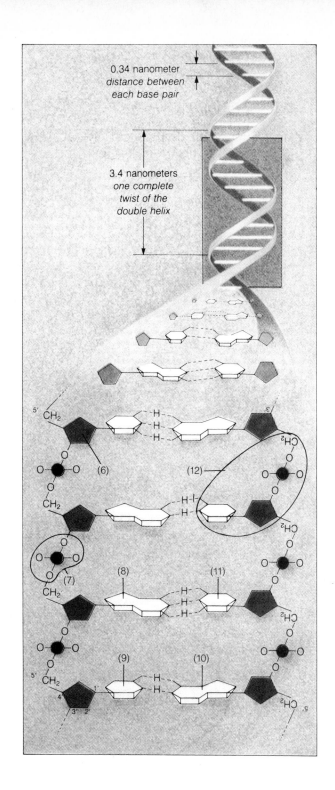

0.34 nanometer
*distance between
each base pair*

3.4 nanometers
*one complete
twist of the
double helix*

(2) In DNA, complementary base pairing occurs between _____.
   (a) cytosine and uracil
   (b) adenine and guanine
   (c) adenine and uracil
   (d) adenine and thymine

_a_ (3) Adenine and guanine are _____ .
  (a) double-ringed purines
  (b) single-ringed purines
  (c) double-ringed pyrimidines
  (d) single-ringed pyrimidines

_d_ (4) Franklin used the technique known as _____ to determine many
  of the physical characteristics of DNA.
  (a) transformation
  (b) transmission electron microscopy
  (c) density-gradient centrifugation
  (d) X-ray diffraction

___ (5) The significance of Griffith's experiment that used two strains of
  Diplococcus pneumoniae is that _____ .
  (a) the semiconservative nature of DNA replication was finally demon-
    strated
  (b) it demonstrated that harmless cells had become permanently trans-
    formed through a change in the bacterial hereditary system
  (c) it established that pure DNA extracted from disease-causing bacteria
    transformed harmless strains into "killer strains"
  (d) it demonstrated that radioactively labeled bacteriophage transfer their
    DNA but not their protein coats to their host bacteria

___ (6) The significance of the experiments in which ^{32}P and ^{35}S were used is
  that _____ .
  (a) the semiconservative nature of DNA replication was finally demon-
    strated
  (b) it demonstrated that harmless cells had become permanently trans-
    formed through a change in the bacterial hereditary system
  (c) it established that pure DNA extracted from disease-causing bacteria
    transformed harmless strains into "killer strains"
  (d) it demonstrated that radioactively labeled bacteriophage transfer their
    DNA but not their protein coats to their host bacteria

___ (7) Franklin's research contribution was essential in _____ .
  (a) establishing the double-stranded nature of DNA
  (b) establishing the principle of base pairing
  (c) establishing most of the principal structural features of DNA
  (d) all of the above

___ (8) Density-gradient centrifugation is a technique used by two biochemical
  geneticists to demonstrate that _____ .
  (a) two separated "parental" strands of DNA act as a pair of templates
    as they assemble new complementary "companion" strands
  (b) a variety of enzymes participate in DNA replication
  (c) the plasmids of certain bacteria can be opened up and gene sequences
    taken from other bacteria can be spliced into the opened plasmids
  (d) a variety of mutagenic agents can cause point mutations in DNA
    strands

___ (9) When Griffith injected mice with a mixture of dead pathogenic—encapsulated S cells and living, unencapsulated R cells of *Diplococcus pneumoniae*—he discovered that _____.
  (a) the previously harmless strain had permanently inherited the capacity to build protective capsules
  (b) the dead mice teemed with living pathogenic (R) cells
  (c) the killer strain R was encased in a protective capsule
  (d) all of the above

CHAPTER TEST

INTEGRATING AND APPLYING KEY CONCEPTS

Mention three uses of recombinant DNA that could be immensely beneficial to humans or other life on Earth. Then mention a possible scenario for recombinant DNA that could be disastrous to humans.

16

PROTEIN SYNTHESIS

General Objectives

1. Understand how earlier experimentation led to our current understanding of biochemical genetics.
2. Know how the structure and behavior of DNA determine the structure and behavior of the three forms of RNA during transcription.
3. Know how the structure and behavior of the three forms of RNA determine the primary structure of polypeptide chains during translation.
4. Know that there are exceptions to the universality of the genetic code and know how geneticists try to account for their existence.

16-I
(pp. 208–210)

ONE GENE, ONE POLYPEPTIDE

THE PATH FROM GENES TO PROTEINS
    Enter RNA

Summary

Today, we know that genes are portions of DNA that do not code directly for the production of proteins; that is, proteins are not assembled on the DNA. Rather, certain kinds of RNA are intermediaries constructed from ribose-containing nucleotides that are assembled in a complementary fashion on DNA, yet leave the DNA molecules intact in the nuclear region even as the RNA bears genetic messages to the sites of protein synthesis—the ribosomes. The linear sequence of nucleotide triplets becomes translated into a linear sequence of amino acids in a polypeptide chain, which is the basic structure of proteins. In the early 1900s, Garrod studied several examples of inherited metabolic disorders and linked them to the absence or deficiency of an essential enzyme. He suggested that each enzyme was specified by a single unit of inheritance and that, if the unit itself was defective, the enzyme also would be defective. About thirty-five years later, Beadle and Tatum bombarded the spores of red bread mold with X rays and isolated mutant strains that could grow only on culture media to which specific substances had been added. Each inherited mutation corresponded to a different defective enzyme, so Beadle and Tatum assumed that each gene controls the synthesis of one enzyme. Their assumption became known as the one gene, one enzyme concept.

Studies using *electrophoresis* to measure the rate and direction of movement of normal and abnormal types of hemoglobin provided a better understanding of the link between genes and proteins. Sickle-cell hemoglobin differs from normal hemoglobin because *one* amino acid (valine) is substituted for another (glutamate). Only one of the 150 amino acids in the beta-polypeptide chains has gone astray in HbS, but that is enough to cause a severe form of anemia. On the basis of such experiments, Beadle and Tatum's hypothesis has been amended as the *one gene, one polypeptide hypothesis*; a protein consists of one or more chemically different polypeptide chains, and a single gene codes for each kind. RNA is usually a single strand of ribose-containing nucleotides in which the base uracil replaces thymine. Uracil base-pairs with adenine just as thymine does. Ribonucleotides are joined together by using DNA as a template.

Key Terms

Garrod
Beadle and Tatum
Neurospora crassa
"nutritional mutants"
"one gene, one enzyme"
   hypothesis
HbS, sickle-cell
   hemoglobin

HbA, normal hemoglobin
electrophoresis
Ingram
"one gene, one
   polypeptide"
   hypothesis

Watson and Crick
template
RNA, ribonucleic acid
uracil

Objectives

1. List an inherited metabolic disorder studied by Garrod, state what Garrod thought caused the disorder, and define *inborn error of metabolism*.
2. Describe Beadle and Tatum's experiments with *Neurospora crassa* and explain how their experiments led to the one gene, one enzyme concept.
3. State the one gene, one enzyme concept and explain how it has changed since Beadle and Tatum first suggested it in the early 1940s.
4. State how RNA differs from DNA in structure and function, and indicate what features RNA has in common with DNA.
5. Explain how the process of electrophoresis is used to study proteins.

Self-Quiz Questions

Fill-in-the-Blanks

In (1) _____, urine samples exposed to air turn a dark color because of concentrations of the intermediate alkapton. (2) _____ linked this disease to the absence or deficiency of an essential (3) _____, which is specified by a single (4) _____, which, if defective, will code for a defective

(5) _____. (6) _____ and (7) _____ worked with *Neurospora crassa*, which is a kind of (8) _____ _____ _____. *N. crassa* was a useful organism because it is (9) _____: it has only one gene for each trait, it can reproduce (10) _____ through spore formation and give rise to many offspring, and it can grow on a simple, chemically defined medium containing (11) _____and biotin (one of the B vitamins). From their work with *N. crassa* mutants, they stated the one (12) _____, one (13) _____ concept, which has since been modified to recognize that there are intermediary (14) _____ _____ between the genes and the enzyme products for which they contain assembly instructions. Today we know that (15) _____ molecules serve as the message carriers. Unlike DNA, (16) _____ contains (17) _____ instead of deoxyribose, and, in place of the base thymine, it has (18) _____, which can form hydrogen bonds with the purine (19) _____.

16-II
(pp. 210–212)

THE PATH FROM GENES TO PROTEINS (cont.)
    Overview of Protein Synthesis
TRANSCRIPTION
    Synthesis of RNA
    RNA Transcripts

Summary

Three kinds of RNA govern the assembly of amino acids into proteins. All three are assembled in a complementary fashion from the template of DNA by the process of *transcription*. Messenger RNA (*mRNA*) molecules are linear, unbranched chains of at least sixty nucleotides with leader sequences (promoters) that control the function of mRNA. Messenger RNA alone carries the instructions for assembling a specific protein from the DNA in the nuclear region to the ribosomes. *Ribosomal RNA* (*rRNA*) becomes attached to proteins to form ribosomes, which are the actual protein construction sites in the cytoplasm. rRNA is synthesized largely in the nucleolus of eukaryotic cells. *Transfer RNA* (*tRNA*) is a shuttle molecule that picks up whatever amino acid is dictated by the anticodon at one end, attaches the amino acid to the opposite end, and moves over to the ribosome, where it awaits its chance to dump its load into the growing amino acid chain.

Of the original DNA template (which guided the formation of RNA), the parts that actually dictate the sequence of amino acids in proteins are called *exons*; the intervening DNA sequences are called *introns* and will be snipped out as the exons are spliced together to form the mature mRNA transcript.

Key Terms

transcription	ribosomal RNA, rRNA	transfer RNA, tRNA
RNA transcripts	ribosome	code
translation	messenger RNA, mRNA	complementary to

transcript   introns   "cap"
RNA polymerase   exons   "poly-A tail"
intervening DNA

Objectives

1. Distinguish among the three types of RNA in terms of structure, function, where each is made, and where each carries out its function.
2. Describe the process of transcription and indicate how it differs from replication.
3. Tell what RNA code would be formed from the following DNA code: TAC-CTC-GTT-CCC-GAA.
4. Explain how terms such as *introns, exons, cap,* and *poly-A tails* are related to RNA transcript processing.

Self-Quiz Questions

Fill-in-the-Blanks

During the process of (1) _Transcription_, the two strands of the (2) _DNA_ double helix are separated; (3) _RNA polymerase_, an enzyme, attaches to and moves along (4) _____ [choose one] ☐ one of the strands, ☐ both strands, all the while speeding up the assembly of RNA nucleotide subunits onto (5) _complementary_ regions of the exposed DNA. Then the (6) _mRNA_ strand detaches from the DNA and moves into the (7) _cytoplasm_. Along the way, (8) _introns_ are snipped from future mRNA transcripts and (9) _exons_ are spliced together. At its 3' end, the transcript acquires a (10) _poly-A tail_ before it reaches the ribosome as a mature mRNA transcript.

16-III
(pp. 213, 214)

TRANSLATION
    The Genetic Code
    Commentary: **Mitochondria: Exceptions to the Rule**

Summary

The DNA message is carried by RNA molecules to ribosomes in the cytoplasm, where amino acid sequences are assembled into linear strings. A string of amino acids represents a protein's primary structure, which in turn dictates the protein's final three-dimensional form. This form determines the substances the protein will interact with in helping to build and maintain the cell. There are only four different kinds of nucleotides in DNA and RNA, but there are twenty different amino acids commonly found in the proteins of living things. If each nucleotide called forth a single amino acid, that would give us only four different amino acids. Two nucleotide letters in a row give us only sixteen different amino acids, not twenty. The genetic code consists of nucleotide bases that are read linearly, three at a time, with the sequence of each triplet signifying an amino acid. Any of sixty-one such *triplets* specifies an amino acid; some different triplets specify the same amino acid. Three other triplets serve to terminate the protein chain. Any nucleotide triplet in messenger RNA (RNA that carries protein assembly instructions from the nuclear region to the ribosomes in the cytoplasm) is now known as a *codon*. In all living organisms, essentially the same genetic code is the basic language of protein synthesis—compelling evidence for the fundamental unity of life at the molecular level.

Mitochondrial DNA, with its slightly variant codons, is genetically isolated from the rest of the eukaryotic cell. Many biologists suspect that free-living microorganisms that resembled some modern bacteria were the original ancestors of mitochondria. Somehow the ancestral organisms established an intracellular symbiosis with their host cell. The descendants lost the ability to live independently.

Key Terms

Crick and Brenner
genetic code
"reading frame"
base triplet

Khorana, Nirenberg,
    Ochoa, Holley, and
    others
codon

Margulis
symbiotic
frameshift mutation

Objectives

1. State the relationship between the DNA genetic code and a protein's primary structure.
2. Explain why singlets or doublets of nucleotide bases are insufficient to specify all of the amino acids in the proteins of living forms.
3. Describe Brenner's base-insertion experiments and explain how they led Crick to suggest that the code was read in triplets.
4. Scrutinize Figure 16.8 and decide whether the genetic code in this instance applies to DNA, to messenger RNA, or to some other type of RNA.
   Write your answer in big letters above Figure 16.8 in your text and in this blank _____ only. Explain how you determined your answer.
5. Tell what RNA code would be formed from the following DNA code: TAC-CTC-GTT-CCC-GAA.
6. Explain how the DNA message TAC-CTC-GTT-CCC-GAA would be used to code for a segment of protein and state what its amino acid sequence would be.

Self-Quiz Questions

Fill-in-the-Blanks

A protein's (1) _____primary_____ structure is a string of amino acids, each of which is specified by a sequence of (2) _mRNA or 3_ nucleotide bases. If the code were read in units of two bases instead of three, only (3) __16__ amino acids could be coded for. In the table that showed which triplet specified a particular amino acid, the triplet code was in (4) _mRNA_ molecules. Each of these triplets is referred to as a(n) (5) _codon_. The (6) _third_ [choose one] ☐ first, ☐ second, ☑ or third base in a codon is much less critical in specifying the intended amino acid than is either of the other two.

TRANSLATION (cont.)
    Codon-Anticodon Interactions
    Ribosome Structure
    Stages of Translation

Summary

A *codon* (base triplet on mRNA) is a hydrogen-bonding site for an anticodon. An *anticodon* is a complementary base triplet on tRNA; at the opposite end of the same tRNA molecule is a specific amino acid. Codon–anticodon recognition is precise between their first two pairs. At the third base, some tRNA molecules can form hydrogen bonds with more than one kind of base; hence, they can recognize more than one codon. When the slot for its particular amino acid load clicks into place on the ribosome, the complementary anticodon hydrogen bonds to the codon of mRNA, and this temporary bonding positions the amino acid perfectly in order for the appropriate enzymes to form peptide bonds by dehydration synthesis and to incorporate the amino acid into the growing protein. That accomplished, the tRNA detaches from the codon and chugs off, searching for another amino acid like the first to bring back again. The mRNA strand continues to advance its codons through the groove in the ribosome until protein assembly is complete. In this way, protein chains of as many as 3000 amino acid subunits are assembled from a collection of twenty different amino acids. In summary, DNA transcribes its message into three forms of RNA, which cooperate to translate the message into a three-dimensional protein.

There are some exceptions to the universality of the genetic code. Most organisms use the same code (Figure 16.8 in main text), but mitochondria (organelles in eukaryotic cells) have some variant codons. Margulis and others believe that mitochondria once had free-living ancestors that later established symbiotic relationships with other cells.

Key Terms

anticodon	nanometer	chain elongation
"hook"	polysome	chain termination
"wobble" effect	initiation	release factor
ribosome	P site	A site

Objectives

1. Describe how the three types of RNA participate in the process of translation.
2. Summarize the steps involved in the transformation of genetic messages into proteins, using a diagram that shows the relationships among DNA, RNA, and proteins.

Self-Quiz Questions

Fill-in-the-Blanks

(1) _____ RNA alone carries the instructions for assembling a particular sequence of amino acids from the DNA to the ribosomes, where (2) _____ of the polypeptide occurs. (3) _____ RNA makes up much of the body of a ribosome, and (4) _____ RNA acts as a shuttle molecule as each type brings its particular (5) _____ _____ to the ribosome where

it is to be incorporated into the growing (6) _____. A(n) (7) _____ is a triplet on mRNA that forms hydrogen bonds with a(n) (8) _____, which is a triplet on tRNA.

Problems

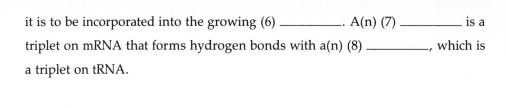

___ (9) Given the following DNA sequence, deduce the composition of the mRNA transcript.

TAC-AAG-ATA-ACA-TTA-TTT-CCT-ACC-GTC-ATC
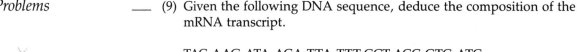

___ (10) From the mRNA transcript above, deduce the composition of the polypeptide sequence.

16-V
(pp. 217–221)

MUTATION AND PROTEIN SYNTHESIS
    Mutation at the Molecular Level
    Mutation Rates
    Mutation and Evolution
PERSPECTIVE
SUMMARY

Summary

Occasional errors occur during replication or during the separation of chromosomes during mitosis. In addition, high-energy radiation and mutagenic chemicals that enter cells can damage strands. Repair enzymes are limited in just how well they can fix damage to DNA, so *deletions* (in which one or more base pairs are lost), *insertions* (in which one or more extra base pairs are incorporated into the original sequence), and base substitutions occasionally occur. The overall precision of DNA replication and repair is the source of life's fundamental unity; the rare "mistakes" account for much of its diversity. Although most mutations are harmful, some may confer upon their bearers a selective advantage to survive in changing environments and to leave more offspring even as others perish. The mutations that are perpetuated and the increasingly abundant phenotypes that they foster are central to the evolutionary process.

    Asexually reproducing haploid organisms are finely tuned to operate under a given range of environmental conditions, but if a mutation occurs in one of their important genes or if there are radical changes in the environment, the outlook for their survival is grim.

    Most eukaryotic cells are diploid, with *two* of each type of chromosome characteristic of the species—one of each type of chromosome from each parent. For diploid organisms, even if a gene in one chromosome mutates, the equivalent gene on the other chromosome remains functional and may help the organism to survive. With these double sets of directions, a diploid organism can store a variety of instructions, which may be called into play at a later date should environmental conditions change.

Key Terms

gene mutation	sequence of base pairs	duplication
mutagens	crossing over	inversion
"base-pair substitution"	recombination	translocation
"frameshift mutation"	chromosomal aberration	aneuploidy
transposition	deletion	polyploidy
mutation rate		

Objectives

1. List some of the agents that can cause mutations.
2. Explain how repair enzymes act to minimize replication errors.
3. Describe the types of defects that occur in DNA from time to time and indicate whether repair enzymes can correct them.
4. Explain the relationship between gene mutations and organism diversity.
5. Explain why frameshift mutations that occur near the beginning of a DNA sequence of bases almost always result in the production of an unusable protein.

Self-Quiz Questions

Fill-in-the-Blanks

(1) _____-_____ _____ and (2) _____ _____ that enter cells can damage strands of DNA. (3) _____ occur when one or more extra base pairs are incorporated into the original sequence; (4) _____ occur when one or more base pairs are left out of the original sequence. (5) _____ mutations generally produce unusable proteins.

CHAPTER TEST

UNDERSTANDING AND INTERPRETING KEY CONCEPTS

____ (1) Transcription _____.
   (a) occurs on the surface of the ribosome
   (b) is the final process in the assembly of a protein
   (c) occurs during the synthesis of any type of RNA by use of a DNA template
   (d) is catalyzed by DNA polymerase

____ (2) _____ carries amino acids to ribosomes, where amino acids are linked into the primary structure of a polypeptide.
   (a) mRNA
   (b) tRNA
   (c) Introns
   (d) rRNA

____ (3) Transfer RNA differs from other types of RNA because it _____.
   (a) transfers genetic instructions from cell nucleus to cytoplasm
   (b) specifies the amino acid sequence of a particular protein
   (c) carries an amino acid at one end
   (d) contains codons

____ (4) _____ dominates the process of transcription.
   (a) RNA polymerase
   (b) DNA polymerase
   (c) Phenylketonuria
   (d) Transfer RNA

____ (5) _____ and _____ are found in RNA but not in DNA.
   (a) Deoxyribose, thymine   (c) Uracil, ribose
   (b) Deoxyribose, uracil   (d) Thymine, ribose

_a_ (6) Each "word" in the mRNA language consists of _____ letters.
   (a) three
   (b) four
   (c) five
   (d) more than five

_a_ (7) If each nucleotide is coded for only one amino acid, how many different types of amino acids could be selected?
   (a) Four
   (b) Sixteen
   (c) Twenty
   (d) Sixty-four

_d_ (8) The genetic code is composed of _____ codons.
   (a) three
   (b) twenty
   (c) sixteen
   (d) sixty-four

_a_ (9) Inborn errors of metabolism _____.
   (a) arise from the absence or deficiency of a vital enzyme
   (b) were first discovered by studies of sickle-cell anemia
   (c) can be corrected by a technique known as electrophoresis
   (d) are caused by misplaced promoters on segments of DNA

_d_ (10) The information content of genes is used to _____.
   (a) specify, in a overlapping reading frame, each amino acid in a polypeptide sequence
   (b) turn on and off specific chemical reactions by activating or inactivating specific enzymes
   (c) generate complementary templates on which carbohydrates and lipids are synthesized
   (d) build molecules of RNA

___ (11) Beadle and Tatum determined the particular substance(s) that could not be synthesized by the mutant forms of *Neurospora* by _____.
   (a) subjecting mutant cell extracts to density-gradient centrifugation
   (b) an elaborate series of electrophoresis studies
   (c) splicing normal genes into the *Neurospora* DNA and comparing the resulting activity with that of the mutant forms
   (d) keeping careful records of the kinds of nutritionally supplemented media on which the mutants could survive

___ (12) Crick and Brenner discovered that the presence of three extra nucleotides inserted in the middle of a gene caused far fewer problems than if only one or two extra nucleotides were inserted. They interpreted this result to mean that _____.
   (a) the genetic code consists of nonoverlapping triplets of nucleotide bases
   (b) the longer the sequence of nucleotides that is added to a gene, the more chemically stable the resulting DNA is
   (c) there had been significant experimental error in their electrophoresis studies
   (d) the "wobble effect" accounts for the unpredictability in codon-anticodon pairing at the third base

___ (13) Electrophoresis is a technique that has been used to study _____.
  (a) nutritional disorders in *Neurospora*
  (b) differences between the ways that mutant and normal forms of proteins behave in electric fields
  (c) how nitrogen bases are inserted into mRNA molecules
  (d) how mitochondrial DNA differs from nuclear DNA

___ (14) Mitochondrial DNA _____.
  (a) contains a few codons that specify amino acids other than what the codons of nuclear DNA specify
  (b) uses the same assortment of codons as does the DNA in the nucleus
  (c) can never replicate itself because DNA polymerases are not present in the mitochondria
  (d) can never be transcribed or translated because RNA polymerases are not in mitochondria

CHAPTER TEST

INTEGRATING AND APPLYING KEY CONCEPTS

Genes code for specific polypeptide sequences. Not every substance in living cells is a polypeptide. Explain how genes might be involved in the production of a storage starch (such as glycogen) that is constructed from simple sugars.

17

CONTROL OF GENE EXPRESSION

GENE REGULATION IN PROKARYOTES
    Mechanisms of Regulation
    Negative Controls
    Positive Controls

GENE REGULATION IN EUKARYOTES
    Selective Gene Expression
    Levels of Gene Control
    Chromosome Organization and Gene
    Activity
    Transcript-Processing Controls
    Gene Control of Cell Division
    Commentary: Altered Regulatory Genes
    and Cancer
    SUMMARY

General Objectives

1. Know the various ways that gene activity (replication and transcription) are turned on (activated) and off (inactivated).
2. Understand how operon controls regulate gene expression in prokaryotes.
3. Be able to visualize the organization of DNA in eukaryote chromosomes and understand how it affects gene expression.
4. Understand how differentiation proceeds by selective gene expression during development.
5. Understand how cancer may result when regulatory genes are altered.

17-I
(pp. 222–224)

GENE REGULATION IN PROKARYOTES
    Mechanisms of Regulation
    Negative Controls
    Positive Controls

Summary

Cells require internal controls over which proteins are assembled at any given moment of the life cycle. These controls govern transcription, translation, and enzyme activity. Genetic information encoded in DNA first becomes transcribed as mRNA and then is translated (with the help of tRNA and rRNA) into polypeptide chains. In prokaryotes, controls over gene expression are basically short-range responses to changing environmental conditions.

    Prokaryotes can respond to changes in the environment by activating or inactivating an *operon*—a set of genes that transcribes the appropriate pieces of mRNA, which in turn translate the code into a series of enzymes constituting a particular metabolic pathway. The term *operon* was coined to signify any set of protein-coding genes that operates as a coordinated unit under the direction of

DNA control elements. The lactose operon of *Escherichia coli* examined by Jacob and Monod is an example. Jacob and Monod identified all the genes in the gene system that regulated the lactose pathway. The lactose operon permits *E. coli* to metabolize lactose as an energy source when it is present in the bacterial environment by *repressing* transcription; a repressor protein bound to the operator portion of the DNA prevents RNA polymerase from binding to the promoter portion of the DNA. This interferes with transcription and is a type of negative control mechanism. An activator protein bound to the promoter portion of the DNA helps RNA polymersase to bind there also. The stronger the bond, the more mRNA molecules will be transcribed in a given period; the rate of transcription is increased. Transcription rates are decreased only when the activator protein is removed from the promoter by some agent.

Key Terms

Escherichia coli	negative control	Jacob and Monod
promoter	positive control	operon
repressor protein	regulator gene	RNA polymerase
activator protein	lactose operon	tryptophan operon
operator		

Objectives

1. Describe the studies of Jacob and Monod with *E. coli* nutrition and explain how their findings led them to develop the operon model.
2. Diagram an operon, locating the regulator, the promoter, the operator, and the structural genes. Indicate where the repressor protein, activator protein, and RNA polymerase bind to the operon.
3. Define the function of each of the components listed in Objective 2.
4. Describe how the rate of transcription can change from time to time.
5. Explain why it is advantageous to a cell to be able to activate and inactivate its enzymes in response to changes in the environment.

Self-Quiz Questions

Fill-in-the-Blanks

In prokaryotes, controls over gene expression are basically (1) _short_-range responses to changing environmental conditions; in eukaryotes, both types of controls govern events of (2) _development_ and (3) _differentiation_. (4) _Repression_ is the process by which genes are prevented from being transcribed and gene function becomes increasingly restricted. If not blocked by a (5) _repressor_ protein, (6) _RNA polymerase_ attaches to the promoter, and transcription begins. A(n) (7) _operon_ is any set of structural genes that operates as a coordinated unit under the direction of DNA control elements. Transcription is reactivated by inducing a change in the shape of the (8) _repressor_ molecule so that it can no longer inhibit transcription. The (9) _regulator_ codes for the formation of mRNA, which assembles a repressor protein. The affinity of the (10) _promoter_ for RNA polymerase dictates the rate at which a particular operon will be transcribed. Each differentiated cell

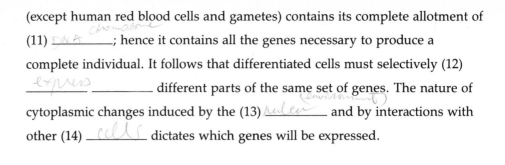

(except human red blood cells and gametes) contains its complete allotment of
(11) _DNA_ [chromosome]; hence it contains all the genes necessary to produce a
complete individual. It follows that differentiated cells must selectively (12)
express _____ different parts of the same set of genes. The nature of
cytoplasmic changes induced by the (13) _nucleus_ [(environment)] and by interactions with
other (14) _cells_ dictates which genes will be expressed.

17-II
(pp. 224–226)

GENE REGULATION IN EUKARYOTES
Selective Gene Expression
Levels of Gene Control

Summary

In eukaryotes, both short-range and long-range controls over gene expression
govern development and differentiation. Along the journey from gene to protein, transcriptional, transcript processing, translational, and posttranslational
controls exert their influence. In eukaryotes, the nuclear membrane acts as a
selective barrier that retains some mRNA and lets other mRNA pass into the
cytoplasm. Sometimes mRNA is changed once it reaches the cytoplasm. Sometimes the different amounts of enzymes and types of tRNA influence which
mRNA is translated into protein. And even after translation, proteins can be
activated, inactivated, or destroyed in a selective manner that can vary from cell
to cell or throughout the life of any given cell. Extracellular hormonal signals,
changing pH levels, and other events influence the amount and activity of the
proteins being synthesized.

Genetic controls direct the unfolding of a single-celled zygote into all the
specialized cells and structures of the adult form; in eukaryotes, the controls
span longer periods of time and involve more intricate cellular arrangements in
space than the controls in prokaryotes do. During _development_ of all complex
eukaryotes, cells come to differ in their positions in the body, in their developmental potential, and in their appearance, composition, and function. These
differences are brought about by mitosis and _differentiation_, during which cells of
the same genetic makeup become more or less abundant and become structurally
and functionally different from one another according to the prescribed program
for the species. Although all the cells in the body inherit the same genes, they
activate or suppress some fraction of these genes in different ways to produce
pronounced differences in their structure or functioning. Selective gene expression is controlled by agents that act within cells, between cells, and between cells
and the environment. Several studies using the slime mold _Dictyostelium discoi-
deum_ demonstrate this. In mammals, most of the regulatory proteins are activator
rather than repressor proteins. More than 90 percent of mammalian DNA is never
transcribed.

Key Terms

differentiate	_Dictyostelium discoideum_	transcript processing
selective gene expression	cyclic AMP	transport controls
signaling molecule	fruiting body	translational controls
target cell	transcriptional controls	posttranslational controls
hormone		

Objectives

1. Define *translational control*, indicate how it differs from *transcriptional control*, and give two examples of translational controls.
2. Define and distinguish between the terms *development* and *differentiation*. State which sorts of organisms are highly differentiated and which show very little differentiation.
3. Describe the life cycle of *Dictyostelium discoideum* and summarize it by means of a diagram with arrows.
4. Give an example of differentiation occurring in *D. discoideum*. Then state why its life cycle leads biologists to say that environmental change *and* interactions between cells influence the activity of genes necessary for differentiation.
5. Define and distinguish among the five levels that control the expression of genes.

Self-Quiz Questions

Fill-in-the-Blanks

Development in all complex eukaryotes consists of cell division and (1) _____, in which genetically identical cells become structurally and functionally distinct from one another. When food is scarce, *Dictyostelium discoideum* amoebas swarm together to form a (2) _____ that crawls around for a while and then differentiates to form a (3) _____ that eventually discharges spores. During much of this activity, some cells start producing the nucleotide (4) _____ _____ _____, which serves as a signaling agent that eventually stimulates differentiation. The life cycle of *D. discoideum* tells us that both (5) _____ _____ _____ and intercellular interactions influence genes to transcribe their products in a differential manner.

Short-Answer Essay

___ (6) If human males (XY) have no Barr bodies and human females have one Barr body per somatic cell, how many Barr bodies would you expect to find in each somatic cell of an XXX individual?

17-III
(pp. 227–230)

GENE REGULATION IN EUKARYOTES (cont.)
    Chromosome Organization and Gene Activity
    Transcript-Processing Controls

Summary

Most of the DNA and RNA in the nucleus is used to direct the activities of the control elements. Proteins called *histones*, which make up about half of all proteins in the nucleus, are tightly linked to DNA and form a nucleoprotein fiber that resembles a chain of beads. Each bead is a *nucleosome*, a complex packing arrangement that seems to be involved in controlling when transcription occurs.

During mitosis or meiosis, the entire chromatin network of nucleoprotein fibers—with their nucleosomes—condenses into chromosomes. Transcription apparently never occurs on any chromatin that has condensed into the chromosome state. Evidence that gene activity varies among cells within the same embryo comes from studies of the giant chromosomes in the salivary glands of midges; the transcription rate in cells with these chromosomes has been correlated with how large and diffuse the regional puffs appear. Puffing varies from one DNA region to the next during different developmental stages. In mammalian females, one of each pair of X chromosomes always remains condensed so that, of any pair of X chromosomes, only one X chromosome is being transcribed in a given cell. The condensed X chromosome is called a *Barr body*. The process of differential transcription is known as *Lyonization*, which ensures that any XX female is a mosaic of X-linked traits.

Transcriptional gene controls are most prevalent. In prokaryotes, operon controls influence the rate of transcription. In eukaryotes, chromosome organization influences the timing and rate of transcription. All cells depend on controls that can respond to short-term changes in cellular or extracellular conditions. Only eukaryotic cells depend on long-term controls over development and differentiation, as well.

Key Terms

histones	polytene chromosomes	calico cats
nucleosome	ecdysone	anhydrotic ectodermal
looped domains	Barr body	dysplasia
lampbrush chromosomes	"mosaic"	introns
chromosome puffs	Lyonization	exons

Objectives

1. Describe how eukaryotic DNA is organized in the nucleus.
2. Define *histones* and explain their relationship to DNA and nucleosomes. Comment on the supposed function of histones.

Self-Quiz Questions

Fill-in-the-Blanks

Cells depend on (1) _____, which govern transcription, translation, and enzyme activity. (2) _____ controls and (3) _____ _____ are two principle control categories that operate in the nucleus.

A (4) _____ _____ is a condensed X chromosome; the process by which an XX embryonic cell chooses which X chromosome will remain active is called (5) _____.

(6) _____ and DNA are tightly linked into a (7) _____ fiber, which looks like a beaded chain. Each bead is called a(n) (8) _____, which is thought to be involved in shutting down transcription.

True-False

If false, explain why.

____ (9) In the polytene chromosomes of *Drosophila*, each band represents the location of one or more genes.

___ (10) Polytene chromosomes develop puffing patterns that vary at different times during development; specific puff patterns correlate strongly with the transcription and translation of specific assemblages of proteins.

17-IV
(pp. 230–233)

GENE REGULATION IN EUKARYOTES (cont.)
    Gene Control of Cell Division
      Commentary: **Altered Regulatory Genes and Cancer**
SUMMARY

Summary

All cells have interacting control elements (such as enzymes, regulatory proteins, and noncoding elements of DNA) that determine which gene products appear at what times and in what amounts. In complex eukaryotes, differentiation of cells arises through selective gene expression; in a given cell type, some genes are activated and others are repressed.

Nuclear-cytoplasmic interactions that maintain variable gene expression do break down occasionally. Sometimes a single cell loses the controls that indicate when to stop cell division; it divides endlessly, crowding certain cells and interfering with vital cell functions. The mass that results will either be *benign*, with only its controls over cell division and growth gone awry, or *malignant*, with the recognition factors on its cell surfaces also disturbed. Benign tumors generally can be removed surgically. Malignant growth can invade and destroy surrounding tissues. Some malignancies do not stay as one solid tumor, but *metastasize*, or disperse to other sites in the body. Such cancers are generally treated with chemical agents, radiation, or both to destroy uncontrollably dividing cells selectively wherever they are in the body. Such traumatic changes in cellular controls seem to have a variety of causes; specific viruses, chemicals, continual prolonged exposure to sunlight, and constant physical irritation of some tissue all may increase susceptibility to some forms of cancer.

It now appears that a small number of altered regulatory genes may give rise to at least some kinds of cancer. Proto-oncogenes are inherent parts of vertebrate DNA, and they encode for proteins necessary in normal cell functioning. They become cancer-causing genes (oncogenes) only on those rare occasions when specific mutations alter their structure or their expression.

Key Terms

tumor	metastasis	adenoviruses
"clonal"	oncogenes	herpesvirus
benign tumor	Rous sarcoma virus, RSV	proto-oncogenes
malignant	papovaviruses	carcinogens
cancer cells		

Objectives

1. Give an example of cells that normally cease dividing. Then state how one such cell could be made to begin dividing again.
2. Define *tumor* and distinguish between *benign* and *malignant* tumors.
3. Explain what happens when a tumor metastasizes. Use leukemia as an example. Then describe how the treatment of a metastasized cancer differs from the treatment of benign tumors.
4. Describe the relationship of proto-oncogenes, environmental irritants, and oncogenes.

Self-Quiz
Questions

True-False

If false, explain why.

___ (1) When cells become cancerous, cell populations decrease to very low densities and stop dividing.

___ (2) All abnormal growths and massings of new tissue in any region of the body are called *tumors*.

___ (3) Malignant tumors have lost only their normal surface markers, but benign tumors have lost both their surface markers and the mechanism that stops mitosis.

___ (4) The term *metastasis* means the acquiring of the shape or form typical of that species.

Multiple Choice

Choose the best answer from the following five lettered choices.

(5) ___ cause fever blisters and cold sores.

(6) ___ cause AIDS.

(7) ___ cause cancer in chickens.

(8) ___ cause lung infections that are curable.

(9) ___ cause genital infections and cancer.

(10) ___ cause warts.

A. Adenoviruses
B. Herpesviruses
C. Papovaviruses
D. Rous sarcoma viruses
E. None of the above

CHAPTER TEST

UNDERSTANDING AND INTERPRETING KEY CONCEPTS

___ (1) _____ refers to the processes by which cells with identical genotypes become structurally and functionally distinct from one another according to the genetically controlled developmental program of the species.
(a) Metamorphosis
(b) Metastasis
(c) Cleavage
(d) Differentiation

___ (2) The _____ determines the rate at which a certain mRNA chain is to be synthesized.
(a) intron
(b) repressor gene
(c) promoter sequence
(d) operator sequence

___ (3) Lampbrush chromosomes are generally seen _____.
(a) in the salivary glands of fruit flies
(b) in many animals' eggs during oogenesis
(c) in embryonic cells during morphogenesis
(d) in cells that have signs of advanced deterioration

___ (4) Jacob and Monod's model of the operon explains the regulation of _____
    in prokaryotes.
    (a) replication
    (b) transcription
    (c) induction
    (d) Lyonization

___ (5) Which of the following characteristics seems to be most uniquely
    correlated with metastasis?
    (a) Loss of nuclear-cytoplasmic controls governing cell growth and division
    (b) Changes in recognition factors on membrane surfaces
    (c) "Puffing" in the polytene chromosomes
    (d) The massive production of cyclic adenosine monophosphate and its
        secretion into the environment

___ (6) The polytene chromosomes of *Drosophila* salivary glands provide
    evidence that _____.
    (a) the transcription rate of a particular region on a polytene chromo-
        some increases as this region enlarges and becomes more diffused
    (b) maternal messages are formed during gametogenesis
    (c) the RNA assembled on lampbrush chromosomes apparently directs
        initial developmental events
    (d) "puffing" apparently is responsible for the forming of recognition
        factors in the surface membranes of most organisms

Matching

(7) ___ Control over enzyme activity

(8) ___ Posttranscriptional control

(9) ___ Posttranslational control

(10) ___ Transcriptional control

(11) ___ Translational control

A. Influences the amount and kinds
   of mRNA assembled from
   structural genes
B. Snipping out of transcript regions;
   chemical modification of transcript
   before it arrives at ribosome
C. Nuclear envelope selectively
   regulates passage of transcripts
D. Phosphorylation of translated
   protein
E. For example, the interior
   environmental pH determines
   whether a protein can act as a cell
   catalyst

CHAPTER TEST

INTEGRATING AND APPLYING KEY CONCEPTS

Suppose that you have been restricting yourself to a completely vegetarian diet
for the past six months. Quite unexpectedly, you find yourself in a social situation
that requires you to eat a half-pound sirloin steak. Would you expect to digest
the steak as easily as you digest soybean burgers? Explain your yes or no answer
in terms of transcriptional controls or feedback inhibition.

18

RECOMBINANT DNA AND GENETIC ENGINEERING

General Objectives

1. Know how genetic recombination occurs naturally.
2. Understand what plasmids are, how they are instrumental in conferring resistance to drugs, and how they may be used to insert new genes into recombinant DNA molecules.
3. Know how DNA can be cleaved, spliced, cloned, used as a probe, and extracted.
4. Be aware of several limits and possibilities for future research in genetic engineering.

18-I
(pp. 235–238)

NATURAL RECOMBINATION MECHANISMS
    Transposable Elements
    Plasmids
    Viruses as Transposable Elements

Summary

Prokaryotes ordinarily reproduce by prokaryotic fission, an asexual process in which the partitioning of a parental cell leads to two separate cells having basically similar parts. *Conjugation* is an additional process that occurs occasionally in prokaryotes and results in *genetic recombination*—the production of individuals having new assortments of genes that were acquired, in one way or another, from more than one parent cell—and is regarded by many as a possible method of exchanging genetic information that helped diversify early prokaryotes. Recombination also occurs naturally through the activity of transposable elements including transposons, plasmids, and viral DNA. These elements have the means to insert themselves into another DNA molecule (or another location in the same molecule) through site-specific recombination. *Site-specific recombination* is a common mechanism by which genes can be transferred from one DNA molecule to another.

Plasmids are small circles of DNA that are found in some bacteria in addition to the main circular DNA molecule. *Integrated F plasmids* help transfer a portion of the main DNA molecule into an F$^-$ bacterial cell during *conjugation*; it will then undergo crossing over and general recombination. *R plasmids* confer to recipient bacteria resistance to antibacterial drugs; thus, resistant bacteria responsible for intestinal tract disorders, gonorrhea, typhoid, and meningitis are increasingly encountered.

Genetic experiments have been occurring in nature for billions of years. Mutations, homologous recombination at meiosis, the novel assortments of alleles brought together at fertilization, and hybridizations between species have all contributed to the current diversity among organisms. *Recombinant DNA technology* is founded on procedures by which DNA molecules can be cut into fragments, then joined with similarly cut fragments from any organism to form recombinant DNA molecules that can be propagated in a line of dividing cells.

Key Terms

recombinant DNA technology	insertion sequence	conjugation
homologous recombination	transposase	site-specific recombination
reciprocal	plasmids	antibiotic
transposable elements	F plasmids	lambda bacteriophage
transposons	R plasmids	lytic pathway
	pilus	lysogenic pathway

Objectives

1. Explain why anyone should care about McClintock's discovery of transposons. Indicate how the tag-along promoter is involved in *jumping genes*.
2. Distinguish *integrated F plasmids* from *R plasmids*.
3. State which organisms carry out site-specific recombination and describe the process using terms such as *lytic pathway* and *lysogenic pathway*.

Self-Quiz Questions

True-False

If false, explain why.

_____ (1) *Plasmids* are organelles on the surfaces of which amino acids are assembled into polypeptides.

Matching

Match the most appropriate letter with its numbered partner.

(2) ___ integrated F plasmid

(3) ___ DNA ligase

(4) ___ lambda bacteriophages

(5) ___ lysogenic pathways

(6) ___ lytic pathways

(7) ___ Barbara McClintock

(8) ___ pilus

(9) ___ R plasmids

(10) ___ restriction enzymes

A. Kept telling everyone about transposons forty years ago, but no one paid attention
B. Serves in pair formation; may also be a bridging tube
C. Infection proceeds by integrating viral DNA into host DNA; both DNAs are then replicated in a latent state.
D. Pasting DNA fragments
E. Confers resistance to antibacterial drugs
F. Helps transfer a portion of the main DNA molecule into an F⁻ cell
G. Experiments with these organisms shed light on the process of site-specific recombination
H. Cutting DNA
I. Viral DNA free in cytoplasm fosters the assembly of many viral particles that burst open the bacterial cell

---

18-II
(pp. 238–246)

RECOMBINANT DNA TECHNOLOGY
    Producing Restriction Fragments
    Preparing and Cloning a DNA Library
    Identifying the Cloned DNA of Interest
    Selected Gene Amplification
    Expressing the Cloned Gene
    Gene Sequencing
GENETIC ENGINEERING: RISKS AND PROSPECTS
    Genetically Engineered Bacteria
    Genetically Engineered Plants
    Genetically Engineered Animals
    Human Gene Therapy
SUMMARY

Summary

Recombinant DNA is a hybrid molecule that results from breaking apart the DNA from one organism and combining it with DNA from another organism. *Restriction enzymes* can be used to cut open plasmids at a particular site, and *DNA ligase*, an enzyme, can be used to attach some foreign DNA to the opened plasmid; then the plasmid is closed again in its circle. A *cloning vector* is a plasmid, virus, or any other self-replicating genetic element that can be inserted into a host cell for propagation. A collection of DNA fragments produced by restriction enzymes and incorporated into cloning vectors is called a *DNA library*. Under certain conditions, the hybrid DNA plasmid can be introduced into new *E. coli* bacterial cells, and it will then be reproduced every time the bacterium undergoes cell division. A *DNA clone* is any DNA library that has been amplified in a line of dividing cells or in a cell-free system (such as a test tube in which polymerase chain reactions proceed). The bacterium's metabolic machinery can churn out many multiple copies, or *clones*, of the genes carried on the hybrid plasmid ring;

with repeated doublings of a bacterial population, millions of these genes can be manufactured overnight. Three approaches to DNA cloning include *shotgun cloning*, *cDNA cloning*, and *cloning of synthesized genes*. Recombinant DNA techniques hold great promise for transferring genes from one organism to another for manufacturing proteins such as insulin in a pure form for organisms that cannot make the required substances themselves. Entirely new species with entirely new assemblages of characteristics are now being synthesized in specially equipped laboratories following very strict safety rules.

Key Terms

genome
restriction enzymes
recognition sequence
DNA ligase
cloning vector
DNA library
recombinant plasmid
cloned DNA
"shotgun cloning"

reverse transcriptase
complementary DNA,
    cDNA
nucleic acid hybridization
    techniques
cDNA probe
autoradiography
polymerase chain reaction
primers

promoter
gel electrophoresis
ice-minus bacteria
halophytes
glycophytes
Brinster, Palmiter, and
    others
gene therapy
eugenic engineering

Objectives

1. Describe the process by which recombinant DNA is produced. Use terms such as *plasmid* and *restriction enzymes* in ways that make their meanings clear.
2. Describe the process known as a *probe*.
3. Define *clone* and explain how clones of recombinant DNA can be useful to people suffering from diabetes.
4. Describe how recombinant DNA techniques theoretically could transfer to crop plants the ability to harvest nitrogen (N_2) from the air in order to make proteins, thereby reducing modern agriculture's demand for expensive nitrogen-containing fertilizers.
5. Name one problem that some scientists fear will result from recombinant DNA research. Then state what research geneticists are doing to guard against such a possibility.

Self-Quiz Questions

Fill-in-the-Blanks

(1) _____ protect bacteria against viral infection by cleaving the viral DNA molecule at specific nucleotide sequences. (2) _____ _____ seal the DNA and the plasmid into a small circle; the resulting circle of recombinant DNA is a (3) _____ _____. A cloned gene coding for the protein (4) _____ is now being mass-produced. A(n) (5) _____ detects the specific sequence coding for the desired protein among many thousands of fragments. Mammalian cells can be made to incorporate foreign DNA in a process called (6) _____. (7) _____, which are plants adapted to conventional nonsalty conditions, may be alterable through genetic engineering into more salt-tolerant, high-yield strains of those plants.

UNDERSTANDING AND INTERPRETING KEY CONCEPTS

___ (1) In the lysogenic pathway of the life cycle of the lambda bacteriophage, _____.
   (a) bacterial DNA becomes incorporated in viral DNA
   (b) viral DNA becomes incorporated in bacterial DNA
   (c) viral DNA manufactures bacterial DNA
   (d) viral DNA causes the bacteria to manufacture enzymes that split the bacteria apart
   (e) viral DNA forces the bacteria to manufacture more viral particles

___ (2) Site-specific recombination is characteristic of _____.
   (a) alpha viruses
   (b) beta bacteriophages
   (c) phi particles
   (d) lambda bacteriophages
   (e) gamma viruses

___ (3) Site-specific recombination is involved in _____.
   (a) production of red blood cells
   (b) embryonic development
   (c) coding for antibodies
   (d) gamete formation
   (e) all of the above

___ (4) Small circular molecules of DNA in bacteria are called _____.
   (a) plasmids
   (b) desmids
   (c) pili
   (d) F particles
   (e) transferins

___ (5) Bacteria reproduce sexually by _____.
   (a) fission
   (b) gametic fusion
   (c) conjugation
   (d) lysis
   (e) none of the above because bacteria only reproduce asexually

___ (6) The _____ plasmids protect bacteria from antibiotic drugs.
   (a) B
   (b) F
   (c) H
   (d) L
   (e) R

___ (7) Enzymes used to cut genes in recombinant DNA research are _____.
   (a) ligases
   (b) restriction enzymes
   (c) transcriptases
   (d) DNA polymerases
   (e) replicases

___ (8) "Male" bacteria have _____ that allow two bacteria to join together to transfer genes.
 (a) flagella
 (b) pores
 (c) connecting channels
 (d) pili
 (e) stylets

___ (9) Probes for cloned genes use _____.
 (a) complementary nucleotide sequences labeled with radioactive isotopes
 (b) specific media that contain specific antibodies
 (c) specific enzymes
 (d) certain bacteria sensitive to the genes
 (e) all of the above

___ (10) Recombinant DNA research uses plasmids and _____ as cloning vectors.
 (a) *E. coli*
 (b) viruses
 (c) plants
 (d) fungi
 (e) any type of host cell

CHAPTER TEST

INTEGRATING AND APPLYING KEY CONCEPTS

How could scientists guarantee that *Escherichia coli* will not be transformed into a severely pathogenic form and released into the environment, if researchers use the bacterium in recombinant DNA experiments?

19

PLANT CELLS, TISSUES, AND SYSTEMS

General Objectives

1. Describe the generalized body plan of a flowering plant.
2. Define and distinguish among the various types of ground tissues, vascular tissues, and dermal tissues.
3. Explain how plant tissues develop from meristems.
4. Know the functions of stems, leaves, and roots.
5. Explain what is meant by *secondary growth* and describe how it occurs in woody dicot roots and stems.

19-I
(pp. 248–251)

THE PLANT BODY: AN OVERVIEW
    Monocots and Dicots
    Shoot and Root Systems
PLANT TISSUES
    Ground Tissues

Summary

There are more than 275,000 species of plants. *Nonvascular* plants (certain algae, mosses, and liverworts) lack all but the simplest of internal transport systems; *vascular* plants (ferns, gymnosperms, and angiosperms) have well-developed internal tissues that conduct food and water through the multicellular plant body. *Gymnosperms* include pines, junipers, and redwoods; *angiosperms* are represented by plants as diverse as roses, cherry trees, corn, and dandelions. The plant body differentiates first into three kinds of vegetative organs: root, stem, and leaf. *Primary growth* causes roots, stems, and leaves to increase in length; *secondary growth* causes roots and stems to increase in thickness.

Among the flowering plants (angiosperms), *monocots* differ from *dicots* in body structure and in amount of secondary growth.

Three tissue systems—dermal, vascular, and ground—extend continuously throughout the plant body. Tubes and fibers of *vascular* tissue are embedded in *ground* tissue, which in turn is covered and protected by *dermal* tissue. The *ground* system comprises most of the plant body. It consists mainly of three types of tissue: *parenchyma*, which participates in photosynthesis, food storage, wound healing, and regeneration of plant parts; *collenchyma*, a thin-walled flexible tissue that helps support young plant parts; and *sclerenchyma*, a rigid-walled strengthening tissue that consists mainly of dead cells.

The vegetative body of vascular plants grows continuously throughout its lifetime, producing the same tissue systems that were derived from apical meristems. The structure and function of these tissue systems depend on their locations in the mature plant body.

Key Terms

vascular plants	root system	vascular tissues
nonvascular plants	tissue	dermal tissues
gymnosperms	primary growth	parenchyma
angiosperms	"primary" tissues	collenchyma
monocots	secondary growth	sclerenchyma
dicots	"secondary" tissues	sclereids
"cotyledon"	ground tissues	fibers
shoot system		

Objectives

1. Define and give examples of *nonvascular plants* and *vascular plants*.
2. Distinguish *angiosperms* from *gymnosperms* and give examples of each.
3. Compare the characteristics of monocots with those of dicots.
4. Name the three tissue systems that extend throughout the plant body.
5. Define and distinguish from each other the three principal tissue types that compose most of the ground system of the plant body.
6. State the function(s) of the three types of ground tissue and tell where each is located in the plant body

Self-Quiz Questions

Fill-in-the-Blanks

Root and shoot tips have dome-shaped (1) _____ meristems where new cells form rapidly through mitotic divisions. These meristems give rise to three other kinds of embryonic tissues: (2) _____, which will form the epidermis; (3) _____, which will form the vascular tissue; and the (4) _____ _____, which will form most of the plant body's substance.

(5) _____ is an example of ground tissue that lies centrally inside the rings of vascular bundles in plant stems. (6) _____ is ground tissue that lies between the vascular ring and the dermal surface layer of plant stems and

roots. The main photosynthetic area of a leaf is composed of (7) _____, which lies between the leaf's upper and lower (8) _____.

(9) _____ cells form continuous storage tissues in the stem and root cortex and in stem (10) _____. (11) _____ cells are responsible for healing wounds, regenerating plant parts, and ensuring successful grafts. (12) _____ cells contain cellulose and (13) _____, which are hydrophilic substances promoting pliable cell walls that support young plant parts. (14) _____ tissues strengthen mature plant parts with (15) _____ and fibers.

19-II
(pp. 251–253)

PLANT TISSUES (cont.)
Vascular Tissues
Dermal Tissues
How Plant Tissues Arise: The Meristems

Summary

The *vascular system* contains food-conducting tissues (*phloem*) and water-conducting tissues (*xylem*) that form a network throughout the entire plant. *Xylem* contains parenchyma cells that store water and food, sclereids and fibers that provide mechanical support, and *tracheids* and *vessel* members that passively conduct water and dissolved salts. Like xylem, *phloem* also contains parenchyma, sclereids, and fibers. Unlike the tracheids and vessel members of xylem, which die as soon as they reach maturity, the food-conducting cells of phloem (called *sieve-tube members* or sieve cells) are alive when mature, as are their adjacent parenchyma cells (called *companion cells*), which help load and unload the sieve element pipelines.

In general, *epidermis*, the outermost layer of cells, covers the plant as a single, compact, and continuous layer of cells. Root hairs in root epidermis promote water absorption from the surrounding soil. *Cuticle*, the surface coat on aerial plant parts, restricts water loss and discourages microbial attack. Certain secretory structures that aid in attracting insects, discouraging predators, and ridding the plant body of excess salts are also associated with the epidermis. Periderm is a protective covering that replaces the epidermis of gymnosperms and angiosperms when they are undergoing secondary growth.

Many parts of the mature plant body contain clusters of cells that are undifferentiated; these "embryonic" zones (where mitosis and differentiation occur) are known as *meristems*. Newly formed cells differentiate first into three more specialized embryonic tissues: *protoderm*, which develops into the plant's surface layers; the *ground meristem*, which forms the bulk of the plant body; and *procambium*, which differentiates still further to become the primary vascular tissues.

Secondary growth, which increases the diameter of older stems and roots, originates with two types of lateral meristems; *vascular cambium* forms secondary xylem and phloem, and *cork cambium* forms periderm.

Key Terms

xylem	vessel members	"perforation plates"
phloem	"pit-pairs"	sieve-tube members
tracheids	vessel	sieve cells

"sieve plates"   cork   lateral meristems
companion cells   apical meristem   vascular cambium
epidermis   primary meristematic   cork cambium
cuticle   tissues   primary xylem and
root   protoderm   phloem
hair cells   ground meristem   secondary xylem and
guard cells   procambium   phloem
periderm

Objectives

1. Identify the three main functions of xylem and name the types of cells that participate in each function.
2. Distinguish the structural features of tracheids from those of vessel members.
3. List the principal differences in xylem and phloem tissues and identify the undifferentiated tissue that both arise from.
4. Identify the three main functions of phloem and name the types of cells that participate in each function.
5. Explain how sieve cells differ from companion cells.
6. Distinguish between *epidermis* and *periderm* in terms of structure.
7. Describe how root epidermis differs in function and in structure from shoot epidermis.
8. List some of the structures commonly found in epidermis that change its function from protection to some other role.

Self-Quiz Questions

Fill-in-the-Blanks

Vascular tissues consist of primary (1) _____, which conducts water and ions from the roots to the photosynthetic areas, and primary (2) _____, which conducts the products of photosynthesis away to storage areas and helps support the plant. The highly specialized cells of xylem that passively conduct water and (3) _____ are called (4) _____ _____, which are perforated at one end, and (5) _____, which are not perforated but have (6) _____ that allow water to move from cell to cell. The specialized cells of phloem that conduct and store (7) _____ are called (8) _____-_____; the end walls of these cells are called (9) _____ _____ because they have larger pores than do the side walls. (10) _____ _____ are thought to help load and unload the phloem pipelines.

In a young plant, (11) _____, the outermost layer of cells, is a single, continuous protective covering of cells. Root epidermis tends to promote (12) _____ _____ from the surrounding soil; shoot epidermis tends to restrict (13) _____ _____ and discourage (14) _____ _____.

THE PRIMARY SHOOT SYSTEM
Stem Primary Structure
Formation of Leaves and Buds
Leaf Structure

Summary

Stems provide a structure that (1) exposes leaves to light; (2) provides routes for food and water movement between roots and leaves; (3) stores food in the parenchyma cells of the cortex, pith, xylem, and phloem; and (4) displays reproductive structures. The stem and its leaves constitute the shoot. At specific points along the stem (nodes), leaves are differentiated and buds develop. Flowering plants have two general patterns of stem structure. Most *monocot* stems usually are uniformly thick along their length; they generally do not undergo secondary growth, and cross sections of their stems reveal that strands of vascular tissue are arranged in many bundles scattered throughout the ground tissue. Most *dicot* stems are usually tapered; they often undergo secondary growth, and their vascular bundles are arranged in a cylinder that separates the ground tissue into pith and cortex. Most leaves are the site for photosynthesis; their structure promotes the absorption of sunlight and the rapid uptake of carbon dioxide.

Generally, dicot leaves have a stalklike petiole attached to the stem; joined to the petiole may be one leaf or several leaflets. Most monocot leaves lack a petiole; the base of the blade forms a sheath around the stem, as in corn. Veins that contain vascular bundles connect the stem's vascular tissue with photosynthetic mesophyll, which is sandwiched between the upper and lower epidermis. *Stomata* are gates to the interior of leaves and stems that may be opened or closed to regulate the inward movement of carbon dioxide and the outward movement of water.

Key Terms

vascular bundle	axillary bud primordia	"deciduous"
cortex	axil	"evergreen"
pith	leaf, leaves	veins
transverse	blade	palisade mesophyll
radial	petiole	spongy mesophyll
tangential	sheath	stoma, stomata
leaf primordium, -dia	"simple" leaf	guard cells
node	"compound" leaf	turgor pressure
internode	leaflet	

Objectives

1. Describe the primary structure of a generalized stem. Distinguish *nodes* from *internodes*, tell where buds are located, and indicate how cross sections of monocot stems differ from those of dicot stems.
2. Show how guard cells are related to a stoma and tell what general roles are performed by guard cells.
3. Describe the primary structure of a leaf. List the principal functions of leaves and tell how leaf structure enables these functions to be carried out.
4. Describe some modifications in leaf structure that enable plants to exploit exceedingly dry or exceedingly wet environments.

Self-Quiz Questions

Fill-in-the-Blanks

All cells of a dicot stem are derived from the (1) _____ _____ at its tip; lateral outgrowths of this structure are called (2) _____ _____, and they develop into the mature leaves of the stem. (3) _____ are parts of the stem where one or more leaves are attached. (4) _____ _____ are strands of xylem and phloem in the ground tissue; ground tissue located within a ring of these is called (5) _____; ground tissue located outside such a ring is called (6) _____.

Leaves are differentiated and buds develop at specific points along the stem called (7) _____. (8) _____ stems generally do not undergo secondary growth, and cross sections of their stems reveal (9) _____ _____ scattered throughout the ground tissue. Most leaves absorb (10) _____, take up carbon dioxide from the air, and use (11) _____ delivered by the xylem in carrying out the complex series of chemical reactions known as (12) _____. (13) _____ are gates in the stem and leaf epidermis that are flanked by (14) _____ _____. (15) _____ replaces epidermis in older gymnosperms and angiosperms.

Labeling

Identify each indicated part of the accompanying illustration.

(16) _____   (18) _____   (20) _____

(17) _____   (19) _____

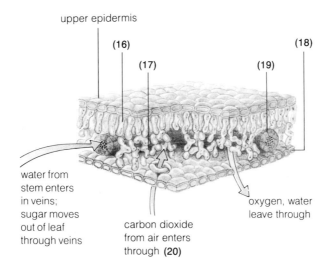

upper epidermis

(16)

(17)

(18)

(19)

water from stem enters in veins; sugar moves out of leaf through veins

carbon dioxide from air enters through (20)

oxygen, water leave through

Plant Cells, Tissues, and Systems   *161*

THE PRIMARY ROOT SYSTEM
    Taproot and Fibrous Root Systems
    Root Primary Structure

Summary

In gymnosperms and flowering plants, the first (primary) root develops from the apical meristem at the tip of the embryo's root. Both taproot and fibrous root systems anchor the plant and store photosynthetically derived food; their most important function is to absorb water and dissolved ions from the soil. The meristematic tip of the primary root, shielded by its root cap as it pushes its way through the soil, gives rise to cells that differentiate into the vascular, dermal, and ground systems. In a mature root, epidermis with root hairs surrounds the cortex, which surrounds a vascular cylinder. Immediately outside the vascular cylinder lies the *endodermis*, a specialized, innermost layer of cortex; this single layer of cells, through active and passive transport mechanisms, helps control the movement of water and dissolved salts into the xylem pipeline to the stem and leaves.

Secondary growth in many gymnosperms and flowering plants increases the diameter of roots and stems. *Procambium,* which lies between primary xylem and phloem in each vascular strand, develops into *vascular cambium*, which gives rise to secondary xylem and phloem. As the volume of new phloem increases, it pushes the vascular cambium toward the root's periphery. Part of the ground tissue between the phloem and the endodermis begins dividing soon afterward and becomes the *pericycle*; pericycle cells line up and merge with the vascular cambium to produce layer after layer of secondary xylem and phloem. Cork cambium arises on the phloem tissue's outer ridges and produces *periderm*, the corky covering that replaces the epidermis.

Key Terms

primary root	root cap	endodermis
lateral roots	root epidermis	Casparian strip
taproot system	pericycle	cortex
adventitious root	root apical meristem	vascular column
fibrous root system	vascular column	pith

Objectives

1. Distinguish between the taproot system and the fibrous root system in terms of structure and developmental origins.
2. State the principal functions of roots and explain how the structural components of roots allow those functions to be performed.
3. Describe how secondary growth increases the diameter of a root. Define all the structural features that participate in the secondary growth of roots.

**Self-Quiz
Questions**

Fill-in-the-Blanks

(1) _____ systems are based on a primary root and its lateral branches;

(2) _____ root systems arise from the young stem. The root (3) _____

protects the delicate young root from being torn apart during growth through

the soil; (4) _____ _____ gives rise to cells that differentiate into the

three basic tissue systems. (5) _____ with root hairs surrounds the

(6) _____, which surrounds the (7) _____ _____. Immediately

outside the vascular column lies the innermost layer of cortex, the
(8) _____, which regulates water and solute movements into the xylem.
During secondary growth, the procambium develops into (9) _____
_____, a thin layer of lateral meristem that gives rise to secondary xylem
and phloem. Divisions in the (10) _____ _____ produce periderm,
which replaces the epidermis.

Labeling   Identify each indicated part of the following illustration.

(11) _____ _____   (15) _____ _____   (18) _____

(12) _____   (16) _____   (19) _____

(13) _____   (17) _____   (20) _____ _____

(14) _____ _____ _____

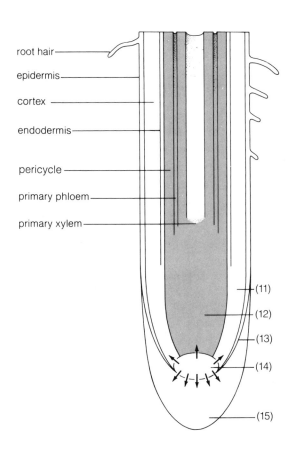

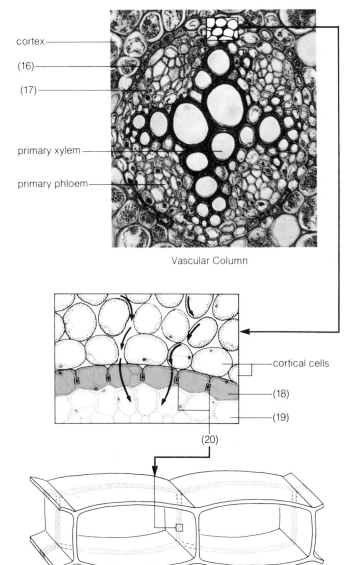

Vascular Column

SECONDARY GROWTH
    Seasonal Growth Cycles
    Vascular Cambium Activity
    Cork Cambium Activity
    Early and Late Wood
SUMMARY

Summary

Secondary growth occurs in stems much as it does in roots. The fusiform initials and the ray initials—two kinds of cells in the vascular cambium—produce the vertical system and the ray system, respectively, in woody stems. The *vertical system* conducts the food and water up and down the stem; the *ray system* consists mostly of parenchyma cells that act as lateral conduits and food-storage centers. Through the ray system, the vascular cambium and phloem are fed water, and the food from the secondary phloem enters the vascular cambium and the still-living cells of secondary xylem. Growth rings are produced by alternate periods of activity and inactivity in the vascular cambium.

Key Terms

herbaceous plants	fusiform initials	bark
woody plants	vertical system	heartwood
annuals	ray initials	sapwood
biennials	ray system	early wood
perennials	cork	late wood

Objectives

1. Explain how some stems increase their diameters each year. Contrast the functions of the vertical and ray systems of xylem.
2. Tell how the xylem cells of woody plants indicate the relative amounts of water available to the plant from season to season.
3. Define and distinguish among *annuals*, *biennials*, and *perennials*.

Self-Quiz Questions

Fill-in-the-Blanks

Two kinds of cells in the vascular cambium produce the (1) _____ and (2) _____ systems in woody stems; the (3) _____ system consists mostly of (4) _____ cells that act as lateral conduits and food storage centers. Growth rings are produced by fluctuating levels of activity in the (5) _____ _____.

Labeling   Identify each indicated part of the accompanying illustration.

(6) _____ _____   (8) _____ _____   (10) _____

(7) _____ _____   (9) _____

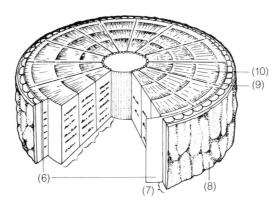

UNDERSTANDING AND INTERPRETING KEY CONCEPTS

___ (1) _____ develops into the plant's surface layers.
   (a) Procambium
   (b) Protoderm
   (c) Ground meristem
   (d) Pericycle

___ (2) Which of the following is *not* considered part of the ground system?
   (a) Endodermis
   (b) Pericycle
   (c) Leaf mesophyll
   (d) Periderm

___ (3) The main photosynthetic area of a leaf is composed of _____.
   (a) mesophyll
   (b) cortex
   (c) xylem
   (d) epidermis

___ (4) Annual growth layers are formed in woody stems principally through the activities of _____.
   (a) the pericycle
   (b) the pith
   (c) vascular cambium
   (d) mesophyll

___ (5) Leaves are differentiated and buds develop at specific points along the stem called _____.
  (a) nodes
  (b) internodes
  (c) vascular bundles
  (d) cotyledons

___ (6) Which of the following structures is *not* considered to be meristematic?
  (a) Vascular cambium
  (b) Procambium
  (c) Protoderm
  (d) Endodermis

___ (7) Which of the following is *not* generally true of monocot stems?
  (a) They do not undergo secondary growth.
  (b) They are not tapered along their length.
  (c) Their vascular bundles are scattered throughout the ground tissue.
  (d) Monocot stems have a single central vascular cylinder.

___ (8) Of the following tissues, the ground meristem would *not* give rise to _____.
  (a) epidermis
  (b) leaf mesophyll
  (c) cortex
  (d) pith

___ (9) If a mutation prevented the formation of lignin, which tissue would be most affected?
  (a) Vascular cambium
  (b) Parenchyma
  (c) Collenchyma
  (d) Sclerenchyma

___ (10) Tissue noted for storage, healing wounds, and regenerating plant parts is _____.
  (a) Vascular cambium
  (b) Parenchyma
  (c) Collenchyma
  (d) Sclerenchyma

___ (11) If cell walls did not vary in thickness or were not perforated, _____.
  (a) xylem and phloem would not be able to function as vascular tissue
  (b) gas exchange could not occur across the stomata
  (c) water could not enter the epidermal cells of the root
  (d) the terminal bud could not cause the stem to elongate

___ (12) The ray system of a tree consists mostly of _____ cells that act as horizontal conduits and food-storage centers.
 (a) parenchyma
 (b) collenchyma
 (c) sclerenchyma
 (d) spongy mesophyll

CHAPTER TEST

INTEGRATING AND APPLYING KEY CONCEPTS

Try to imagine the specific behavioral restrictions that might be imposed if the human body resembled the plant body in having (1) open growth with apical meristematic regions, (2) stomata in the epidermis, (3) cells with chloroplasts, (4) excess carbohydrates stored primarily as starch rather than as fat, and (5) dependence on the soil as a source of water and inorganic compounds.

20

WATER, SOLUTES, AND PLANT FUNCTIONING

General Objectives	
	1. Know which elements are essential to plant health.
	2. Explain how water is absorbed, transported, used, and lost by a plant.
	3. Describe how the intake of CO_2 is connected with water loss.
	4. Explain how essential mineral ions are taken up by a plant.
	5. Know how translocation of organic substances occurs, according to the pressure flow theory.

20-I
(pp. 268–270)

ESSENTIAL ELEMENTS AND THEIR FUNCTIONS
    Oxygen, Carbon, and Hydrogen
    Mineral Elements

Summary

Unlike most animals, land plants generally absorb water, carbon dioxide, and necessary ions from sources in which concentrations of these substances are low. About 90 percent of the volume of a mature plant cell is occupied by the vacuole, which is mostly water; thus, the vacuole requires minimal energy outlay for assembly and maintenance. Nutrients can be stockpiled in vacuoles as they become available and dispensed as they become scarce. Extensive systems of cylindrically shaped roots allow efficient absorption of dilute substances and efficient burrowing through the soil. Broad, thin leaves present a large exterior surface area for sunlight absorption and a large interior surface area in the spongy parenchyma for gas absorption.

Most plants require sixteen essential elements in order to grow and reproduce. *Oxygen, hydrogen,* and *carbon* account for 96 percent of the dry weight of plants. Oxygen is incorporated into organic compounds that make up the plant's dry weight; it comes from water, gaseous oxygen (O_2), and carbon dioxide (CO_2) in the air. Nitrogen is a necessary constituent of amino acids and nitrogen bases from which plants synthesize their proteins and nucleic acids. Ions such as potassium, calcium, magnesium, phosphorus, chlorine, iron, boron, manganese, zinc, copper, and molybdenum are necessary to form and activate plant enzymes and other substances; these ions are absorbed from the soil dissolved in water. They are also important in establishing solute concentration gradients that distribute water necessary for plant growth and for sustaining a nonwilted plant form.

Key Terms

plant physiology	micronutrients	legumes
mineral ions	trace elements	nitrogen-fixing bacteria
macronutrients	nodules	turgor pressure

Objectives

1. Explain why a central vacuole that contains mostly water is useful for plant cells but is not an advantage for animal cells.
2. Explain why cylindrically shaped roots are better suited to carry out the functions of roots and why broad, thin leaves are better suited to carry out the functions of leaves, than vice versa.
3. List the sources of oxygen, hydrogen, carbon that make up 96 percent of the dry weight of plants. Also indicate the source of nitrogen that plants use.
4. Describe how plants obtain the important ions they need for enzyme formation and growth.
5. Explain how turgor pressure participates in the growth of cells.

Self-Quiz Questions

Fill-in-the-Blanks

(1) _____, which composes most of the contents of the central (2) _____ in a plant cell, is energetically (3) [choose one] ☐ more expensive ☐ less expensive to accumulate compared with other cytoplasmic substances that might otherwise be stored. In plants, nitrogen is a necessary component of amino acids and (4) _____ _____. Nitrogen is available to plants in the form of either (5) _____ or (6) _____ ions.

(7) _____ atoms enter the plant via water, O_2, and CO_2. (8) _____ atoms also enter the plant via water. (9) _____ is the pressure applied to cell walls as water is absorbed into the cell; it allows the plant to sustain a(n) (10) _____ form.

Matching   Choose at least one and no more than two letters per blank.

(11) ___ boron   A. Macronutrient
                             B. Micronutrient
(12) ___ calcium   C. Component of nucleic acids and cell membranes
(13) ___ chlorine   D. At center of chlorophyll molecule
(14) ___ copper   E. Helps to establish osmotic gradients; as CO_2
                                dwindles in guard cells, this is pumped in
(15) ___ iron   F. Part of a coenzyme needed in enzyme-mediated
                                reactions that reduce nitrate
(16) ___ magnesium   G. In ammonium, nitrate, and nitrite ions; needed
                                for protein and nucleotide synthesis
(17) ___ manganese   H. Involved in electron transport during
                                photosynthesis and aerobic respiration
(18) ___ molybdenum   I. Needed in cell walls, in cell growth, and in cell
                                division
(19) ___ nitrogen

(20) ___ phosphorus

(21) ___ potassium

(22) ___ sulfur

(23) ___ zinc

20-II   **WATER UPTAKE, TRANSPORT, AND LOSS**
(pp. 270–273)   **Water Absorption by Roots**
                                 Transpiration and Water Conduction
                                 Cohesion Theory of Water Transport
                                 The Dilemma in Water and Carbon Dioxide Movements

Summary
Water is absorbed into the root through the root hairs of the thin-walled epidermal cells and then is passed along the highly permeable cortex cells. At the endodermis, water is funneled directly into the endodermal cytoplasm and into the vascular columns.

Evaporation from stems and leaves, known as *transpiration*, is caused by the drying power of air. According to the *cohesion theory of water transport*, as long as water molecules vacate the transpiration sites, replacement water is sucked up through the xylem from the roots in continuous columns as a result of hydrogen bonding between the water molecules. The cuticle prevents excessive water loss from plants; stomata in the epidermis open and close according to the amounts of water and carbon dioxide in their guard cells. High levels of potassium in guard cells are established by active transport as carbon dioxide levels dwindle during photosynthesis; high levels of potassium cause water to move in osmotically, and the increased turgor pressure causes stomata to open. When stomata are open and carbon dioxide is moving into the plant, water tends to move out. During heat stress, the plan produces *abscisic acid*, which causes potassium ions to leave the guard cells and causes turgor pressure to drop. As a result, stomata close, photosynthesis slows and stops, and so must growth.

Key Terms

fibrous root system	tension	cuticle
taproot system	cohesion theory of water	stoma, -ata
mycorrhiza	transport	abscisic acid
transpiration		

Objectives

1. Explain how water passes from dry soil into the vascular column of the root.
2. Define *transpiration* and list the main points of the cohesion theory of water transport.
3. Compare the amount of water that gets stored or used in metabolism with the amount that is transpired.
4. Explain how land plants regulate water loss as environmental conditions change. List the plant structures that participate in regulating water loss.

Self-Quiz Questions

Fill-in-the-Blanks

(1) _____ is evaporation of water from stems and leaves. The

(2) _____ theory of water transport suggests that (3) _____

_____ allows water molecules to cohere tightly enough to keep from

breaking apart as they are pulled up through the plant body. High levels of

(4) _____ in guard cells are established by (5) _____ _____ as

carbon dioxide levels dwindle during photosynthesis. During (6) _____

stress, the plant produces (7) _____ _____, which causes

(8) _____ ions to leave the guard cells, the turgor pressure to drop, and

the stomata to close.

20-III
(pp. 273–277)

UPTAKE AND ACCUMULATION OF MINERALS
    Active Transport of Mineral Ions
    Controls Over Ion Absorption
TRANSPORT OF ORGANIC SUBSTANCES IN PHLOEM
    Translocation
    Pressure Flow Theory
SUMMARY

Summary

Solute absorption and accumulation occur as energy from ATP drives the membrane pumps involved in active transport to move substances into cells against concentration gradients. Mineral ion absorption and accumulation are coordinated throughout the plant body in ways that affect growth. Insoluble storage starches and fats are converted into soluble forms (most often sucrose) for transport from one plant organ to another via the sieve-tube system of phloem tissue (*translocation*). According to the *pressure flow theory*, translocation is driven by differences in water pressure from one region of the phloem pipeline to another. Sucrose and other organic molecules are actively transported into sieve-tube cells in the leaf, water follows osmotically due to the increased solute concentration, and pressure builds up on the leaf sieve-tube regions. Stems, fruits, and roots are all regions of lower pressure because cells there convert the food molecules delivered to them into substances they need and use these substances in cellular respiration. Food (sucrose) therefore tends to flow from the leaves (the *source*) to the fruits, seeds, and roots (*sink* regions).

Key Terms

solute
membrane pumps
translocation

sieve-tube members
source regions
sink regions

pressure flow theory
pressure gradients

Objectives

1. Explain how plants absorb and accumulate solutes. Then explain how the rate at which a plant absorbs and accumulates solutes affects its rate of photosynthesis, its storage of foods, its rates of cellular respiration, and its rate of growth.
2. Discuss the evidence that leads us to believe in the pressure flow theory.
3. Describe the process of translocation and list the key points of the pressure flow theory of translocation. Explain what causes sucrose to be transported from one plant organ to another.

Self-Quiz Questions

Fill-in-the-Blanks

When soil is sufficiently moist, dissolved ions are carried rapidly into roots, where (1) _____ _____ moves them through the cortex and into the (2) _____ for transport upward. (3) _____, the dominant food storage product in plants, generally is converted into (4) _____ for transport through the plant body to another plant organ; this transport process is known as (5) _____. The (6) _____ _____ theory suggests a mechanism by which food can be transported from one plant region to another. Sucrose is actively transported into (7) _____-_____ cells, (8) _____ follows osmotically, and (9) _____ builds up in the leaves. Sucrose flows from the leaves to the stems, fruits, and (10) _____ because all three are regions of lower pressure.

UNDERSTANDING AND INTERPRETING KEY CONCEPTS

___ (1) The _____ theory of water transport states that hydrogen bonding allows water molecules to maintain a continuous fluid column as water is pulled from roots to leaves.
(a) pressure flow
(b) evaporation
(c) cohesion
(d) abscission

___ (2) _____ control(s) movement of nutrients to and from the cytoplasm of plant cells.
(a) The nuclear membrane
(b) The vacuole membrane
(c) Chromosomes
(d) Chloroplasts

___ (3) Of all the water moving into a leaf, about _____ percent is stored or used in metabolism.
(a) 1
(b) 2
(c) 10
(d) 70

___ (4) Without _____, plants would rapidly wilt and die during hot, dry spells.
(a) a cuticle
(b) a mycorrhiza
(c) phloem
(d) cotyledons

___ (5) Water inside all of the xylem cells is being pulled upward by _____.
(a) turgor pressure
(b) tension and negative pressure
(c) osmotic gradients
(d) pressure-flow theory

___ (6) Most plants require sixteen essential elements to grow and reproduce. Which of the following is *not* an essential element for most land plants?
(a) Sodium, Na
(b) Potassium, K
(c) Copper, Cu
(d) Molybdenum, Mo

___ (7) Which of the following substances does *not* require nitrogen-containing ions for its synthesis?
(a) Nucleic acids
(b) Amino acids
(c) Proteins
(d) Cholesterol

___ (8) Most of the water moving into a leaf is lost through _____.
   (a) osmotic gradients being established
   (b) transpiration
   (c) pressure-flow forces
   (d) translocation

___ (9) _____ causes transpiration.
   (a) Hydrogen bonding
   (b) The drying power of air
   (c) Cohesion
   (d) Turgor pressure

___ (10) _____ activates enzymes involved in protein synthesis and sets up an osmotic gradient across the plasma membrane.
   (a) Carbon dioxide
   (b) Hydrogen
   (c) Oxygen
   (d) Potassium

___ (11) By control of _____ levels inside the guard cells of stomata, the activity of stomata is controlled when leaves are losing more water than roots can absorb.
   (a) oxygen
   (b) potassium
   (c) carbon dioxide
   (d) ATP

___ (12) Large pressure gradients arise in sieve-tube systems by means of _____.
   (a) vernalization
   (b) abscission
   (c) osmosis
   (d) transpiration

CHAPTER TEST

INTEGRATING AND APPLYING KEY CONCEPTS

How do you think maple syrup is made from maple trees? Which specific systems of the plant are involved, and why are maple trees only tapped at certain times of the year?

21

PLANT REPRODUCTION AND EMBRYONIC DEVELOPMENT

General Objectives

1. Describe the typical patterns of life cycles in flowering plants.
2. Draw and label the parts of a perfect flower. Explain where gamete formation occurs in the male and female structures.
3. Define and distinguish between *pollination* and *fertilization*.
4. Trace embryonic development from zygote to seedling.
5. Describe the various styles presented by asexual reproduction in flowering plants.

21-I
(pp. 278–279)

SEXUAL REPRODUCTION OF FLOWERING PLANTS
   Life Cycles of Flowering Plants
   Floral Structure

Summary

Flowering plants (angiosperms) can reproduce asexually (sex cells and fertilization are not involved) as well as sexually. In flowering plant sporophytes, the vegetative diploid growth phase of roots, stems, and leaves gives way to a reproductive haploid phase that begins with flower production; a sort of *alternation of generations* exists as, in time, the flowers produce seeds, which begin a new vegetative phase. A *flower* is a cluster of reproductive and nonreproductive organs that are thought to originate from highly modified leaves. The outermost green *sepals* protect the buds; *petals* attract bird and insect pollinators and enclose the stamen and one or more carpels in a perfect flower. *Stamens* are male reproductive organs, each of which consists of an *anther* (in which microspores develop into pollen grains) perched on its supportive stalk (the filament). *Carpels* are female reproductive organs, each of which consists of a swollen base (the ovary) that produces at least one *ovule*, within which megaspores are produced; megaspores develop into egg-bearing gametophytes.

Angiosperms form flowers that have *microspore-* and *megaspore*-producing structures, as well as accessory parts that have both direct and indirect roles in sexual reproduction. Sexual reproduction depends on pollen grains being transferred from anthers (male reproductive structures) to *stigmas* (female reproductive structures). Once a pollen grain has been deposited on the stigma, a pollen tube forms and grows down to the ovarian chamber below. In flowering plants, *double fertilization* occurs: One sperm nucleus fuses with the nucleus of the egg, and the other sperm nucleus fuses with two other nuclei in the ovule to form a single triploid (3n) nucleus. Following double fertilization, the *ovule* expands and develops into a *seed*. The ovarian wall expands and ripens into a *fruit*, which may be an edible structure, a small capsule, or one of a variety of other forms.

Key Terms

pollinators	ovary	imperfect flower
sporophyte	stigma	anther
flower	style	petal
gametophyte	stamen	corolla
alternation of generations	filament	sepals
receptacle	perfect flower	calyx
carpel		

Objectives

1. List in sequence the major events that constitute the reproductive portion of the flowering plant life cycle.
2. Explain how the inclusion of a reproductive phase in their life cycles is advantageous to flowering plants. Explain why reproduction by budding or other asexual means is less advantageous to plant populations than reproduction by sexual means.
3. List the parts of a flower and state the functions of each part.
4. Distinguish between *perfect* and *imperfect flowers*.

Self-Quiz Questions

Fill-in-the-Blanks

In a flower, the outermost (1) _____ protect the buds; (2) _____ attract bird and insect pollinators. (3) _____ are male reproductive organs, each of which consists of two parts: a(n) (4) _____, in which (5) _____ develop into pollen grains, and its supportive stalk, the filament. (6) _____ are female reproductive organs, each of which consists of a swollen base, the (7) _____, that produces at least one (8) _____; inside this, (9) _____ develop into egg-bearing gametophytes.

GAMETE FORMATION
    Microspores to Pollen Grains
    Megaspores to Eggs

Summary

Reproductive cells in flowers develop and undergo meiosis, forming *microspores* in the male reproductive organs (anthers). Haploid microspores develop into *pollen grains*, which are immature male gametophytes. Pollen grains later develop into mature sperm-bearing gametophytes. Meanwhile, in the ovary lie one or more ovules, within which reproductive cells develop and undergo meiosis, forming *megaspores*. Megaspores develop into egg-bearing female gametophytes. An ovule consists of a *nucellus* (the site of megaspore production), one or more integuments, and a stalk attached to the ovary wall. A female gametophyte (*embryo sac*) is a seven-celled eight-nucleate body. One cell of the sac is the *egg*; another cell has two nuclei and helps give rise to *endosperm*.

Key Terms

microspores	integuments	microspore mother cell
pollen sac	megaspores	ovule
pollination	micropyle	seed
pollen tube	embryo sac	megaspore mother cell
pollen grain	endosperm mother cell	female gametophyte
nucellus	endosperm	

Objectives

1. Explain where male sex organs form in a flowering plant. Then describe, beginning with microspore production, all of the steps and structures formed along the way to producing sperm nuclei.
2. Distinguish between *pollination* and *fertilization*.
3. Explain where female sex organs form in a flowering plant. Then describe, beginning with megaspore production, all of the steps and structures formed along the way to producing egg and endosperm nuclei.

**Self-Quiz
Questions**

Matching

(1) ___ carpel

(2) ___ generative cell

(3) ___ micropyle

(4) ___ microspore mother cell

(5) ___ ovule

(6) ___ pollen grain

(7) ___ stamen

A. A tiny opening through the integuments
B. Will become a seed, especially if fertilization occurs
C. Female sex organs
D. Male sex organs
E. Undergo meiosis
F. Will divide mitotically to form two sperm nuclei
G. Immature male gametophyte

POLLINATION AND FERTILIZATION
     Pollination and Pollen Tube Growth
     Fertilization and Endosperm Formation
     Case Study: Coevolution of Flowering Plants and Pollinators

Summary

Pollen grains (the immature male gametophytes) are transferred from the anthers to the female gametophytes, where they develop into mature male gametophytes called pollen tubes and deliver their mature sperms. In *double fertilization*, one sperm nucleus fuses with the egg to form a *zygote*, which, through mitotic division, eventually develops into the plant *embryo*. The other sperm nucleus fuses with the two nuclei of the endosperm mother cell in the ovule and develops into *endosperm*, a mass of triploid tissue that surrounds and nurtures the developing embryo.

Many successful plants (grasses, oaks, birches, and maples) are wind-pollinated; their flowers typically do not have nectar or perfume and tend to lack colorful petals. Many flowering plants, however, coevolved with insects, birds, or bats and depend on them to disperse their pollen. Colors, patterns, and odors are important attractants for many pollinators. Once a pollinator locates a flower, color patterns and petal shapes guide it to the nectar and channel the pollinator's movements in ways that aid pollination; for example, petals of hummingbird-pollinated flowers often form a long tube, which corresponds to length and bill shape. The more refined the "fit" between plant and pollinator, the more efficient pollination can be. But the more specialized the plant–pollinator relationship, the greater the chance that the plant may face extinction if its pollinator should happen to disappear. Regardless of how these interactions developed, diverse species of pollinators and flowering plants have been interdependent since about the dawn of the Cenozoic, about 65 million years ago. The first seed-bearing plants evolved in the Devonian period, some 350 million years ago; ovules and pollen sacs were borne on modified leaves or scales, often in conelike structures.

Key Terms

pollination	triploid	coevolution
double fertilization	primary endosperm cell	nectary
diploid zygote	vector	nectar

Objectives

1. Describe fertilization as it occurs in a flowering plant.
2. Identify the parts of the flower in Figure 21.6 of your main text and locate the sites of the (a) megaspore mother cell, (b) megaspore, (c) female gametophyte, (d) egg cell, (e) pollen grain, (f) pollen tube, (g) microspore, (h) immature male gametophyte, and (i) mature male gametophyte.
3. Define *double fertilization* and *endosperm*.
4. Explain how seed formation and fruit development are related.
5. State when during geologic history the flowering plants became established and tell which other groups of organisms paralleled their expansion.
6. Explain how some insect (or bird or bat) populations have helped some plant populations to compete more effectively for nutrients, living space, and water. Then tell how the same plant populations have helped the same insect (or bird or bat) populations.
7. Summarize the nature of the plant-pollinator coevolutionary relationship.

Self-Quiz Questions

Fill-in-the-Blanks

Pollen grains are transferred from (1) _____ to (2) _____. The mature male gametophyte is the (3) _____ _____, which grows into the ovule and delivers two (4) _____ _____. One of the (5) _____ fuses with the egg and forms the (6) _____; the other fuses with nuclear material and develops into (7) _____, which surrounds and nourishes the developing embryo. Fully mature ovules are (8) _____; the ripened ovary of one or more carpels constitutes most or all of a (9) _____.

The diversity of flowering plants increased dramatically between (10) _____ and _____ million years ago; by about 65 million years ago, they had become abundant and were found in diverse environments. (11) _____-pollinated plants have little or no color and odor and no nectar at all; many do not even have petals. Bird-pollinated flowers tend to be (12) _____. Bee-pollinated flowers tend to be blue or (13) _____, with prominent (14) _____ components.

Labeling

Identify each indicated part of the illustration on page 180.

(15) _____	(21) _____ _____ _____	(26) _____
(16) _____	(22) _____ _____ _____	(27) _____
(17) _____	(23) _____	
(18) _____	(24) _____ _____	
(19) _____	(25) _____ _____ _____	
(20) _____		

21-IV
(pp. 287–289)

EMBRYONIC DEVELOPMENT
From Zygote to Plant Embryo
Seed and Fruit Formation

Summary

Following double fertilization, the *ovule* of an angiosperm expands. Integuments thicken and harden, forming the *seed coat*. Inside, the zygote develops into an embryo that consists of diploid cells. Nutritive endosperm is formed by repeated mitotic divisions and consists of triploid cells. Fully mature ovules are called seeds. The ripened ovary of one or more carpels constitutes most or all of a *fruit*. Fruits are specialized structures that protect seeds and enhance their dispersal in specific environments.

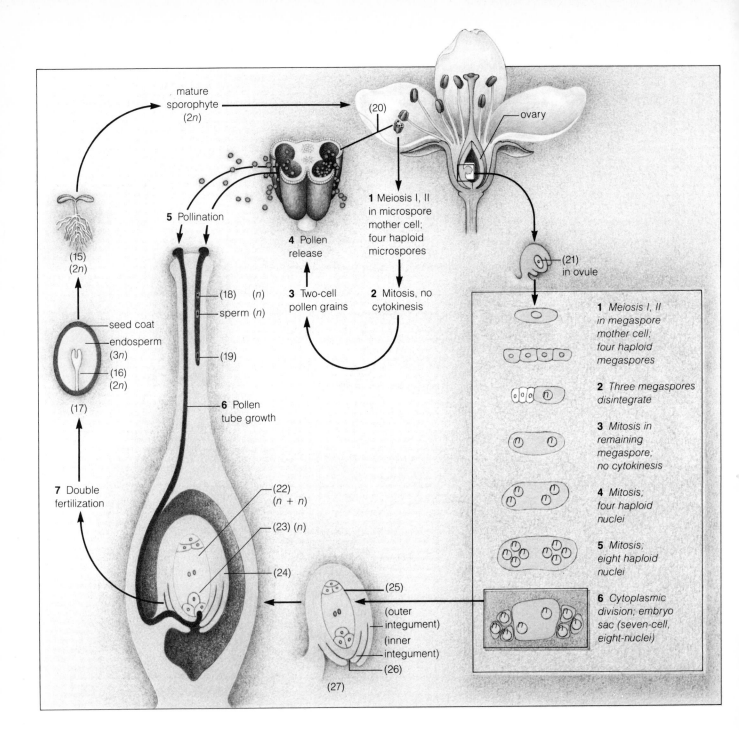

mature
sporophyte
(2n)

(20)

ovary

5 Pollination

4 Pollen
release

1 Meiosis I, II
in microspore
mother cell;
four haploid
microspores

(21)
in ovule

(15)
(2n)

(18) (n)
— sperm (n)

3 Two-cell
pollen grains

(19)

2 Mitosis, no
cytokinesis

seed coat
endosperm
(3n)

(16)
(2n)

6 Pollen
tube growth

(17)

1 *Meiosis I, II
in megaspore
mother cell;
four haploid
megaspores*

2 *Three megaspores
disintegrate*

3 *Mitosis in
remaining
megaspore;
no cytokinesis*

4 *Mitosis;
four haploid
nuclei*

5 *Mitosis;
eight haploid
nuclei*

7 Double
fertilization

(22)
(n + n)

(23) (n)

(24)

(25)

(outer
integument)

(inner
integument)

(26)

(27)

6 *Cytoplasmic
division; embryo
sac (seven-cell,
eight-nuclei)*

Key Terms

Capsella, shepherd's purse
suspensor
embryo
ovule

endosperm
seed
seed coat
cotyledons

fruit
simple fruit
aggregate fruit
multiple fruit

Objectives

1. Explain how a flower is related to a fruit and how an ovule is related to a
   seed.
2. Describe the role(s) played by the cotyledon(s) of flowering plants.
3. Distinguish among *simple, aggregate,* and *multiple fruits.* Provide an example
   of each.

4. Describe some of the strategies that have evolved in plants that decrease the competition between parent plants and offspring.

Self-Quiz Questions

Fill-in-the-Blanks

A fully matured ovule is a seed; the (1) _____ surrounding it develops into most or all of a fruit. (2) _____, which is triploid tissue, results when a single sperm fuses with two other nuclei in the (3) _____ during double fertilization in flowering plants. Grasses, oaks, birches, and maples are (4) _____-pollinated; their (5) _____ typically do not have perfume or nectar and are not colorful. (6) _____ and nuts are dry fruits; apples and tomatoes are (7) _____ fruits. A raspberry is a(n) (8) _____ of many separate carpels from one flower. (9) _____ matured ovaries remain clustered together and grow into a mass in a pineapple.

21-V
(pp. 289–291)

ASEXUAL REPRODUCTION OF FLOWERING PLANTS
SUMMARY

Summary

Asexual reproduction in plants can occur by any of several mechanisms: (1) reproduction on modified stems (runners, rhizomes, corms, tubers, and bulbs); (2) *parthenogenesis*; (3) vegetative reproduction; and (4) tissue culture propagation. By means of *tissue culture*, thousands of identical plants can be propagated from a single specimen.

Key Terms

runners	clone	tuber
parthenogenesis	rhizome	bulb
vegetative reproduction	corm	tissue culture propagation

Objectives

1. Distinguish between *parthenogenesis* and *vegetative reproduction*.
2. Name and give one example each of four asexual modes of reproduction by modified stems. Describe each of the four modified stems.
3. Outline the steps involved in tissue culture and the production of plant clones.

Self-Quiz Questions

Fill-in-the-Blanks

An embryo that develops from an unfertilized egg is an example of (1) _____. (2) _____ along the stems of many plants will develop new roots and shoots if cultivated properly. Onions and lilies can grow from

(3) _____, which are examples of underground stems. Asexually

produced offspring are genetically identical to their parents and are called

(4) _____.

UNDERSTANDING AND INTERPRETING KEY CONCEPTS

____ (1) A stamen is _____.
   (a) composed of a stigma
   (b) the mature male gametophyte
   (c) the site where microspores are produced
   (d) part of the vegetative phase of an angiosperm

____ (2) A gametophyte is _____.
   (a) a gamete-producing plant
   (b) haploid
   (c) both a and b
   (d) the plant produced by the fusion of gametes

____ (3) The process during which the diploid set of chromosomes become haploid is _____.
   (a) metastasis
   (b) fertilization
   (c) cleavage
   (d) meiosis

____ (4) A fruit is formed from carpels of several associated flowers is referred to as a(n) _____ fruit.
   (a) aggregate
   (b) simple
   (c) multiple
   (d) fleshy

____ (5) The significance of polarization is that _____.
   (a) cell systems near the anterior (head) end of an organism are aligned differently from cell systems at the posterior end
   (b) controlled cell death occurs in a gradient along the body axis in most organisms
   (c) malignant cells metastasize more readily in polarized regions
   (d) cytoplasmic substances located at one end of the zygote differ from cytoplasmic substances located at the other end

____ (6) Flowering plants and insects have been interdependent and coevolutionary since _____.
   (a) Cretaceous Period of the Mesozoic Era
   (b) Triassic Period of the Mesozoic Era
   (c) Carboniferous Period of the Paleozoic Era
   (d) Paleocene Epoch of the Cenozoic Era

___ (7) Which of the following is *not* primarily related to asexual reproduction?
   (a) Mitosis
   (b) Pollination
   (c) Runner formation
   (d) Cloning and tissue culture

___ (8) If a plant is said to propagate vegetatively, it means that _____.
   (a) part of a leaf, a stem, or a root, when torn away from the parent plant and planted under proper conditions, can develop into a new plant
   (b) it cannot reproduce by forming flowers, fruits, and seeds
   (c) the leafy part of the gametophyte can grow into a new plant if planted and grown similar to the way most vegetables are grown
   (d) flowers and fruits from one plant can be grafted onto another closely related plant to produce hybrids

___ (9) In flowering plants, one sperm nucleus fuses with that of an egg, and a zygote forms that develops into an embryo. Another sperm nucleus fuses with _____.
   (a) a primary endosperm cell to produce three cells, each with one nucleus
   (b) a primary endosperm cell to produce one cell with one triploid nucleus
   (c) the diploid endosperm mother cell, forming a primary endosperm cell with a single triploid nucleus
   (d) one of the smaller megaspores to produce what will eventually become the seed coat

Matching

Choose the most appropriate answer for each.

(10) ___ bulb

(11) ___ corm

(12) ___ rhizome

(13) ___ runner

(14) ___ tuber

A. New plants arise at nodes of underground horizontal stem, as in Bermuda grass
B. Onions and lilies
C. New plant arises from axillary bud on short, thick vertical, underground stem, as in gladiolus
D. New shoots arise from buds on enlarged tips of slender, underground rhizomes
E. New plants arise at nodes of an above-ground horizontal stem

CHAPTER TEST

INTEGRATING AND APPLYING KEY CONCEPTS

Unlike the growth pattern of animals, the vegetative body of vascular plants continues growing throughout its life by means of meristematic activity. Explain why most representatives of the plant kingdom probably would not have survived and grown as well as they do if they had followed the growth pattern of animals.

22

PLANT GROWTH AND DEVELOPMENT

General Objectives

1. Describe the general pattern of plant growth and list the factors that cause plants to germinate.
2. List the various chemical messengers that regulate growth and metabolism in plants. Explain how plants respond to changes in their environment.
3. Know the factors that cause a plant to flower, to age, and to enter dormancy. Describe each process.

22-I
(pp. 292–293)

SEED GERMINATION
PATTERNS OF GROWTH

Summary

The whole seed swells with incoming water molecules, which hydrogen bonding has attracted to proteins and polysaccharides stored in the seed. In the seeds of most species, the first cells to grow are in the embryonic root (the radicle), and they grow longitudinally. A slender primary root results; when it protrudes from the seed, germination is complete. Plants grow because their cells grow. On average, about half of the daughter cells formed enlarge—often by twenty times. The other half remain meristematic; they grow only as large as the mother cell and then probably divide again. Cell growth is driven by water uptake and the ensuing turgor pressure against the cell wall. To a large extent, the arrangement of polysaccharides in cell walls dictates the direction in which the wall yields under turgor pressure. The expansive properties of primary cell walls permit cell enlargement. Therefore, how the wall polysaccharides are oriented governs whether a cell becomes long and slender or short and broad. If most cells in a plant organ are elongated, so is the organ; this is true of many fruits.

Key Terms

seed germination	dicot	elastic
imbibition	radicle	plastic
monocot	primary root	true growth

Objectives

1. Compare the growth of corn (a monocot) with that of a soybean plant (a dicot).
2. Explain how the arrangement of cellulose molecules in a cell wall controls the nature and direction of plant growth.
3. State the mechanism by which some cells remain meristematic and others enlarge and differentiate.
4. Distinguish *elastic* from *plastic*.

Self-Quiz Questions

Fill-in-the-Blanks

During (1) _____, water molecules enter the seed and are attracted to hydrophilic groups of the stored proteins. (2) _____ is a process in which a plant embryo divides and daughter cells grow to form first roots and then a stem and leaves. On the average, about half of the daughter cells (3) _____, often by twenty times; the other half remain (4) _____. Cell growth is driven by (5) _____ uptake and the ensuing (6) _____ _____ against the cell wall. To a large extent, the arrangement of (7) _____ in cell walls dictates the direction in which the wall yields under pressure. Each soybean plant develops a single embryonic root called the (8) _____, which develops into the primary root that later sends out lateral branches.

22-II
(pp. 293–295)

PLANT HORMONES
  Types of Plant Hormones
  Examples of Hormonal Action

Summary

Plant hormones are information-carrying messenger molecules that cause certain cell types to change their activities in response to environmental conditions. Plant hormones are less specific than animal hormones with respect to the physiological responses they activate, and the target tissues themselves may be less specific. Five hormones (or groups of hormones) have been identified in vascular plants. *Auxins* are best known for their ability to stimulate cell elongation in coleoptiles and stems. 2,4-D is a synthetic auxin that is used to kill dicot weeds.

Gibberellins, like auxins, promote stem elongation in plants; they have been identified in plants and fungi. *Cytokinins* stimulate cell division, promote leaf cell expansion, and retard leaf aging. *Abscisic acid* stimulates stomate closure, may bring about seed and bud dormancy, and may promote resistance to water stress in many species. *Ethylene* is a simple hydrocarbon gas that stimulates fruit ripening.

The growth of a coleoptile (a hollow, cylindrical organ that protects tender young leaves growing within it) is controlled by IAA (an auxin) synthesized in its tip. The hormone moves down the coleoptile and increases cell wall plasticity, which encourages elongation. Auxin synthesized at a dicot stem tip is essential for the growth of cells below. The growth of stems that elongate slowly is especially influenced by gibberellin. In contrast to coleoptiles and dicot stems, root and leaf cells themselves synthesize most or all of the hormones they require.

Key Terms

hormone	gibberellic acid	florigen
target cell	cytokinins	coleoptile
auxins	zeatin	apical dominance
IAA, indoleacetic acid	abscisic acid, ABA	cotyledon
herbicides	abscission	hypocotyl
2,4-D	ethylene	
gibberellins		

Objectives

1. State what plant hormones are and, in general, what they do.
2. Identify the five principal plant hormones (or hormone groups) and state the known or suspected effects of each.
3. List three examples of specific plant hormones and describe how each hormone affects the different parts of a plant.
4. Define the relationship between the terms *radicle*, *embryonic root*, *primary root*, and *hypocotyl*.

Self-Quiz Questions

Fill-in-the-Blanks

Plant (1) _____ are information-carrying messenger molecules that cause certain cell types to change their activities in response to environmental conditions. (2) _____ and (3) _____ both promote elongation of cells in stems. (4) _____ stimulate cells to divide. (5) _____ promotes the closure of stomata and may bring about the end of (6) _____ dormancy. (7) _____ promotes the ripening of fruit. The growth of a coleoptile is controlled by (8) _____ synthesized in its tip. Corn plants develop (9) _____ roots.

Identification

Name the hormone that is primarily responsible for the effects described after each blank.

(10) _____ Promote stem elongation (especially in dwarf plants); might help break dormancy of seeds and buds

(11) _____ Arbitrary designation for as yet unidentified hormone (or hormones) thought to cause flowering

(12) _____ Promote cell elongation in coleoptiles and stems; long thought to be involved in phototropism and gravitropism

(13) _____ Promotes stomatal closure; might trigger bud and seed dormancy

(14) _____ Promotes fruit ripening; promotes abscission of leaves, flowers, and fruits

(15) _____ Promote cell division; promote leaf expansion and retard leaf aging

---

22-III
(pp. 295–299)

PLANT RESPONSES TO THE ENVIRONMENT
    The Many and Puzzling Tropisms
    Response to Mechanical Stress

Summary

Plant tropisms are familiar but as yet unexplained growth responses in which an environmental stimulus causes the plant to respond with a faster rate of cell elongation on one side than on the other. *Phototropism* is caused by light coming in mainly from one side. IAA moves to the shaded side and downward, where it causes stems to curve toward the light, and the flat surface of a leaf blade becomes perpendicular to the light. In *gravitropism* (a response to Earth's gravitational force), the root cap synthesizes a growth inhibitor that is transported to the bottom of a horizontally positioned root and causes the downward curvature of the growing root tip. *Thigmotropism* is a positive growth response to contact.

Key Terms

tropism	flavoprotein	geotropism
phototropism	gravitropism	thigmotropism

Objectives

1. Define *geotropism*, state the cellular and hormonal events that bring it about, and indicate how it promotes the survival of a plant.
2. Define *phototropism*, state the cellular and hormonal events that bring it about, and indicate how it promotes the survival of a plant.
3. Define *thigmotropism* and indicate how it promotes the survival of specific plant types.

**Self-Quiz
Questions**

Matching

Choose the one most appropriate answer for each.

(1) ___ gravitropism

(2) ___ phototropism

(3) ___ thigmotropism

A. A response to electric shock caused by potassium redistribution
B. Negative response in shoot; positive response in root due to growth inhibitor produced by cap
C. A response to physical contact evident in climbing vines
D. Known to be controlled by indoleacetic acid and blue light

PLANT RESPONSES TO THE ENVIRONMENT (cont.)
  Biological Clocks in Plants
  Photoperiodism
THE FLOWERING PROCESS

Summary

Time-measuring devices that allow all eukaryotes, including plants, to anticipate and adjust to environmental changes are called biological clocks. *Circadian* rhythms occur daily; *lunar* rhythms every 28 days; *annual* rhythms every 365 days.

Both temperature and sunlight influence the growth of coleoptiles, stems, roots, and leaves. Environmental temperature affects cell metabolism and transport of water, ions, and sucrose. Especially for dicots, exposure to sunlight promotes leaf expansion and stem branching but inhibits stem elongation; *phytochrome*, a blue-green pigment, receives light energy and seems to be an on-off switch for expansion, branching, and elongating activities—and in many plants, for seed germination and flowering.

As a flowering plant matures, its physiological processes become directed toward flower, fruit, and seed production. Plants such as corn, soybeans, and peas live only one growing season; they are known as *annuals*. Plants that live for many growing seasons are called *perennials*. Still other species produce only roots, stems, and leaves the first growing season; then they die back to soil level in autumn and then grow a new flower—forming stem from a bud. These plants, which typically live for two growing seasons, are called *biennials*; they include cabbages, carrots, and turnips.

Day length and low temperature are the strongest stimuli for the flowering of land plants. Low-temperature stimulation of flowering is called *vernalization*. Day length cues flowering and reproduction in many species. *Long-day* plants are adapted to long days and reproduce early in summer. *Short-day* plants are attuned to the shorter day lengths of late summer and flower in autumn. *Day-neutral* plants flower almost independently of day length. A relationship between the pigment phytochrome and an elusive hormone, florigen, is supposed to exist but has never been demonstrated conclusively.

Key Terms

biological clock	Pfr, active phytochrome	vernalization
circadian rhythms	Pr, inactive phytochrome	long-day plants
seasonal adjustments	annuals	short-day plants
photoperiodism	perennials	day-neutral plants
phytochrome	biennials	florigen

Objectives

1. Explain how temperature influences the growth of coleoptiles, stems, roots, and leaves.
2. Describe how sunlight influences plant growth, by explaining how phytochrome responds to specific colors of light and how seed germination, stem elongation, leaf expansion, stem branching, and flowering are affected by the activated form of phytochrome.
3. Define *annual*, *perennial*, and *biennial*.
4. Identify factors that stimulate flowering in land plants. Present the reasons some botanists believe there must be a flowering hormone named *florigen*.
5. Distinguish among *long-day*, *short-day*, and *day-neutral* plants and give an example of each.

Self-Quiz Questions

Fill-in-the-Blanks

For dicots, exposure to sunlight (1) _____ leaf expansion and stem branching but (2) _____ stem elongation. (3) _____ seems to be an on-off switch for hormone activities that govern leaf expansion, stem branching, stem length, and (in many plants) seed (4) _____ and flowering. Before the sun rises, phytochrome exists mainly in an (5) _____ [choose one] ☐ active ☐ inactive form that absorbs (6) [choose one] ☐ red ☐ far-red light and is converted to the (7) _____ [choose one] ☐ active ☐ inactive form (Pfr). In the shade, at sunset, or at night, (8) [choose one] ☐ Pr is converted to Pfr ☐ Pfr is converted to Pr.

Plants that typically live two growing seasons are called (9) _____, whereas plants that live year after year are known as (10) _____. (11) _____ is a highly reliable cue for flowering and reproduction of many species. (12) _____ organs withdraw nutrients from vegetative organs through connecting (13) _____ _____. In deciduous trees, nutrients are transported to reproductive organs and to (14) _____ _____ cells in twigs, stems, and roots prior to abscission.

22-V
(pp. 302–306)

SENESCENCE
DORMANCY
    Case Study: From Embryogenesis to the Mature Oak
SUMMARY

Summary

All the processes that lead to the death of a plant or any of its organs are called senescence. One stimulus for senescence could be the drain of nutrients during the growth of reproductive organs, but some sort of death signal that forms during short days is also suspected. When any plant stops growing under physical conditions that are actually quite suitable for growth, it is said to have entered a period of *dormancy*. Abscisic acid movement from leaves to buds in late summer may trigger dormancy, and abscisic acid breakdown along with gibberellin accumulation in buds during late autumn and winter may end dormancy.

An organism such as the coast live oak develops and grows in an ecosystem according to interactions between environmental cues and its own genotype. Humans, in their ignorance of the requirements of other organisms, often sentence their intended neighbors to death.

Key Terms

abscission   senescence   dormancy

Objectives

1. Identify factors that bring about senescence in plants.
2. Distinguish between *senescence* and *dormancy* and identify the factors that promote dormancy.
3. Explain how dormancy can promote the survival of plants.
4. Identify factors that help a plant to break out of dormancy.

Self-Quiz Questions

Fill-in-the-Blanks

(1) _____ is the principal substance that stimulates abscission. Any plant that stops growing under physical conditions that are actually quite suitable for growth is said to have entered a period of (2) _____. (3) _____ _____ movement from leaves to buds in late summer may trigger dormancy; its breakdown, along with gibberellin accumulation in buds in late autumn and winter, may end dormancy.

CHAPTER TEST

UNDERSTANDING AND INTERPRETING KEY CONCEPTS

___ (1) Exposure to wind or rain causes some plants to develop shorter, thicker stems than similar plants grown indoors; this is a growth response known as _____.
  (a) geotropism
  (b) phototropism
  (c) thigmotropism
  (d) thigmomorphogenesis

___ (2) In comparison with young trees growing out in the open, young trees growing in the darker forest understory tend to have longer, thinner trunks with less branching; this developmental pattern is principally caused by _____.
  (a) phototropism
  (b) thigmotropism
  (c) activated phytochrome being converted to inactive phytochrome
  (d) inactive phytochrome being converted to active phytochrome

___ (3) _____ is the principal substance that regulates thigmotropism.
  (a) Indoleacetic acid
  (b) Gibberellic acid
  (c) Abscisic acid
  (d) Ethylene

___ (4) _____ is the principal substance that causes phototropism in stems or leaves.
  (a) Indoleacetic acid
  (b) Gibberellic acid
  (c) Abscisic acid
  (d) Ethylene

___ (5) _____ controls the direction of cell growth.
    (a) Ethylene
    (b) Mycorrhizae
    (c) The carbon dioxide level in a leaf
    (d) The arrangement of cellulose molecules

___ (6) 2,4-D, a potent dicot weed killer, is a synthetic _____.
    (a) auxin
    (b) gibberellin
    (c) cytokinin
    (d) phytochrome

___ (7) All the processes that lead to the death of a plant or any of its organs are called _____.
    (a) dormancy
    (b) vernalization
    (c) abscission
    (d) senescence

___ (8) Zeatin is a(n) _____, a substance that stimulates cell division, promotes leaf expansion, and retards leaf aging.
    (a) auxin
    (b) gibberellin
    (c) cytokinin
    (d) phytochrome

___ (9) Which of the following is *not* promoted by the active form of phytochrome?
    (a) Seed germination
    (b) Stem elongation
    (c) Leaf expansion
    (d) Stem branching

___ (10) Which of the following has *not* been implicated as a factor in breaking dormancy?
    (a) Abscisic acid movement from leaves to buds in late summer
    (b) The breakdown of abscisic acid in late autumn and winter
    (c) Gibberellin accumulation in buds in late autumn and winter
    (d) An exposure to low temperature for hundreds of hours

CHAPTER TEST

INTEGRATING AND APPLYING KEY CONCEPTS

An oak tree has grown up in the middle of a forest. A lumber company has just cut down all of the surrounding trees except for a narrow strip of woods that includes the oak. How will the oak be likely to respond as it adjusts to its changed environment? To what new stresses will it be exposed? Which hormones will most probably be involved in the adjustment?

Crossword
Number Three

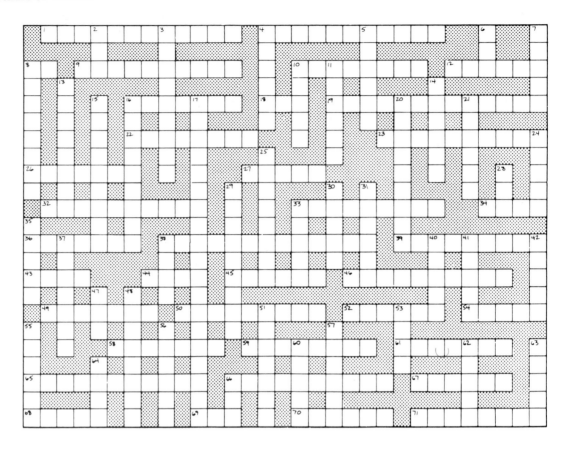

The terms in the puzzle are not necessarily biology terms.

Across

1. not vegetative
4. the transfer of the immature male gametophyte
8. inactive phytochrome
9. chiefly conifers in this group
10. slope
12. ground tissue located between the vascular ring and the dermal surface
16. a reversible form of growth
18. the most widely encountered auxin
19. evaporation from stems and leaves
22. fusiform _____ produce xylem and phloem cells that are arranged parallel with the stem's long axis
23. the portions of the stem between the nodes
26. embryonic root
27. one of the two groups of angiosperms
32. gives rise to vascular tissue
33. the exterior region of any body
34. a straight line about which a body or geometrical object rotates
36. elements of xylem
38. low-temperature stimulation of flowering
39. components of xylem
43. the below-ground portion of a plant

44. ground tissue located within a ring of vascular bundles
45. functions in food conduction, storage, and support
46. stimulate cell division; promote leaf cell expansion and retard leaf aging
49. meristem that runs parallel with the sides of roots and stems
50. _____ fruit: formed from numerous but separate carpels of a single flower
52. female reproductive organ in a flower
54. stalk that connects the ovary with the stigma
58. male reproductive organs in flowers
59. circulatory
61. a _____ system has a primary root and its various lateral branchings
65. flowering plants
66. seed leaves
67. at the peak of development; generally able to reproduce
68. the above-ground part of a plant
69. plant fluid
70. a group of interacting elements
71. sticking together

Down

2. the _____ system acts as a system of conduits and food storage centers in wood
3. the _____ strip is a continuous band of fatty suberin deposits
4. a hydrophilic substance found in the cell walls of collenchyma
5. of, or related to, the apex
6. active phytochrome
7. a class of plant growth hormones
8. the _____ flow theory says that translocation depends on pressure gradients in the sieve-tube system
10. a small, hard seed or fruit, especially that produced by a grass such as rice, corn, or wheat
11. a pollen-bearing structure
13. the vascular _____ is surrounded by cortex in a plant root
14. long cells that have somewhat flexible walls
15. ground tissue between the phloem and endodermis in roots and stems that becomes part of the active cambium
16. surface layer of cells
17. generally, overall plant growth is inhibited as a response to contact
20. leading to the death of a plant or any of its organs
21. a plant that lives for one growing season
24. fertilized ovules
25. plant that typically lives for two growing seasons
28. any of a particular class of fatty derivatives that are insoluble in water
29. will develop into an egg-bearing female gametophyte
30. the ripened ovary of one or more carpels
31. _____ parenchyma lies between the palisade tissue and the lower cuticle-covered epidermis
33. a sieve _____ has larger pores than do the side walls
35. a multicellular organ that produces eggs
37. gates across the epidermis of leaves and stems
38. a vascular structure
40. the _____ system of secondary vascular tissues conducts food and water up and down the stem
41. root _____ are absorptive structures
42. _____ elements and companion cells are components of phloem

47. abscisic acid
48. a _____ fruit is formed from the carpels of several associated flowers
51. a part of one plant is joined to another at an incision made in a stem or root
53. _____ are on the side walls of vessel elements
55. conspicuously colored leaves
56. an organism that is in an undifferentiated state before it has achieved a distinctive recognizable form
57. the flattened portions of a leaf
60. the simplest complete living systems
62. structure within which megaspores are produced
63. a waxy waterproofing substance
64. a seed _____ protects against desiccation
66. the root _____ protects against abrasion during growth

23

ANIMAL CELLS, TISSUES, AND ORGAN SYSTEMS

General Objectives

1. Understand the various levels of animal organization (cells, tissues, organs, and organ systems) and be familiar with the anatomical terms provided.
2. Know the characteristics of the various types of tissues. Know the types of cells that compose each tissue type and cite some examples of organs that contain significant amounts of each tissue type.
3. Describe how the four principal tissue types are organized into an organ such as the skin.
4. Explain how the human body maintains a rather constant internal environment despite changing external conditions.

23-I
(pp. 308–310)

ANIMAL STRUCTURE AND FUNCTION: AN OVERVIEW
    Levels of Organization
    Some Anatomical Terms
    Organ Systems and the Internal Environment
    Major Organ Systems

Summary

Individual cells can recognize and adhere to each other because patterns of *recognition proteins* on the plasma membrane serve as adhesion sites. Chemical and structural bridges established at the molecular level unite cells similar in form and function as cohesive *tissue*. Animal cells become differentiated into four main tissue types: epithelial, connective, muscle, and nervous tissues. Tissues that work together to perform a task compose an *organ*. Organs acquire their particular character from the tissues of which they are composed and from the ways in which they are joined together. Individual cells carry out the same general kinds of activities as multicellular organisms. However, because most cells in the multicellular animal body lack direct access to the resources of the external environment, the complex of organ systems works together to maintain within the body a stable environment that supplies resources and ensures favorable conditions. Each cell must carry out its own tasks even as it performs some special activity that stabilizes the internal environment of the whole organism. Extracellular fluid bathes the cells and exchanges substances with them. In vertebrates, most extracellular fluid is interstitial (occupying spaces between cells and tissues). The remainder consists of blood plasma (which is contained in blood vessels and the heart) and lymph (which is contained in lymph vessels and lymph nodes).

Key Terms

anatomy	integumentary system	urinary system
physiology	muscular system	reproductive system
tissue	skeletal system	bilateral symmetry
organ	nervous system	midsagittal plane
organ system	endocrine system	transverse plane
invertebrates	circulatory system	anterior
vertebrates	lymphatic system	posterior
extracellular fluid	respiratory system	dorsal
interstitial fluid	digestive system	ventral
plasma		

Objectives

1. Explain how, if each cell can perform all of its basic activities, organ systems contribute to cell survival.
2. Define *extracellular fluid* and *interstitial fluid*.
3. Explain how extracellular fluid helps cells survive.
4. Distinguish *tissues* from *organs* and give an example of each structure.
5. List the various types of tissues that are assembled into animal organs, organ systems, and organisms.
6. List each of the eleven principal organ systems and match each with its main tasks.

Self-Quiz Questions

Fill-in-the-Blanks

(1) _____ fluid fills the spaces between cells and tissues in vertebrates. Groups of *like* cells that work together to perform a task are known as a(n) (2) _____. Groups of *different* types of tissues that interact to carry out a task are known as a(n) (3) _____. The (4) _____ plane divides the body into right and left halves; the (5) _____ plane divides the body into (6) _____ (front) and posterior (back) parts. The (7) _____ plane divides it into dorsal (upper) and (8) _____ (lower) parts. The urinary system of an animal is responsible for the disposal of (9) _____ wastes; fecal material is not considered in the category. The endocrine system is generally responsible for internal (10) _____ control; together with the (11) _____ system, it integrates physiological processes.

Labeling   Identify each indicated part of the accompanying illustration.

(12) _____ _____   (15) _____ _____   (18) _____

(13) _____ _____   (16) _____   (19) _____

(14) _____   (17) _____ _____

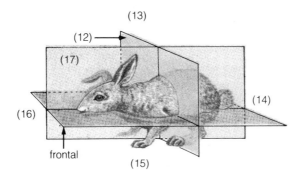

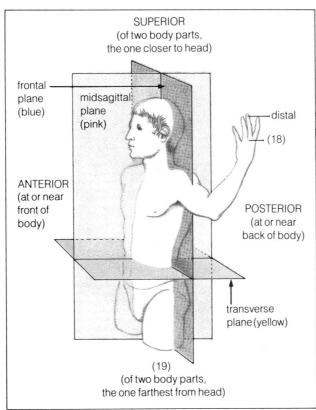

23-II
(pp. 311–313)

ANIMAL TISSUES
Tissue Formation
Epithelial Tissue

Summary   All the diverse body parts found in different animals may be assembled from a few tissue types, simply through variations in how those tissues are combined and arranged. *Somatic* cells constitute the physical structure of the animal body, and *germ* cells develop into gametes. In animal embryos, *ectoderm*, *mesoderm*, and *endoderm* are unspecialized layers of tissues that will differentiate into the various specialized tissues. Ectoderm forms the tissues of the nervous system and outer skin; mesoderm forms muscle, skeletal, and circulatory tissues, in addition to other tissues. Endoderm becomes the gut lining and digestive glands.

Epithelial tissues cover the external body surface of all animals and line internal organs from gut cavities to vertebrate lungs. These tissues are sheets of densely packed cells with little space or intercellular material at intercellular junctions. They form a continuous barrier between the body parts they cover and the surrounding medium. There may be one flat layer of cells, or the tissue may be stratified into different layers. Cells in epithelial tissues may resemble flat floor tiles, cubes, or columns. Functionally, epithelium may serve as a covering tissue or as a glandular tissue. Simple epithelium blocks out some substances and facilitates passage of some other substances across the tissue. Stratified epithelium protects underlying tissues. Glandular epithelium includes endocrine glands that secrete hormones into the bloodstream and exocrine glands that secrete their various products (digestive enzymes, mucus, saliva, wax, oil, milk) through ducts that open out onto an epithelial surface.

Key Terms

germ cells	cell junctions	squamous epithelium
embryo	epithelium, -lia	cuboidal epithelium
primordial tissue layer	tight junctions	columnar epithelium
ectoderm	adhering junctions	gland
mesoderm	communication junctions	endocrine gland
endoderm	simple epithelium	hormone
somatic cells	stratified epithelium	exocrine gland

Objectives

1. Describe Humphreys and Moscona's sponge research in the 1960s and indicate the significance of the results.
2. Explain the nature of three different cell-to-cell junctions and state the types of tissues in which these junctions occur.
3. Define what is meant by *epithelial tissue*.
4. Identify and distinguish each of the three structural types of epithelial tissue.
5. List the functions carried out by epithelial tissue and state the general location of each type.
6. Describe how the general structure of epithelial tissues permits it to carry out its principal functions.
7. Explain the meaning of the term *gland*, cite three examples of glands, and state the extracellular products secreted by each.

Self-Quiz Questions

Fill-in-the-Blanks

All of the diverse body parts found in different animals may be assembled from a few (1) _____ types, through variations in the way they are combined and arranged. (2) _____ cells constitute the physical structure of the animal body; they become differentiated into the components of four main types of tissue: (3) _____, (4) _____, (5) _____, and (6) _____.

If false, explain why.

___ (7) Mammalian skin contains squamous epithelium.

___ (8) The more tight junctions there are in a tissue, the more permeable the tissue will be.

___ (9) Endocrine glands secrete their products through ducts that empty onto an epithelial surface.

___ (10) Endocrine cell products include digestive enzymes, saliva, and mucus.

23-III
(pp. 314–316)

ANIMAL TISSUES (cont.)
Connective Tissue

Summary

Connective tissue cells are usually scattered throughout an extensive extracellular ground substance; these tissues bind together and generally support other animal tissues. *Loose connective tissue* functions as a packing material composed of strong, flexible collagen fibers and a few highly elastic *elastin* (a protein) fibers. *Dense fibrous tissue* attaches and holds movable body parts together; it consists generally of parallel collagen fibers with little ground substance between them. *Tendons* and *ligaments* are made of dense fibrous tissue. *Adipose* tissue consists of large cells, each containing a single fat-filled vacuole. Fat contains stored energy, cushions parts of the body against shocks and blows, and helps retain body heat.

Cartilage and *bone* are supportive connective tissues. Cartilage has small clusters of living cartilage cells suspended in the firm, rubbery ground substance they secreted. *Bone* consists of living bone cells embedded in a dense collagen-rich ground substance they secreted; calcium salts strengthen and make the collagen matrix rigid. The long bones of mammals contain red marrow (a major site of blood cell formation) and yellow marrow (a reserve tissue).

Blood consists of a fluid ground substance (*plasma*) and free cells; it transports substances to and from body cells, regulates pH, and helps to distribute body heat, nutrients, water, wastes, hormones, and infection-fighting cells.

Key Terms

connective tissue	loose connective tissue	spongy bone tissue
connective tissue proper	elastin	red marrow
ground substance	adipose tissue	yellow marrow
fibroblast	cartilage	compact bone tissue
dense connective tissue	bone	matrix
collagen	calcium phosphate	lamellae
tendon	lacunae	Haversian canal
ligament	osteocytes	blood

Objectives

1. Describe the basic features of connective tissue and explain how they enable connective tissue to carry out its various tasks.
2. List each type of connective tissue and characterize each in terms of its component materials, its location in the body, and its function.
3. Explain how bone is shaped and reshaped throughout the life span of a large mammal.
4. State the functions of marrow and adipose tissue.
5. List three functions of blood.

Self-Quiz Questions

Fill-in-the-Blanks

Cells of the connective tissues are scattered throughout an extensive extracellular (1) _____ _____. (2) _____ connective tissue contains a weblike scattering of strong, flexible protein fibers (3) (_____) and a few highly elastic protein fibers (4) (_____) and serves as a packing material that holds in place blood vessels, nerves, and internal organs. (5) _____ and (6) _____ are examples of dense fibrous tissue that help connect elements of the skeletal and muscular systems. (7) _____ and (8) _____ are examples of supportive connective tissue. (9) _____ systems are found in the long bones of mammals and contain living bone cells that receive their nutrients from the blood. (10) _____ _____ is a major site of blood cell formation. The fluid matrix of blood is called (11) _____: (12) _____ tissue serves to store fat as a food reserve.

Labeling

Identify each indicated part of the accompanying illustration.

(13) _____ _____   (15) _____ _____   (17) _____ _____

(14) _____ _____   (16) _____ _____   (18) _____

23-IV
(pp. 317–318)

ANIMAL TISSUES (cont.)
   Muscle Tissue
   Nervous Tissue

Summary

Nerve cells have highly excitable plasma membranes. In most animals, nerve cells (neurons) are arranged in organized networks concerned with receiving signals and conducting nerve impulses to other neurons, muscles, or glands, causing them to respond by contracting or secreting substances. Muscle cells enable an animal to move in place or through the environment; contractile thread-like structures shorten upon chemical stimulation and move body parts to which the muscle cells are attached. *Striated* (striped) muscle forms skeletal and cardiac (heart) muscle tissues. *Smooth* muscle tissue is not banded, contracts slowly, and is responsible for involuntary movements such as those associated with digestion, circulation, and respiration.

Key Terms

nervous tissue	striated	muscle bundle
neurons	smooth muscle tissue	cardiac muscle tissue
muscle tissue	involuntary	intercalated disks
contract	skeletal muscle tissue	

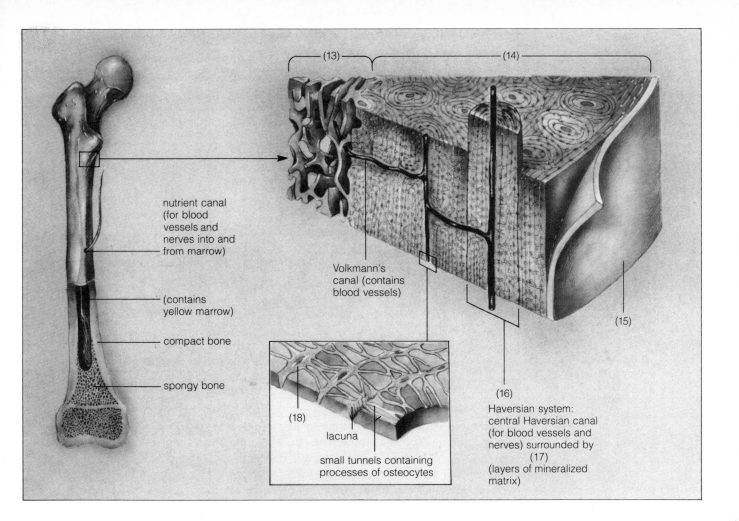

nutrient canal (for blood vessels and nerves into and from marrow)

(contains yellow marrow)

compact bone

spongy bone

(13)

(14)

(15)

Volkmann's canal (contains blood vessels)

lacuna

(18)

small tunnels containing processes of osteocytes

(16)
Haversian system: central Haversian canal (for blood vessels and nerves) surrounded by (17) (layers of mineralized matrix)

Objectives

1. Explain how changes in plasma membranes provide the basis for receiving and responding to signals in the environment.
2. Define *neuron* and describe the basic features of a true nervous system.
3. Describe the relationships among muscle cells, muscle bundles, and muscles.
4. Distinguish among *skeletal*, *cardiac*, and *smooth muscle tissues* in terms of location, structure, and function.

Self-Quiz Questions

True-False

If false, explain why.

___ (1) Muscle bundles are identical to skeletal muscle cells.

___ (2) Both skeletal and cardiac muscle tissues are striated.

___ (3) Cardiac muscle cells are fused, end-to-end, at regions called muscle bundles.

___ (4) Smooth muscle tissue contracts more slowly than does striated muscle tissue.

___ (5) Smooth muscle tissue contracts voluntarily.

___ (6) Neurons conduct messages to other neurons or to muscles or glands.

23-V
(pp. 318–321)

ANIMAL TISSUES (cont.)
    Case Study: The Tissues of Skin
HOMEOSTASIS AND SYSTEMS CONTROL
SUMMARY

Summary

Human skin is an example of an organ system composed of organs such as sebaceous glands, blood vessels, sweat glands, and so on. All four basic types of tissues are included in the skin, which is arranged in three principal layers: the epidermis, the dermis, and the subcutaneous layer.

Organisms maintain internal conditions through systems of homeostatic control, which try to preserve physical and chemical aspects of the body's interior within specific ranges of tolerance. In *negative feedback mechanisms*, deviations from stable conditions activate reactions that work to counteract or reverse the change. *Positive feedback mechanisms*, in which some disturbance to the homeostatic state activates a chain of events that throws things even further off balance, are also called into play in such situations as the formation of a blood clot or the birth process. Homeostatic feedback mechanisms have three basic components: receptors, integrators, and effectors. Receptors detect specific types of stimuli (energy changes in the environment). Integrators are control points where responses to the stimulus are selected and signals are sent to the organism's effectors (the muscles and glands), which respond to the stimuli. *Feedforward mechanisms* detect small environmental changes at their surface receptors and send messages to organs in the body that can begin corrective measures before the change in external environment significantly alters the internal environment.

Key Terms

epidermis	follicle	stimulus
dermis	collagen	integrators
hypodermis	Bernard	effectors
keratinization	homeostatic mechanisms	"set point"
keratin	"feedback"	positive feedback
melanocytes	negative feedback	mechanisms
basement membrane	mechanisms	feedforward mechanisms
sebaceous glands	receptors	

Objectives

1. Define *epidermis, dermis,* and *subcutaneous layer.*
2. List all the various components found in each of the three skin layers.
3. Draw a diagram that illustrates the mechanism of homeostatic control.
4. Distinguish between *negative* and *positive feedback mechanisms.* Give an example of each in human physiology.
5. Describe the relationships among receptors, integrators, and effectors in a negative feedback system.
6. Explain how feedforward mechanisms refine homeostasis.

Self-Quiz

Fill-in-the-Blanks

Human skin is an organ system that consists of two layers: the outermost (1) _____, which contains mostly dead cells, and the middle (2) _____, which contains hair follicles, nerves, tiny muscles associated with the hairs, and various types of glands. The (3) _____ layer, with its loose connective tissue and store of fat in (4) _____ tissue, lies beneath the skin.

Bernard discovered that the (5) _____ absorbs many of the nutrients carried to it by blood and converts them into complex storage forms; he also showed that the amount of blood being supplied to different body regions can be varied by the (6) _____ and (7) _____ of small blood vessels. Blood, with its cargo of (8) _____ and oxygen, can be diverted to regions where it is needed most.

True-False

If false, explain why.

____ (9) The process of childbirth is an example of a negative feedback mechanism.

____ (10) The human body's effectors are, for the most part, muscles and glands.

____ (11) An integrator is constructed in such a way that it detects specific energy changes in the environment and relays messages about them to a receptor.

CHAPTER TEST

UNDERSTANDING AND INTERPRETING KEY CONCEPTS

____ (1) Which of the following is *not* included in connective tissues?
(a) Bone
(b) Lymph
(c) Cartilage
(d) Collagen
(e) Skeletal muscle

____ (2) A surrounding substance within which something originates, develops, or is contained is known as a _____.
(a) lamella
(b) ground substance
(c) plasma
(d) lymph

____ (3) Blood is considered to be a(n) _____ tissue.
(a) epithelial
(b) muscular
(c) connective
(d) none of the above

___ (4) From smallest structure to largest, which group is arranged correctly?
(a) Muscle cells, muscle bundle, muscle
(b) Muscle cells, muscle, muscle bundle
(c) Muscle bundle, muscle cells, muscle
(d) None of the above

___ (5) Muscle that is not striped and is involuntary is _____.
(a) cardiac
(b) skeletal
(c) striated
(d) smooth

___ (6) Haversian systems _____.
(a) are composed of receptors, modulators, and effectors
(b) are negative feedback mechanisms
(c) are positive feedback mechanisms
(d) supply bone cells with nutrients, oxygen, and signals from else-where in the body

___ (7) Humphreys and Moscona's work with sponges demonstrated that _____.
(a) germ cells develop into gametes
(b) sponges are composed of epithelial, connective, muscular, and nervous tissues
(c) the body covering of sponges is composed of epidermis, dermis, and a subcutaneous layer
(d) cells from each species of sponge can distinguish "own-species" cells from "other-species" cells

___ (8) Chemical and structural bridges link groups or layers of like cells, uniting them in structure and function as a cohesive _____.
(a) organ
(b) organ system
(c) tissue
(d) cuticle

___ (9) A fish embryo was accidently stabbed by a graduate student in developmental biology. Later, the embryo developed into a creature that could not move and had no supportive or circulatory systems. Which embryonic tissue had suffered the damage?
(a) Ectoderm
(b) Endoderm
(c) Mesoderm
(d) Protoderm

___ (10) If its cells are striated and fused at the ends by intercalated disks so that the cells are not autonomous, the tissue is _____.
(a) smooth muscle
(b) dense fibrous connective
(c) supportive connective
(d) cardiac muscle

___ (11) The secretion of tears, milk, sweat, and oil are functions of _____ tissue.
   (a) epithelial
   (b) loose connective
   (c) lymphoid
   (d) nervous

CHAPTER TEST

INTEGRATING AND APPLYING KEY CONCEPTS

(1) Explain why, of all places in the body, marrow is located on the interior of long bones.
(2) Explain why your bones are remodeled after you reach maturity. Why does your body not keep the mature skeleton throughout life?

24

INFORMATION FLOW
AND THE NEURON

General Objectives

1. Describe the visible structure of neurons, neuroglia, nerves, and ganglia, both separately and together as a system.
2. Describe the distribution of the invisible array of large proteins, ions, and other molecules in a neuron, both at rest and as a neuron experiences a change in potential.
3. Understand how a nerve impulse is received by a neuron, conducted along a neuron, and transmitted across a synapse to a neighboring neuron, muscle, or gland.
4. Outline some of the ways by which information flow is regulated and integrated in the human body.

NEURONS: FUNCTIONAL UNITS OF NERVOUS SYSTEMS
Classes of Neurons
Structure of Neurons
Neuroglia
Nerves and Ganglia

Summary

Throughout the life cycle, each body part is constantly monitored and evaluated, not only for its own sake but also for its contribution to overall working patterns. Whole constellations of *neurons* (nerve cells arranged in precise message-conducting and information-processing pathways) and *endocrine cells* (which secrete chemical messages that affect target cells some distance away) form the framework of animal integration. Most neurons have three message-handling zones that occur on the cell body and on its cytoplasmic extensions (collectively called processes): an *input zone* (the dendrites), where the neuron receives most incoming messages; a *conducting zone* (the axon), along which information travels rapidly, without alteration, from one region to another; and an *output zone* (the axon terminals), where chemical or electrical signals are released that affect another neuron, muscle cells, or gland cells. In some animals, billions of neurons are organized into an interconnected system. Some are *sensory* neurons, which are receptors for environmental stimuli. *Interneurons* connect and integrate different neurons in the system. *Motor* neurons connect with muscles or glands (effectors) that increase or decrease their activity according to incoming signals. Neuroglial cells, which make up at least half the cells in a nervous system, nurture the neurons in various ways.

Key Terms

neuron	effectors	output zone
system	innervate	neuroglia
sensory neurons	dendrites	macrophages
receptors	axon	nerves
sensory stimuli	axon terminals	nerve pathways
integrators	input zone	nerve tracts
interneurons	trigger zone	nuclei
motor neurons	conducting zone	ganglion, -glia

Objectives

1. Describe the three classes of neurons and tell how each is related to the other.
2. Draw a neuron and label it according to its three general zones, its specific structures, and the specific function(s) of each structure.
3. Define *neuroglia* and state how they are related to neurons and the nervous system.

**Self-Quiz
Questions**

Fill-in-the-Blanks

Nerve cells that conduct messages are called (1) _____. (2) _____ cells, which support and nurture the activities of neurons, make up about half the volume of the nervous system. Most neurons have four zones: an input zone consisting of one or more (3) _____; the (4) _____ _____, which is the site on the neuronal membrane where an electrical signal is produced; the (5) _____, which consists of a single threadlike process called the (6) _____; and an output zone, the axonal (7) _____, where chemical or electrical signals are released. (8) _____ neurons are receptors for environmental stimuli, (9) _____ connect different neurons in the nervous system, and (10) _____ neurons are linked with muscles or glands.

24-II
(pp. 324–325)

ON MEMBRANE POTENTIALS
Membrane Excitability
The Neuron "At Rest"
Changes in Membrane Potential

Summary

The neuronal plasma membrane is permeable in different ways to different ions. Two steep concentration gradients—one for potassium ions, the other for sodium ions—exist across the neuronal membrane. In addition, an electrical gradient also exists across the neuronal membrane; the interior of a neuron at rest has an overall negative charge relative to the exterior. The difference in charge represents a potential to do work called the *resting membrane potential*. The potential energy from these ion-concentration and electrical gradients can be used to conduct messages whenever the gradients exist. The gradients for potassium and sodium ions are reestablished by membrane-bound enzyme systems called *sodium-potassium pumps*. Energy from a stimulus changes the proteins that span the membrane width at a certain point such that the membrane's permeability, first to sodium and then to potassium ions, is increased, and the ions flow along their natural gradients—sodium in and potassium out. This ion flow increases the permeability of the adjacent regions of the membrane, and the disturbance (now called the *nerve impulse*, or *message*) radiates outward from the initial point of stimulation.

   Graded potentials are short-range messages that can change and vary in size as they travel along short sections of dendrites, sensory receptors, and axonal terminals. Graded potentials arriving at a neuron from different pathways can be added up (summed) to determine the course of further transmissions.

Key Terms

electric gradient	open channels	resting membrane
concentration gradient	gated channels	potential
polarity	sodium-potassium pumps	receptor potential
current	millivolt	synaptic potential
action potentials	voltage difference	graded signal
membrane excitability		

Objectives

1. Explain how the plasma membrane of a neuron differs in function from that of a bone cell.
2. Describe the types of gradients associated with the neuronal membrane at rest.
3. Define *resting membrane potential*; explain what establishes it and how it is used by the cell neuron.

4. Describe what happens to the gradients associated with the neuronal membrane when it is disturbed.
5. Define *sodium-potassium pump* and state how it helps to maintain the resting membrane potential.
6. Explain how alterations in the electrical gradient across the neuron membrane are involved in the transmission of messages carried by the nervous system.

Self-Quiz Questions

Fill-in-the-Blanks

A plasma membrane is (1) _____ permeable; it passively permits some molecules and ions to move across it along their (2) _____ gradient, and it passively prevents others from doing so. Sometimes it actively transports ions so that they move (3) _____ their gradient. A neuronal membrane has far more positively charged (4) _____ ions inside than out and far more positively charged (5) _____ ions outside than inside. An electrical gradient also exists across the neuronal membrane; compared with the outside, the inside of a neuron at rest has an overall (6) _____ charge. For most neurons in most animals, the difference in charge across the neuronal membrane is within the range of 60–90 (7) _____, which represents an ability to do work; it is the (8) _____ _____ _____. A mechanism built into the neuronal membrane, the (9) _____-_____ _____, prevents net diffusion from equalizing concentrations across the membrane.

Alterations in the electrical gradient across the membrane constitute the (10) _____ that travels over a neuronal membrane surface. Short-range neural messages that decay within a short distance from the point of stimulation are called (11) _____ _____; those arriving at a neuron from different pathways can be (12) _____ to determine the course of further transmissions.

24-III
(pp. 325–329)

THE ACTION POTENTIAL
Mechanism of Excitation
Duration of Action Potentials
Propagation of Action Potentials
Saltatory Conduction

Summary

Some messages travel without changing along the neuronal membrane, and others do not. Fast-responding neurons usually have axons with broad diameters and effective insulation. Messages that travel unaltered and with great speed are encoded as *action potentials*. An action potential is an all-or-nothing event. When a *threshold stimulus* (the minimum change in membrane potential needed to generate a nerve impulse) is applied to a resting neuron, the electrical disturbance associated with ion flow across the membrane causes all of the sodium gates to open at once, which causes the membrane potential to reverse abruptly. The sodium gates then close, and the potassium gates open and then partly close when the membrane potential has returned to its resting value. If the threshold is reached, all of the associated membrane changes occur; if it is not reached, they do not occur at all. An action potential lasts for only a few milliseconds and is self-propagating: it triggers the same permeability changes in the adjacent membrane region, and so on away from the stimulation site. The time during which potassium ions are flowing out of the neuron is known as its *refractory period*—the time it is insensitive to stimulation. By the time the resting membrane potential is restored (repolarization) and the refractory period ends, the electrical disturbance has moved far enough away that it no longer can cause sodium gates to open in the region of the original impulse. Some fast-conducting cells have very thin axons that are wrapped in the electrically resistant myelin membranes of Schwann cells.

Key Terms

electrode	"spike"	refractory period
polarized	all-or-nothing event	Schwann cells
depolarized	cotransport mechanism	myelin sheath
repolarized	propagate	node of Ranvier
threshold level	message	saltatory conduction
positive feedback		

Objectives

1. Describe how gated channels through resting membranes participate in moving sodium ions.
2. Define *action potential* by stating its three main characteristics.
3. Explain the chemical basis of the action potential. Look at Figure 24.5 in your main text and determine which part of the curve represents:
   a. the point at which the stimulus was applied;
   b. the events prior to achievement of the threshold value;
   c. the opening of the ion gates and the diffusing of the ions;
   d. the change from net negative charge inside the neuron to net positive charge and back again to net negative charge; and
   e. the active transport of sodium ions out of and of potassium ions into the neuron.
4. Define *refractory period* and state what causes it.
5. Define *Schwann cell, nodes of Ranvier,* and *myelin sheath* and explain how each helps narrow-diameter neurons conduct nerve impulses quickly.
6. State the characteristics of neurons that help a message travel without changing along the neuronal membrane.

Self-Quiz Questions

Fill-in-the-Blanks

A(n) (1) _____ _____ is an all-or-nothing, brief reversal in membrane potential; it is also known as a(n) (2) _____ _____. Once an action potential has been achieved, it is (3) _____-_____-_____; its amplitude will not change even if the strength of the stimulus changes. The minimum change in membrane potential needed to achieve an action potential is the (4) _____ value. Each action potential is followed by a (5) _____ period: a time of insensitivity to stimulation. Some narrow-diameter neurons are wrapped in the lipid-rich (6) _____ of Schwann cells; each of these is separated from the next by a (7) _____ _____ _____—a small gap where the axon is exposed to extracellular fluid. An action potential jumps from one node to the next in line and is called (8) _____ conduction.

24-IV
(pp. 329–330)

SYNAPTIC POTENTIALS
    Chemical Synapses
    Excitatory and Inhibitory Postsynaptic Potentials
    Synaptic Potentials in Muscle Cells

Summary

Chemical synapses are junctions between two neurons or between a neuron and a muscle or gland cell. Only a small gap (the *synaptic cleft*) separates them. At a chemical synapse, only one of the cells releases transmitter substance into the synaptic cleft. Its action identifies it as the *presynaptic* cell, which relays signals to the other, *postsynaptic* cell.

    Transmitter molecules bind briefly to receptors on the postsynaptic cell membrane. Depending on the type of receptors activated on the postsynaptic cell, it may respond with excitatory or inhibitory potentials. An *excitatory postsynaptic potential* (EPSP) is depolarizing; it brings the membrane closer to the threshold of an action potential. An *inhibitory postsynaptic potential* (IPSP) is usually hyperpolarizing and drives the membrane away from the threshold.

    Many EPSPs acting together are usually needed to initiate a series of action potentials, and they must be strong enough to overcome the effects of inhibitory potentials at nearby synapses.

    Synaptic potentials are graded membrane responses at chemical synapses. *Graded* means they can vary in magnitude, depending on the stimulus energy. They are not all-or-nothing, and they are not self-propagating.

Key Terms

chemical synapse
synaptic cleft
transmitter substance
acetylcholine (ACh)
gamma aminobutyrate
  (GABA)
presynaptic cell
postsynaptic cell
vesicles

gated calcium channels
graded signals
synaptic potentials
excitatory postsynaptic
  potential (EPSP)
depolarizing
inhibitory postsynaptic
  potential (IPSP)

hyperpolarizing
neuromodulators
neuromuscular
  junctions
motor end plate
acetylcholinesterase

Objectives

1. Name all the structures and substances associated with a synapse and explain the role of each structure or substance in conducting a message across a synapse to a receiving neuron.
2. State the functional features shared by muscle cells and nerve cells.
3. Distinguish the way excitatory synapses function from the way inhibitory synapses function.
4. Explain how graded signals differ from action potentials.

Fill-in-the-Blanks The junction specialized for transmission between a neuron and another cell is
called a (1) _____. Usually, the signal being sent to the receiving cell is
carried by chemical messengers called (2) _____ _____.
(3) _____ is an example of such a chemical messenger that diffuses across
the synaptic cleft, combines with protein receptor molecules on the muscle cell
membrane, and soon thereafter is rapidly broken down by enzymes. At an
(4) _____ synapse, the membrane potential is driven toward the threshold
value and increases the likelihood that an action potential will occur. At an
(5) _____ synapse, the membrane potential is driven away from the
threshold value, and the receiving neuron is less likely to achieve an action
potential.

24-V
(pp. 330–334)

FROM SYNAPSE TO NEURAL CIRCUIT
  Synaptic Integration
  Circuit Organization
  The Stretch Reflex
SUMMARY

Summary *Integration* is the moment-by-moment combining of all excitatory and inhibitory
inputs acting at different synapses on a neuron. *Synaptic integration* means that
signals arriving at a neuron can be reinforced or dampened, sent or suppressed.
The direction of information flow through the body depends on the organization
of neurons into interconnecting circuits. In the brain and spinal cord, *local circuits*
are sets of interacting neurons confined to a single region. The circuits called
nerve pathways extend from neurons in one region to neurons in different regions.
    A *reflex* is a sequence of events elicited by a stimulus. In a monosynaptic
reflex pathway, a sensory receptor is stimulated, resulting in an impulse's being
generated in the sensory axon. The sensory neuron synapses in the spinal cord
or brain with a motor neuron, which generates an impulse in the motor neuron
that is carried to a muscle or gland (effectors), where the response will be a
contraction or secretion if the impulse is sufficiently strong.

Key Terms

synaptic integration	local circuit	muscle spindles
temporal summation	reflex arc	monosynaptic pathways
spatial summation	reflex	withdrawal reflex
tetanus	stretch reflex	polysynaptic pathways
lockjaw		

1. State what is meant by *integration* when the term is used in conjunction with the nervous system.
2. Explain what a *reflex* is, by drawing and labeling a diagram and telling how it functions.
3. Explain what the *stretch reflex* is and tell how it helps an animal survive.

Self-Quiz Questions

Fill-in-the-Blanks

(1) _____ at the cellular level is the moment-by-moment summation of all excitatory and inhibitory signals acting on a neuron. Incoming information is (2) _____ by cell bodies, and the charge differences across the membranes are either enhanced or inhibited; this summation of messages at the synapses is referred to as (3) _____. In the brain and spinal cord (= central nervous system), (4) _____ _____ are sets of interacting neurons confined to a single region. Circuits called (5) _____ _____ connect neurons from one region to neurons in different regions.

A (6) _____ is an involuntary sequence of events elicited by a stimulus. During a (7) _____ _____, a muscle contracts involuntarily whenever conditions cause a stretch in length; many of these help you to maintain an upright posture despite small shifts in balance. Located within skeletal muscles are length-sensitive organs called (8) _____ _____, which generate action potentials when stretched beyond a critical point; these potentials are conducted rapidly to the (9) _____ _____, where they are communicated to motor neurons leading right back to the muscle that was stretched.

CHAPTER TEST

UNDERSTANDING AND INTERPRETING KEY CONCEPTS

___ (1) Which of the following is *not* true of an action potential?
    (a) It is a short-range message that can vary in size.
    (b) It is an all-or-none brief reversal in membrane potential.
    (c) It doesn't decay with distance.
    (d) It is self-propagating.

___ (2) The conducting zone of a neuron is the _____.
    (a) axon
    (b) axonal terminals
    (c) cell body
    (d) dendrite

___ (3) The integrative zone of a neuron is the _____.
  (a) axon
  (b) axonal terminal
  (c) cell body
  (d) dendrite

___ (4) In the nervous system, a cotransport mechanism involves _____.
  (a) the inactivation of signals along the parasympathetic nerves by signals from the sympathetic division
  (b) acetylcholine being reciprocally inactivated by cholinesterase in the synaptic zone
  (c) the exchange of signals between the peripheral and central nervous systems
  (d) two different substances being exchanged across a membrane by the same enzyme system

___ (5) An action potential is brought about by _____.
  (a) a sudden membrane impermeability
  (b) the movement of negatively charged proteins through the neuronal membrane
  (c) the movement of lipoproteins to the outer membrane
  (d) a local change in membrane permeability caused by a greater-than-threshold stimulus

___ (6) The resting membrane potential _____.
  (a) exists as long as a charge difference sufficient to do work exists across a membrane
  (b) occurs because there are more potassium ions outside the neuronal membrane than there are inside
  (c) occurs because of the unique distribution of receptor proteins located on the dendrite exterior
  (d) is brought about by a local change in membrane permeability caused by a greater-than-threshold stimulus

___ (7) The phrase "all or none" used in conjunction with discussion about an action potential means that _____.
  (a) a resting membrane potential has been received by the cell
  (b) an impulse does not decay or dissipate as it travels away from the stimulus point
  (c) the membrane either achieves total equilibrium or remains as far from equilibrium as possible
  (d) propagation along the neuron is saltatory

CHAPTER TEST

INTEGRATING AND APPLYING KEY CONCEPTS

What do you think might happen to human behavior if inhibitory postsynaptic potentials did not exist and if the threshold stimulus necessary to provoke an EPSP were much higher?

25

NERVOUS SYSTEMS

General Objectives

1. Contrast invertebrate and vertebrate nervous systems in terms of neural patterns.
2. Describe the organization of peripheral versus central nervous systems.
3. Identify the parts of primitive brains, then tell how the human brain is advanced beyond the primitive types.

25-I
(pp. 335–337)

NEURAL PATTERNING: AN OVERVIEW

INVERTEBRATE NERVOUS SYSTEMS
    Nerve Nets
    Cephalization and Bilateral Symmetry
    Segmentation

Summary

Reflexes are simple, stereotyped movements made in response to sensory stimuli. In the simplest reflex pathways, a sensory neuron directly signals a motor neuron, which acts on muscle cells. Reflex pathways are the basic operating machinery of nervous systems. Nervous systems evolved through accretion: a layering of additional nervous tissues over reflex pathways of more ancient origin. Nervous systems evolved along with sensory organs (such as eyes) and motor structures (such as legs and wings), and together they provided the foundation for more active and intricate life styles. The oldest parts of the vertebrate brain deal with reflex coordination of sensory inputs and motor outputs beyond that afforded by the spinal cord alone. Even the most recent layerings of nerve tissue deal partly with reflex coordination. But they also deal with storing, comparing, and using experiences to initiate novel, nonstereotyped action. These regions are the basis of memory, learning, and reasoning.

No matter how specialized a nervous system has become, the same basic principles apply: (1) Information flows by way of *graded* potentials (in receptors and at synapses) and *action* potentials (along axons). (2) Pathways of divergence allow a message being conducted along a neuron to travel through its branched endings and activate many other neurons. (3) Pathways of convergence enable each neuron to receive both excitatory and inhibitory signals coming from many other neurons. (4) For all bilaterally symmetrical animals, each through-conducting pathway leads from sensory axons to the central nervous system and then to motor axons leading away to muscles, glands, or other effectors. (5) Excitatory and inhibitory signals are summed at each synaptic transfer. (6) Signals of all sorts are sent to the brain, which sends out command signals when a change in response is required. The brain's action represents the highest level of integration yet developed.

Nervous systems presumably evolved from simple systems such as those seen in cnidarians (radially arranged multidirectional networks dispersed throughout the body) to a bilaterally symmetrical, centralized system with interneurons generally arranged along the body axis and the major sense organs in the head. A shift from radial to bilateral symmetry could have led, in some evolutionary lines of animals, to paired nerves and muscles, paired sensory structures such as eyes, and paired brain regions. Differences that do exist among complex animals relate largely to brain size and its degree of control over the rest of the nervous system.

Key Terms

theory of neural patterning
reflex
stereotyped
radial symmetry
nerve net

planula
polyp
cephalization
bilateral symmetry
ganglion, ganglia

nerves
nerve cords
rudimentary
segmentation
segmental ganglion

Objectives

1. Contrast the sensitivity and motor abilities of sponges and cnidarians.
2. List some of the changes that might have occurred in promoting the change from a radially arranged, noncephalized, multidirectional nervous system to a system that is bilaterally arranged, cephalized, and with to a one-directional flow of information.
3. Describe how the shift from radial to bilateral symmetry influenced the number of organs and their location in the body.

Self-Quiz Questions

Fill-in-the-Blanks

The symmetrical arrangement of constituents, especially of radiating parts, about a central point is (1) _____ symmetry. Organized knots of neurons encased in connective tissue that form integrative centers are known as (2) _____. (3) _____ increases the concentration of nervous structures and coordinative functions in the head; organisms with this body pattern tend to have (4) _____ symmetry. A shift from radial to bilateral symmetry could have led, in some evolutionary lines of animals, to paired (5) _____ and (6) _____, paired sensory structures such as eyes, and paired (7) _____ centers. The (8) _____ nervous system consists of the brain and nerve cord (or paired cords). The (9) _____ nervous system includes cell bodies of sensory neurons and all nerves (bundles of axons).

25-II
(pp. 337–340)

THE VERTEBRATE PLAN
Evolution of Vertebrate Nervous Systems
Functional Divisions of the Vertebrate Nervous System
PERIPHERAL NERVOUS SYSTEM
Nerves Serving Autonomic Functions
Cranial Nerves

Summary

In all vertebrates, the *central nervous sytem* is enclosed in protective walls and cushioned and bathed with fluid. Bony fishes, amphibians, reptiles, birds, and mammals all have bony plates joined together to form a *skull* that protects the *brain* and another series of bony segments linked into a *vertebral column* that protects the *spinal cord*. In the more primitive fishlike vertebrates, cartilage never develops into bone for this protection. The spinal cord is a region of local integration and reflex connections and serves as the nerve pathway leading to and from the brain. Nerves leading to and from the central nervous system and associated ganglia lying outside the brain and spinal cord make up the peripheral nervous system.

The skeletal muscles are the effectors of the *somatic* nervous system, which in humans is under conscious control. The *autonomic* nervous system, on the other hand, generally is not under conscious control; parts of the peripheral and central nervous systems cooperate to adjust the body's organs and organ systems to changing conditions. The autonomic nervous system is subdivided into two networks that act antagonistically to each other. *Parasympathetic* nerves generally dominate internal events when environmental conditions permit normal body functioning, and *sympathetic* nerves dominate internal events in times of stress and danger, mobilizing the whole body for rapid response to change.

Key Terms

notochord
vertebra, -ae
vertebral column
central nervous system, CNS
peripheral nervous system, PNS
afferent
nerve cord

efferent
somatic system
autonomic system
visceral
sympathetic
parasympathetic
forebrain
midbrain

hindbrain
cervical
thoracic
lumbar
sacral
coccygeal nerves
cranial nerves
biofeedback

Objectives

1. Describe the basic structural and functional organization of the spinal cord. In your answer, distinguish spinal cord from vertebral column.
2. Distinguish the somatic and autonomic nervous systems with respect to location and chief activities.
3. Explain how parasympathetic nerve activity balances sympathetic nerve activity. List activities of the sympathetic and parasympathetic nerves in regulating pupil diameter, rate of heartbeat, activities of the gut, and elimination of urine.
4. Describe how your autonomic nervous system would act to promote your survival if you were in a serious car accident and rendered unconscious.
5. Define and contrast *central* and *peripheral* nervous systems.

Self-Quiz Questions

Fill-in-the-Blanks

All motor-to-skeletal muscle pathways and all sensory pathways make up the (1) _____ nervous system. The remaining nerve tissue, which generally isn't under conscious control, is collectively known as the (2) _____ nervous system; it is subdivided into two parts: (3) _____ nerves, which respond to emergency situations, and (4) _____ nerves, which oversee normal body functioning. The (5) _____ _____ _____ consists of the brain and spinal cord. A skull encloses the brain, and the (6) _____ _____ encloses and protects the spinal cord in vertebrates. (7) _____ never is replaced by bone in vertebrates that are primitive fishes. In humans, thirty-one pairs of spinal nerves connect with the spinal cord and are grouped by anatomical region; twelve pairs of (8) _____ nerves connect parts of the head and neck with brain centers.

Matching

Choose the most appropriate letter to match with the numbered blank.

(9) ___ cervical
(10) ___ coccygeal
(11) ___ lumbar
(12) ___ sacral
(13) ___ thoracic

A. Chest
B. Neck
C. Pelvic
D. Tail
E. Waist

CENTRAL NERVOUS SYSTEM
   The Spinal Cord
   Divisions of the Brain
   Hindbrain
   Midbrain
   Forebrain

Summary

Direct reflex connections between the sensory and motor neurons controlling the limbs and trunk are made in the spinal cord. Most often, interneurons are interposed between sensory input and motor output. Interneurons connect with one another up, down, and laterally in the *gray matter* of the spinal cord, providing a degree of control over activities within the cord itself. Major nerve tracts (bundles of myelinated axons in the *white matter* of the cord) ascend from specific brain centers that provide more refined control over activities.

At the head end of the spinal cord are three distinct masses of neurons: the hindbrain, midbrain, and forebrain. The *hindbrain* contains the *medulla oblongata*, a center that controls breathing, heart rate, and blood pressure reflexes; and the *cerebellum*, which helps coordinate limb movements concerned with maintaining balance, posture, and spatial orientation, and the *pons*. The *midbrain* is the primary "association center" in fishes and amphibians. The optic lobes, which are visual centers associated with the optic nerves, are located dorsally. In reptiles, birds, and mammals, some integration occurs in the midbrain, but messages are sent on to the forebrain for further interpretation. The *forebrain* is divided into two halves—the cerebral hemispheres—which overlie forebrain regions called the thalamus, hypothalamus, and pituitary.

Key Terms

spinal cord	medulla oblongata	olfactory bulbs
white matter	pons	cerebrum
myelin sheath	reticular formation	association center
gray matter	cerebellum	forebrain
meninges	spatial orientation	thalamus
intervertebral disk	tremor	hypothalamus
dorsal root ganglion	midbrain	cerebral cortex
interneurons	tectum	limbic system
hindbrain		

Objectives

1. Compare the structures of the spinal cord and brain concerning white matter and gray matter.
2. Describe the ways the brain is protected and list its three principal divisions.
3. List the parts of the brain found in the hindbrain, midbrain, and forebrain and tell the basic functions of each.
4. Think about each part of the brain in the list you developed for Objective 3 and state how the behavior of a normal person would change if he or she suffered a stroke in that part of the brain.
5. Tell what happened to the importance of the midbrain during the evolution of the vertebrates.

Self-Quiz Questions

Fill-in-the-Blanks

The (1) _____ _____ is a region of local integration and reflex connections with nerve pathways leading to and from the brain; its (2) _____ _____, which contains myelinated sensory and motor axons, is the through-conducting zone. The (3) _____ _____ includes nerve cell bodies, dendrites, and nonmyelinated axon terminals; this is the (4) _____ zone.

The hindbrain is an extension and enlargement of the upper spinal cord; it consists of the (5) _____ _____, which contains the control centers for the heartbeat rate, blood pressure, and breathing reflexes. The hindbrain also includes the (6) _____, which helps coordinate motor responses associated with refined limb movements, maintenance of posture, and spatial orientation.

The (7) _____ contains two cerebral hemispheres composed of gray matter. Underlying the cerebral hemisphere is the (8) _____, a region that links many of the activities of the nervous and (9) _____ systems, and the (10) _____, which is the principal gateway to the cerebral hemispheres.

25-IV
(pp. 344–347)

THE HUMAN BRAIN
The Cerebral Hemispheres
Memory

Summary

The cerebral hemispheres are paired masses that structurally appear to be mirror images of each other. Functionally, however, they differ considerably. Inside each hemisphere is a core of *white matter* that contains axonal pathways connecting the rest of the central nervous system with the brain's surface layer. An expanse of *gray matter* about .6 cm thick, the *cerebral cortex*, covers the white matter. Some cortical regions receive signals from receptors on the body's periphery; others coordinate and process sensory input, and yet others coordinate instructions for motor responses.

The two cerebral hemispheres are connected by the *corpus callosum*. Experiments severing the corpus callosum have revealed how the cerebral hemispheres function in relation to each other. Many of the lines leading into and out of one hemisphere deal with the opposite side of the body. Each cerebral hemisphere can function separately, but it functions in response to signals mainly from one side of the body. The main association regions responsible for language and analytical skills generally reside in the left hemisphere. The main association regions responsible for nonverbal perception (intuition) generally reside in the right hemisphere.

Conscious experience entails a capacity for *memory* (the storage of individual bits of information somewhere in the brain). So far, experiments suggest that at least two stages are involved in establishing memory traces in the brain. One is a short-term formative period lasting only a few minutes; information then becomes spatially and temporally organized in neural pathways (long-term storage). Long-term memory depends on structural changes in the brain. There is also evidence that neuron fine structure is not static, but rather can be modified in several ways, most likely depending on electrical and chemical interactions with neighboring neurons.

Key Terms

optic chiasm	primary somatic sensory cortex	thinking
pituitary gland	primary visual cortex	memory
corpus callosum	primary auditory cortex	memory trace
motor centers	olfactory area	short-term storage
Sperry	primary receiving centers	long-term storage
Broca's area	association centers	retrograde amnesia
motor cortex		

Objectives

1. Describe how the cerebral hemispheres are related to the other parts of the forebrain.
2. List the four lobes of the cerebral cortex (see Figure 25.10) and state the activities that occur in each.
3. Name the parts of the brain shown in Figures 25.10 and 25.11.
4. State the functions that are associated with the brain structures listed in Objective 3.
5. Describe current theories that attempt to explain how neurons might be the repositories of memory.
6. State what the results of the "split-brain" experiments suggest about the functioning of the cerebral hemispheres.

Self-Quiz Questions

Matching

(1) ___ Broca's area
(2) ___ cerebellum
(3) ___ corpus callosum
(4) ___ hypothalamus
(5) ___ limbic system
(6) ___ medulla oblongata
(7) ___ meninges
(8) ___ motor cortex
(9) ___ olfactory lobes
(10) ___ primary auditory cortex
(11) ___ somatic visual cortex
(12) ___ somatic sensory cortex

A. Monitors visceral activities; influences behaviors related to thirst, hunger, reproductive cycles, and temperature control
B. Coordinates muscles required for speech
C. Receives inputs from cochleas of inner ears
D. Issues commands to muscles
E. Relays and coordinates sensory signals to and from the cerebrum
F. Receives inputs from receptors in nasal epithelium
G. Receives and processes input from body-feeling areas
H. Broad channel of white matter that keeps the two cerebral hemispheres communicating with each other

(13) ___ thalamus

    I. Coordinates nerve signals for maintaining balance, posture, and refined limb movements

    J. Receives inputs from retinas of eyeballs

    K. Connects pons and spinal cord; contains reflex centers involved in respiration, stomach secretion, and cardiovascular function

    L. Protective coverings of brain and spinal cord

    M. Contains brain centers that coordinate activities underlying emotional expression

Fill-in-the-Blanks

The storage of individual bits of information somewhere in the brain is called (14) _____; the neural representation of such bits is known as a (15) _____. Experiments suggest that there are at least two stages involved in its formation. One is a (16) _____-_____ _____ period, lasting only a few minutes; then information becomes spatially and temporally organized in neural pathways. The other is a (17) _____-_____ _____; then information is put in a different neural representation and permanently filed in the brain.

Labeling

Identify each indicated part of the accompanying illustration (p. 223).

(18) _____   (21) _____ _____   (24) _____

(19) _____ _____   (22) _____ _____   (25) _____ _____

(20) _____   (23) _____

25-V
(pp. 347–351)

THE HUMAN BRAIN (cont.)
    States of Consciousness
    Commentary: **Drug Action on Integration and Control**
SUMMARY

Summary

Between the mindless drift of coma and total alertness are many levels of conscious experience: sleeping, dozing, meditating, and daydreaming. Neurons chattering among themselves show up as characteristic wavelike patterns in an *electroencephalogram* (EEG), an electrical recording of the frequency and strength of potentials from the brain's surface. Alpha waves predominate during meditation, a relaxed state of wakefulness. A slow-wave sleep pattern composes about eighty percent of the total sleeping time for adults. EEG arousal occurs when individuals make a conscious effort to focus on external stimuli or even on their own thoughts.

    The REM sleep pattern coincides with vivid dreams. The *reticular formation* governs changing levels of consciousness; within this formation are neurons (the

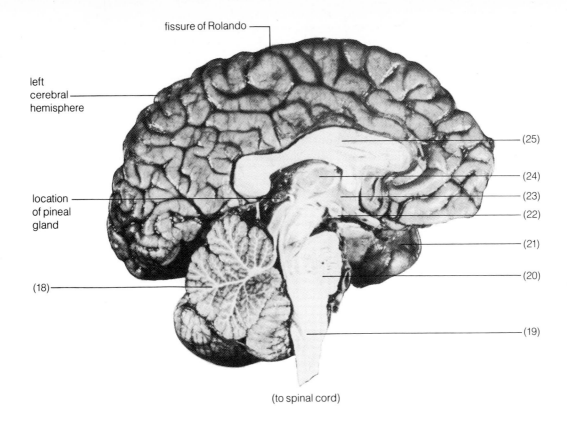

fissure of Rolando

left cerebral hemisphere

location of pineal gland

(18)

(25)
(24)
(23)
(22)
(21)
(20)

(19)

(to spinal cord)

RAS) that connect to the thalamus and arouse the brain and maintain wakefulness. The sleep centers are also in the reticular formation. The alternating states of sleeping and wakefulness are thought to have a physiological basis in neurotransmitters produced in the brain stem and whose interactions influence other brain regions. People deprived of the REM sleep and its accompanying dream state experience irritability and mental instability. Some individuals with brain damage never fall asleep, yet they survive.

Endorphins (which include enkephalins) are *analgesics* (pain relievers that the brain can produce); they bind to neural membranes on the spinal cord and limbic system and thereby inhibit neural activity. Emotional states are expressions of the balance of transmitter substances (such as acetylcholine, serotonin, dopamine, etc.) as they are produced in response to changing conditions in our complex world. Imbalances in the production of transmitter substances can provoke mental illness, the symptoms of which can often be brought under control by drugs that act to balance the imbalance. Tranquilizers, opiates, stimulants, and hallucinogens inhibit, modify, or enhance the behavior of chemical messengers throughout the brain.

Key Terms

consciousness
electroencephalogram, EEG
alpha rhythm
slow-wave sleep
REM sleep
EEG arousal
reticular activating system, RAS

thalamus
sleep centers
psychoactive drug
depressant
hypnotic
stimulant
narcotic analgesic

hallucinogen
psychedelic
antipsychotic
"psychosis"
endorphins
enkephalins

Objectives

1. Tell what an electroencephalogram is and what EEGs can tell us about the levels of conscious experience. Describe three typical EEG patterns and tell which level of consciousness each characterizes.
2. Locate and identify the function of the reticular formation. Explain what happens if a normal RF is stimulated or is damaged.
3. Distinguish the RAS from the sleep centers. Tell which area releases serotonin, and state the effect.
4. Explain the relationship between transmitter substances and analgesics.
5. List the major classes of psychoactive drugs and provide an example of each.

Self-Quiz Questions

Fill-in-the-Blanks

The principle wave pattern for someone who is relaxed, with eyes closed, is a(n) (1) _____ _____ The (2) _____- _____ _____ pattern occupies about eighty percent of the total sleeping time for adults. When individuals shift from sleep to a conscious focus on external stimuli or on their own thoughts, the pattern is called (3) _____ _____.

(4) _____ _____ accompanies vivid dreaming periods. Activities in the (5) _____ _____ determine whether you are awake or asleep.

High (6) _____ levels in the brain stem's core bring about drowsiness and sleep. (7) _____ are analgesics produced by the brain that inhibit regions concerned with our emotions and perception of (8) _____. Imbalances in (9) _____ _____ can produce emotional disturbances.

Matching

Match the most appropriate letter with any appropriate numbered item.

(10) ___ alcohol
(11) ___ amphetamines
(12) ___ caffeine
(13) ___ cocaine
(14) ___ lithium
(15) ___ LSD
(16) ___ marijuana
(17) ___ nicotine
(18) ___ opium
(19) ___ Quaalude
(20) ___ Valium

A. Antipsychotic drug
B. Depressants or hypnotic drugs
C. Narcotic analgesic drug
D. Psychedelic or hallucinogenic drug
E. Stimulant

UNDERSTANDING AND INTERPRETING KEY CONCEPTS

_____ (1) All nerves that lead away from the central nervous system
are _____.
(a) efferent nerves
(b) sensory nerves
(c) afferent nerves
(d) spinal nerves
(e) peripheral nerves

_____ (2) _____ nerves generally dominate internal events when
environmental conditions permit normal body functioning.
(a) Ganglia
(b) Pacemaker
(c) Sympathetic
(d) Parasympathetic
(e) All of the above

_____ (3) The center of consciousness and intelligence is the _____.
(a) medulla
(b) thalamus
(c) hypothalamus
(d) cerebellum
(e) cerebrum

_____ (4) _____ are the protective coverings of the brain.
(a) Ventricles
(b) Meninges
(c) Tectum
(d) Olfactory bulbs
(e) Pineal gland

_____ (5) Broca's area is concerned with _____.
(a) coordination of hands and fingers
(b) speech
(c) memory
(d) vision
(e) sense of taste and smell

_____ (6) The left hemisphere of the brain is responsible for _____.
(a) music
(b) mathematics
(c) language skills
(d) abstract abilities
(e) artistic ability and spatial relationships

_____ (7) The part of the brain that controls the basic responses necessary to
maintain life processes (breathing, heartbeat) is _____.
(a) the cerebral cortex
(b) the cerebellum
(c) the corpus callosum
(d) the medulla
(e) Broca's area

___ (8) To produce a split-brain individual, an operation would need to sever the _____.
   (a) pons
   (b) fissure of Rolando
   (c) hypothalamus
   (d) reticular formation
   (e) corpus callosum

___ (9) The center for balance and coordination in the human brain is the _____.
   (a) cerebrum
   (b) pons
   (c) cerebellum
   (d) hypothalamus
   (e) thalamus

___ (10) The sleep center of the human brain is the _____.
   (a) medulla
   (b) pons
   (c) thalamus
   (d) hypothalamus
   (e) reticular activating system

CHAPTER TEST

INTEGRATING AND APPLYING KEY CONCEPTS

Suppose that anger eventually is determined to be caused by excessive amounts of specific transmitter substances in the brains of angry people. Suppose that an inexpensive antidote to anger that neutralizes these anger-producing transmitter substances is also readily available. Can violent murderers now argue that they have been wrongfully punished because they were victimized by their brain's transmitter substances and could not have acted in any other way? Suppose an antidote is prescribed to curb violent tempers in a person. Suppose also that the easily angered person forgets to take the pill and subsequently murders a family member. Can the murderer still claim to be "victimized" by transmitter substances?

26

INTEGRATION AND CONTROL: ENDOCRINE SYSTEMS

General Objectives

1. Know the general mechanisms by which molecules integrate and control the various metabolic activities in organisms.
2. Understand how the neuroendocrine center controls secretion rates of other endocrine glands and responses in nerves and muscles.
3. Know how sugar and salt distribution is regulated in hormones.
4. Diagram the relationship between the various hormones that control reproduction in the human female and the human male.

26-I
(pp. 352–354)

"THE ENDOCRINE SYSTEM"
HORMONES AND OTHER SIGNALING MOLECULES

Summary

Transmitter substances are secreted from nerve cell endings and travel only a short distance, across a synaptic cleft, to an adjacent cell, in effect saying, "It's time to change your pattern of activity." Some transmitter substances are acetylcholine, epinephrine, norepinephrine, serotonin, and dopamine. They function, for a fleeting moment, as information carriers. *Neurohormones* are produced by *neurosecretory cells*; they travel slowly and farther by way of the bloodstream to many nonadjacent cells. In simple animals, neurohormones are the principal physiological controls over cells and tissues concerned with growth and reproduction.

In more complex animals, secretory endocrine cells function in a variety of ways; some operate individually, while others are organized into tissue patches or organs. All secrete true animal *hormones*, which are transported by the bloodstream and regulate specific cellular reactions in tissues and organs some distance away. Some hormones activate short-term adjustments (heart rate, blood chemical composition); others promote long-term adjustments, such as those involved in growth, differentiation, and reproduction. Generally, through hormonal action, specific enzyme activities (or enzyme formation) are accelerated or slowed down.

Pheromones, which signal other animals of the same species, are secreted by *exocrine glands*.

In addition to all of these chemical signals, *local mediator cells* secrete local signaling molecules that alter chemical conditions in their immediate vicinity and then are quickly degraded.

Key Terms

Bayliss and Starling	hormones	transmitter substances
endocrine system	neurosecretory cells	local mediator cells
signaling molecules	neurohormones	exocrine glands
target cell	synapsing neurons	pheromones
endocrine glands (cells)		

Objectives

1. Define *neurotransmitters, neurohormones,* and *hormones* and list their functions.
2. State what is secreted by nerve cell endings, neurosecretory cells, endocrine cells, and exocrine cells.
3. Name six examples of endocrine glands.

Self-Quiz Questions

Fill-in-the-Blanks

(1) _____ _____ are substances that are secreted from nerve endings and travel only a short distance across a synaptic cleft to an adjacent cell.

(2) _____ cells release information carriers, (3) _____, that travel more slowly and travel farther, by body fluids, to many nonadjacent cells. Animal (4) _____ are produced by endocrine cells and transported by the bloodstream to tissues and organs some distance away, where they regulate specific cellular reactions in (5) _____ cells. Pheromones are produced by specialized (6) _____ glands, which have ducts that lead out to the body surface; pheromones may activate behavioral changes in other animals of the (7) _____ species. The (8) _____ are the female gonads, and the (9) _____ are the male gonads.

NEUROENDOCRINE CONTROL CENTER
   The Hypothalamus-Pituitary Connection
   Posterior Lobe Secretions
   Anterior Lobe Secretions

Summary

Most endocrine elements are controlled either directly by a separate nerve supply or indirectly by secretions from particular centers of the nervous system. All vertebrates have a neuroendocrine control center, which consists of the hypothalamus and pituitary.

Information about changing conditions in the external and internal worlds flows constantly to the *hypothalamus*. Some information arrives along neural pathways from the cerebral cortex. Some endocrine elements arrive in the bloodstream. In the hypothalamus, incoming neural messages are summed, shifts in hormonal concentrations are detected, and the responses are sent out in the form of *releasing hormones* or *inhibiting hormones*, which are routed to one of two pituitary regions. The anterior pituitary is mostly glandular tissue; the posterior lobe and its connecting stalk are nervous tissue.

Key Terms

neuroendocrine control
   center
hypothalamus
pituitary gland
posterior lobe
anterior lobe
intermediate lobe
antidiuretic hormone
oxytocin

"collecting ducts"
vasopressin
releasing hormones
portal vessels
corticotropin-stimulating
   hormone, ACTH
thyrotropin-stimulating
   hormone, TSH

follicle-stimulating
   hormone, FSH
luteinizing hormone, LH
prolactin, PRL
somatotropin, STH (=
   growth hormone, GH)
pituitary dwarfism
gigantism
acromegaly

Objectives

1. Explain why the hypothalamus and pituitary are regarded as the neuroendocrine control centers.
2. State how, even though the anterior and posterior lobes of the pituitary are compounded as one gland, the tissues of each part differ in character.
3. Identify the hormones produced by the anterior lobe of the pituitary and tell which target tissues or organs that each one acts on.
4. Identify the hormones released from the posterior lobe of the pituitary and state their target tissues.

Self-Quiz Questions

Fill-in-the-Blanks

In the (1) _____, incoming neural messages are summed, shifts in hormonal concentrations are detected, and responses are sent out in the form of (2) _____ _____ to one of two regions of the (3) _____. The tissues of the anterior lobe are (4) _____ in nature; the tissues of the posterior lobe are (5) _____. Vessels that connect two distinct capillary beds are called (6) _____ vessels; such vessels link the capillary bed in the

(7) _____ with appropriate capillary beds in either lobe of the pituitary. From the anterior pituitary come several hormones: (8) _____, which stimulates the adrenal cortex; (9) _____, which stimulates the thyroid to produce thyroxin; and (10) _____, which stimulates milk production in mammary glands.

26-III
(pp. 360–362)

ADRENAL GLANDS
Adrenal Cortex
Adrenal Medulla
THYROID GLAND
PARATHYROID GLANDS
GONADS

Summary

The outer layer of the adrenal gland, the *adrenal cortex*, secretes three types of hormones: glucocorticoids, which help regulate food metabolism; mineralocorticoids, which influence salt and water concentrations; and sex hormones (the androgens and estrogens). The inner region, the *adrenal medulla*, secretes epinephrine (adrenalin) and norepinephrine, which help regulate blood circulation and carbohydrate metabolism.

The *thyroid gland* stores and releases hormones—among them thyroxin and calcitonin—that help govern growth, development, and metabolic rates throughout the body. The four parathyroid glands counterbalance the effects of calcitonin by raising the level of calcium in the blood.

The gonads (ovaries and testes) make gametes and secrete hormones that build and maintain ducts, glands, and tissues involved in reproduction. A variety of regulatory hormones produced by the anterior pituitary participate in feedback loops.

Key Terms

adrenal cortex	norepinephrine	parathyroid hormone,
glucocorticoids	thyroid gland	PTH
mineralocorticoids	TSH	rickets
aldosterone	triiodothyronine	gonads
parathyroid glands	thyroxin	testes
cortisol	hypothyroidism	ovaries
homeostatic feedback	hyperthyroidism	androgens
loops	goiter	estrogens
corticotropin, ACTH	calcitonin	progesterone
adrenal medulla	parathyroid glands	testosterone
epinephrine		

Objectives

1. List the major endocrine glands and patches of tissue in this section, tell which substance(s) each secretes, tell what tissues or organs each substance affects, and state how the "target" tissue responds.
2. State which two glands seem to be the principal regulators of the amount of calcium deposited in the bones and distributed in the muscles. Then tell which of the anterior pituitary hormones interact with these two glands.

3. Describe how the endocrine elements of the gonads and adrenal cortex affect aspects of reproduction.

Self-Quiz Questions

Fill-in-the-Blanks

The (1) _____ _____ produces glucocorticoids, such as (2) _____, in response to (3) _____ secreted by the anterior pituitary lobe. The adrenal medulla produces (4) _____ and (5) _____, which help regulate blood circulation and carbohydrate metabolism. When calcium levels in the blood plasma rise, the (6) _____ secretes calcitonin, which (7) [choose one] ☐ promotes ☐ inhibits the release of calcium from bone storage sites. Counterbalancing the effects of calcitonin, the (8) _____ secrete their hormone, which (9) [choose one] ☐ promotes ☐ inhibits the release of calcium into the bloodstream.

26-IV
(pp. 362–363)

OTHER ENDOCRINE ELEMENTS
    Pancreatic Islets
    Thymus Gland
    Pineal Gland
LOCAL CHEMICAL MEDIATORS
    Prostaglandins
    Growth Factors

Summary

Although the *pancreas* serves primarily as a secretor of digestive enzymes, about 2 million pancreatic islets secrete *insulin, glucagon,* and *somatostatin,* which counterbalance each other in regulating aspects of glucose, fat, and protein metabolism.

The *pineal gland* plays a role in reproductive physiology by cueing reproductive cycles to the changing seasons. The *thymus* produces infection-fighting cells. *Growth factors* are signaling molecules that influence growth by regulating the rate at which certain types of cells divide.

Key Terms

pancreatic islets	"type 1 diabetes"	puberty
alpha cells	"type 2 diabetes"	mediator
glucagon	thymus gland	prostaglandins
beta cells	thymosins	corpus luteum
insulin	lymphocytes	ovulation
delta cells	pineal gland	nerve growth factor, NGF
somatostatin	melatonin	epidermal growth factor,
glycogen	diurnal	EGF
diabetes mellitus		

1. List all the various hormones (and their secretory tissues) involved in regulating carbohydrate metabolism.
2. Explain how the thymus gland carries out its function.
3. Discuss the connection between melatonin and puberty.
4. Know what stimulates the production of various prostaglandins and describe their local effects on the female reproductive system.
5. Describe the symptoms of *diabetes mellitus* and distinguish between type 1 and type 2.

Self-Quiz Questions

Fill-in-the-Blanks

In humans, the pineal gland takes part in (1) _____ physiology, although its precise role has not been indentified. (2) _____ is a hormone that promotes the breakdown of glycogen (a storage starch) into glucose subunits; it counterbalances some aspects of (3) _____ action. Both hormones are produced by the pancreatic islets. The (4) _____ is involved in the immune response by producing a hormone that takes part in the production of specific types of infection-fighting cells.

Labeling

Identify each indicated part of the accompanying illustration.

(5) _____ gland

(6) _____ gland

(7) _____ glands

(8) _____ gland

(9) _____

(10) _____ gland

(11) _____

(12) _____ _____

(13) _____ (in female)

(14) _____ (in female during pregnancy)

(15) _____ (in male)

26-V
(pp. 363–365)

SIGNALING MECHANISMS
Steroid Hormone Action
Nonsteroid Hormone Action
SUMMARY

Summary

Steroid hormones pass easily through the plasma membrane of a target cell. Inside they bind to a receptor molecule in the nucleus of target cells and form a complex that "turns on" specific genes in the nucleus to make specific mRNA transcripts that alter the cell's activity. *Nonsteroid hormones* do not pass through the plasma membrane, but they do bind to the receptor sites on the membrane, which activates adenylate cyclase, an enzyme that converts ATP to cyclic AMP. cAMP activates a different enzyme that alters cellular activity.

Steroid hormones trigger gene activation and protein synthesis; nonsteroid hormones alter the activity of proteins already present in target cells. These cellular responses help maintain the internal environment and influence developmental and reproductive programs.

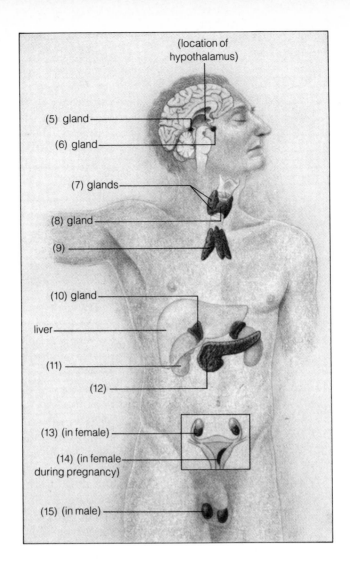

(location of hypothalamus)

(5) gland

(6) gland

(7) glands

(8) gland

(9)

(10) gland

liver

(11)

(12)

(13) (in female)

(14) (in female during pregnancy)

(15) (in male)

Key Terms

steroid hormones
testosterone
testicular feminization
  syndrome

nonsteroid hormones
second messengers
cyclic AMP, cAMP
adenylate cyclase

hormone-receptor
  complex
amplify

Objectives

1. Contrast the behavior of chemical signals that involve intracellular receptors with the behavior of those that involve surface membrane receptors. Identify which types of chemical signals participate in the two different mechanisms.
2. Steroids are produced by the gonads and adrenal cortex. Name all of the hormones involved and envision them participating in a mechanism like that shown in Figure 26.9.
3. A woman reports that she has absolutely no interest in sex. Before you recommend that she consult a psychologist, as a practicing endocrinologist state the items that you would test for first.
4. Contrast the proposed mechanisms of hormone action on target cell activities by (a) steroid hormones and (b) hormones that are proteins or derived from proteins.
5. Name four factors that influence hormone levels in the bloodstream.

Self-Quiz Questions

Matching

(1) ___ ACTH (corticotropin)

(2) ___ ADH

(3) ___ aldosterone

(4) ___ calcitonin

(5) ___ cortisol

(6) ___ epinephrine

(7) ___ estrogen

(8) ___ FSH

(9) ___ insulin

(10) ___ melatonin

(11) ___ parathyroid hormone

(12) ___ STH (= GH)

(13) ___ testosterone

(14) ___ thyroxine

(15) ___ TSH

A. Produced by anterior pituitary; essential for egg maturation

B. Influences daily biorhythms, sexual activity, and sexual development

C. Essential for sperm production; secreted by gonad

D. Increases heart rate and force of contraction; the main "emergency hormone"

E. Produced by gonad; essential for egg maturation and maintenance of secondary sex characteristics in the female

F. The water conservation hormone; released from posterior pituitary

G. Lowers blood sugar by encouraging cells to take in glucose; protein synthesis; fat storage

H. Stimulates adrenal cortex to secrete hormones involved in responses to stress

I. Elevates calcium levels in blood by stimulating calcium reabsorption from bone, kidneys, and absorption from gut

J. Influences overall metabolic rate, growth, and development and sensitivity to temperature extremes

K. Promotes sodium reabsorption; salt, water balance; produced by adrenal cortex

L. Lowers calcium levels in blood by inhibiting reabsorption from bone

M. Raises blood sugar by stimulating glucose production

N. Secreted by anterior pituitary; stimulates release of thyroid hormones

O. Secreted by anterior pituitary; enhances growth in young animals; stimulates release of somatomedins

Fill-in-the-Blank

(16) _____ is one kind of second messenger that communicates information from cell surfaces to specific inner regions of the cell.

UNDERSTANDING AND INTERPRETING KEY CONCEPTS

___ (1) The _____ is often called the master gland, but it is controlled by
    the _____.
    (a) pituitary, hypothalamus
    (b) pancreas, hypothalamus
    (c) thyroid, parathyroid glands
    (d) hypothalamus, pituitary
    (e) pituitary, thalamus

___ (2) Although the _____ is nervous in embryonic origin, structure,
    and behavior, it also secretes substances into the bloodstream.
    (a) anterior pituitary
    (b) adrenal cortex
    (c) pancreas
    (d) thyroid
    (e) posterior pituitary

___ (3) If you were lost in the desert and had no fresh water to drink, the
    level of _____ would increase in your blood as a means to
    conserve water.
    (a) insulin
    (b) erythropoietin
    (c) oxytocin
    (d) antidiuretic hormone
    (e) salt

For Questions 4–6, choose from these answers:
    (a) estrogen
    (b) progesterone
    (c) FSH
    (d) LH
    (e) prolactin

___ (4) _____ prepares and maintains the uterine lining for pregnancy.

___ (5) Sudden high levels of _____ bring about ovulation.

___ (6) _____ is a hormone produced by the pituitary and stimulates
    female gametes to mature in the ovary.

For Questions 7–9, choose from these answers:
    (a) adrenal medulla
    (b) adrenal cortex
    (c) thyroid
    (d) anterior pituitary
    (e) posterior pituitary

___ (7) The _____ produces steroid hormones that exert their effects on a
    target nucleus.

___ (8) The gland that is most closely associated with emergency situations is
    the _____.

___ (9) The _____ gland regulates the basic metabolic rate.

___ (10) If all sources of calcium were eliminated from your diet, your body would secrete more _____ in an effort to release calcium stored in your body to the tissues that require it.
(a) parathyroid hormone
(b) aldosterone
(c) calcitonin
(d) mineralocorticoids
(e) None of the above

INTEGRATING AND APPLYING KEY CONCEPTS

Suppose that you suddenly quadruple your already high daily consumption of calcium. State which body organs would be affected and tell how they would be affected. Name two hormones, the levels of which would most probably be affected, and tell whether your body's production of them would increase or decrease. Suppose that you continue this high rate of calcium consumption for ten years. Can you predict the organs that would be stressed most?

27

SENSORY SYSTEMS

**General
Objectives**

1. Know what a receptor is and list the various types of receptors.
2. Contrast the mechanism by which the chemical senses work with that by which the somatic senses work.
3. Understand how the senses of balance and hearing function.
4. Describe how the sense of vision has evolved through time.
5. Draw a medial section of the human eyeball through the optic nerve, identify each structure, and tell the function of each.

27-I
(pp. 368–371)

SENSORY PATHWAYS
    Receptors Defined
    Principles of Receptor Function
    Primary Sensory Cortex
CHEMICAL SENSES

Summary

The only windows between the nervous system and events going on within and around the animal body are *receptors*: finely branched peripheral endings of sensory neurons (or specialized cells adjacent to them) that respond to specific kinds of stimuli. A receptor is a *transducer*: it converts one form of energy into another. A particular sensation is not triggered by action potentials themselves but by their travel along a particular nerve pathway. Variations in stimulus intensity are converted into a code that is based on both the *frequency* of action potentials in a single axon and the *number* of axons acting in a given tissue at a particular time. A *stimulus* is any form of energy change in the environment that the body actually detects. Receptor cells translate stimulus energy into electrochemical messages that can be dealt with by the nervous system. *Chemoreceptors* detect impinging chemical energy; they include odor and taste receptors and internal receptors such as those sensitive to levels of specific chemicals in the bloodstream. Taste buds are examples of organs that contain chemoreceptors.

Key Terms

neural programs	thermoreceptors	graded response
stimulus	somatic sense	intensity
receptors	kinesthesia	frequency
sensory organs	acoustical sense	taste receptors
chemoreceptors	carotid bodies	taste buds
mechanoreceptors	transducer	olfaction
photoreceptors	receptor potential	

Objectives

1. Explain how a photoreceptor behaves as a transducer.
2. Define and distinguish among *chemoreceptors, mechanoreceptors, photoreceptors,* and *thermoreceptors*. Name at least one example of each type that appears in an animal.
3. Explain how a taste bud works and distinguish the types of stimuli it detects from those detected by tactile or stretch receptors.

Self-Quiz Questions

Fill-in-the-Blanks

Finely-branched peripheral endings of sensory neurons that detect specific kinds of stimuli are (1) _____. A (2) _____ is any form of energy change in the environment that the body actually detects. (3) _____ detect impinging chemical energy; (4) _____ detect mechanical energy associated with changes in pressure, position, or acceleration; (5) _____ detect the energy of visible and ultraviolet light; (6) _____ detect radiant energy asssociated with temperature changes; (7) _____ receptors sampling odors from food in the mouth are important for our sense of taste.

27-II
(pp. 371–376)

SOMATIC SENSES
THE SENSE OF BALANCE
THE SENSE OF HEARING
  Which Animals Hear?
  Echolocation

Summary

Mechanoreceptors detect mechanical energy associated with changes in pressure, position, or acceleration; they include receptors of stretch, touch, equilibrium, and hearing. Somatic senses are those of touch, pressure, temperature, and pain near the body surface. Somatic receptors are activated when their plasma membranes are mechanically deformed. *Pain* is perceived injury to some region of the body. *Sound* is perceived as a form of vibration. Sound detectors pick up on differences in the rate and density of high- and low-pressure regions in sequences of compressional waves. The sound receptors in the human ear are hair cells in the organ of Corti.

Key Terms

somatic senses	"vibration"	pitch
Pacinian corpuscles	tympanic membrane	outer ear
Meissner corpuscle	oval window	middle ear
Ruffini endings	scala vestibuli	hammer
equilibrium	scala tympani	anvil
hair cells	round window	stirrup
vestibular apparatus	basilar membrane	inner ear
semicircular canals	organ of Corti	cochlea
otolith organ	tectorial membrane	echolocation
statocyst	auditory nerve	
hearing	amplitude	

Objectives

1. Distinguish the types of stimuli detected by touch receptors from those detected by hearing and equilibrium receptors.
2. Follow a sound wave from pinna to organ of Corti; mention the name of each structure it passes and state where the sound wave is amplified and where the pattern of pressure waves is translated into electrochemical impulses.
3. State how low- and high-pitch sounds affect the organ of Corti.
4. State how low- and high-amplitude sounds affect the organ of Corti.
5. Explain how the three semicircular canals of the human ear detect changes of position and acceleration in a variety of directions. Compare semicircular canal function with statocyst function.

Self-Quiz Questions

Fill-in-the-Blanks

The (1) _____ (perceived loudness) of sound depends on the density difference between areas of high and low pressure in wave trains. The (2) _____ (perceived pitch) of sound depends on how fast the wave changes occur. The faster the vibrations, the (3) [choose one] ☐ higher ☐ lower the sound. Hair cells are (4) [choose one] ☐ nocireceptors ☐ mechanoreceptors ☐ electroreceptors that detect vibrations. The hammer, anvil, and stirrup are located in the (5) [choose one] ☐ inner ☐ middle ear. The (6) _____ is a coiled tube that resembles a snail shell and contains the (7) _____ _____—the organ that changes vibrations into electrochemical impulses. Structures that detect rotational acceleration in humans are (8) _____ _____.

Labeling   Identify each indicated part of the accompanying illustrations.

(9) _____ _____   (12) _____ _____ _____   (14) _____ _____

(10) _____   (13) _____ _____   (15) _____ _____

(11) _____ _____

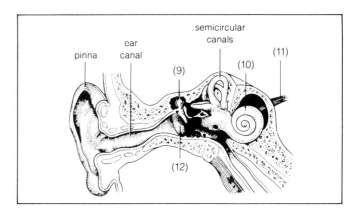

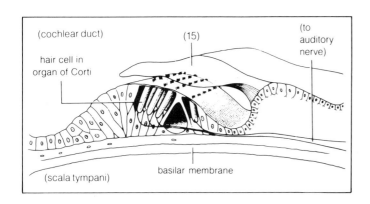

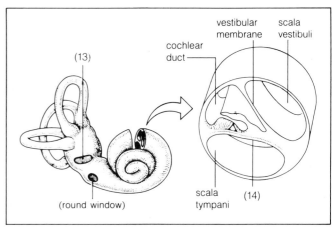

27-III
(pp. 376–383)

SENSE OF VISION
    Invertebrate Photoreception
    Vertebrate Photoreception
    Rods and Cones
    Processing Visual Information
SUMMARY

Summary

Light is a flow of discrete energy packets known as *photons*. Each photon is a bit of energy that has escaped from an excited atom—not only from atoms of an original light source, such as the sun, but from atoms of objects in the environment. Photons travel through the air in a straight line. Photoreceptors detect photon energy of visible and ultraviolet light. As the *pigment molecules* of photoreceptors selectively absorb photons, their molecular structures are altered; eventually, the alteration causes the incoming energy of photons to be transformed into the electrochemical energy of a nerve signal. *Vision* requires precise focusing of light onto a layer of photoreceptive cells that are sufficiently densely packed to sample details concerning the light stimulus, followed by interpretation of the signal pattern into an image (which occurs in the brain). In the vertebrate eye, adjustments by a lens ensure that the focal point for a given batch of light rays will fall on the retina (a tissue containing densely packed photoreceptors). Thermoreceptors detect radiant energy associated with temperature; they include infrared receptors.

Key Terms

photoreception	compound eyes	rod cell
phototaxis	ommatidium, -dia	cone cell
visual system	mosaic theory	fovea
lens	sclera	rhodopsin
eyespots	choroid	cis-retinal
microvilli	aqueous humor	trans-retinal
eyes	vitreous	optic nerve
cornea	focal point	bipolar cells
retina	accommodation	ganglion cells
iris	nearsighted	
pupil	farsighted	

Objectives

1. Define *light* and explain the general principles that affect how light is detected by photoreceptors and changed into electrochemical messages.
2. Give some examples demonstrating that many organisms lack eyes yet respond to light.
3. Explain what a visual system is and list four aspects of a visual stimulus that are detected by different components of a visual system.
4. Describe the main parts of an image-forming eye and contrast the way it functions with the way an ocellus (eyespot) functions.
5. Contrast the structure of compound eyes with the structures of ocelli and of the human eye.
6. Distinguish the way accommodation occurs in fish from the way it occurs in humans.
7. Define *nearsightedness* and *farsightedness* and relate each to eyeball structure.
8. Describe how the human eye perceives color and black and white.
9. Explain how Hubel and Wiesel's experiments suggest that the key to visual perception resides in the organization and synaptic connections between columns of neurons in the brain.

Self-Quiz Questions

Fill-in-the-Blanks

Light is a stream of (1) _____—discrete energy packets. (2) _____ is a process in which photons are absorbed by pigment molecules and photon energy is transformed into the electrochemical energy of a nerve signal.

(3) _____ requires precise light focusing onto a layer of photoreceptive cells that are dense enough to sample details of the light stimulus, followed by image formation in the brain. (4) _____ are simple clusters of photosensitive cells, usually arranged in a cuplike depression in the epidermis.

(5) _____ are well-developed photoreceptor organs that allow at least some degree of image formation. The (6) _____ is a transparent cover of the lens area, and the (7) _____ consists of tissue containing densely packed photoreceptors. Compound eyes contain several thousand photosensitive units known as (8) _____. In the vertebrate eye, lens adjustments assure that the (9) _____ _____ for a specific group of light rays lands on the retina. (10) _____ refers to the lens adjustments that bring about precise focusing onto the retina. (11) _____ people focus light from nearby objects posterior to the retina. (12) _____ cells are concerned with daytime vision and, usually, color perception. A (13) _____ is a funnel-shaped pit on the retina that provides the greatest visual acuity.

Labeling Identify each indicated part of the accompanying illustration.

(14) _____   (18) _____ _____   (22) _____ _____

(15) _____   (19) _____ _____   (23) _____ _____

(16) _____   (20) _____   (24) _____

(17) _____   (21) _____

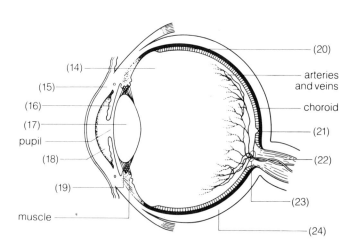

_____   _____

CHAPTER TEST   **UNDERSTANDING AND INTERPRETING KEY CONCEPTS**

___ (1) According to the mosaic theory, _____.
  (a) the basement membrane's pigment molecules prevent the scattering of light

(b) light falling on the inner area of an "on-center" field activates firing of the cells

(c) hair cells in the semicircular canals cooperate to detect rotational acceleration

(d) each ommatidium detects information about only one small region of the visual field; many ommatidia contribute "bits" to the total image

(e) All of the above

___ (2) The principal place in the human ear where sound waves are amplified is _____.
(a) the pinna
(b) the ear canal
(c) the middle ear
(d) the organ of Corti
(e) None of the above

___ (3) The place where vibrations are translated into patterns of nerve impulses is _____.
(a) the pinna
(b) the ear canal
(c) the middle ear
(d) the organ of Corti
(e) None of the above

For Questions 4–8, choose from these answers:
(a) fovea
(b) cornea
(c) iris
(d) retina
(e) sclera

___ (4) The white protective fibrous tissue of the eye is the _____.

___ (5) Rods and cones are located in the _____.

___ (6) The highest concentration of cones is in the _____.

___ (7) The adjustable ring of contractile and connective tissues that controls the amount of light entering the eye is the _____.

___ (8) The outer transparent protective covering of part of the eyeball is the _____.

___ (9) Accommodation involves the ability to _____.
(a) change the sensitivity of the rods and cones by means of transmitters
(b) change the width of the lens by relaxing or contracting certain muscles
(c) change the curvature of the cornea
(d) adapt to large changes in light intensity
(e) All of the above

___ 10. Nearsightedness is caused by _____.
   (a) eye structure that focuses an image in front of the retina
   (b) uneven curvature of the lens
   (c) eye structure that focuses an image posterior to the retina
   (d) uneven curvature of the cornea
   (e) None of the above

CHAPTER TEST

INTEGRATING AND APPLYING KEY CONCEPTS

How might human behavior be changed if human eyes were compound eyes composed of ommatidia and if humans perceived only vibrations—as fish do—rather than sounds?

28

MOTOR SYSTEMS

General Objectives

1. Compare invertebrate and vertebrate motor systems in terms of skeletal and muscular components and their interactions.
2. Identify human bones by name and location.
3. Describe some types of injuries that are commonly encountered by people in exercise programs.
4. Explain in detail the structure of muscles, from the molecular level to the organ systems level. Then explain how biochemical events occur in muscle contractions and how antagonistic muscle action refines movements.

28-I
(pp. 384–390)

INVERTEBRATE MOTOR SYSTEMS

VERTEBRATE MOTOR SYSTEMS

SKELETAL STRUCTURE AND FUNCTION
   Human Skeleton
   Types of Bones
   Development of Bones
   Types of Joints
   Commentary: On Runner's Knee
   Bone Tissue Turnover

Summary

Motor systems, which are adapted to respond to a variety of stimuli, run parallel with sensory systems. Motor systems are based on the ability of certain cells to contract (shorten) and relax (extend in length); they also depend on the presence of some medium or structure against which the contractile force may be applied. Contractile cells exert force against a variety of resistant surfaces (cylinders of fluid encased in membrane, inner plates, exterior plates, inner rods). Connective tissue generally links the contractile cells to the resistant structure, against which they pull in an *antagonistic muscle system*, where the action of one motor element opposes the action of another.

Invertebrates have several types of skeletomuscular systems, but vertebrates generally have rigid endoskeletons with antagonistic muscle sets applying force against them. Long, short, flat, and irregular bones may be linked by fibrous, cartilaginous, or sinovial joints.

Key Terms

motor system	scapula	short bones
muscle fiber	humerus	flat bones
antagonistic system	radius	irregular bones
hydrostatic skeleton	ulna	osteoblasts
cuticle	carpals	osteocytes
exoskeleton	metacarpals	cartilage models
endoskeleton	phalanges	epiphyseal plates
tendons	femur	fibrous joints
ligaments	tibia	cartilaginous joints
axial skeleton	tarsals	sinovial joints
appendicular skeleton	metatarsals	osteoarthritis
pectoral girdle	cranium	rheumatoid arthritis
pelvic girdle	cervical	osteoclasts
cranium	thoracic	runner's knee
vertebrae	lumbar	bursae
sternum	sacral	osteoporosis
clavicles	long bones	

Objectives

1. Distinguish between *contraction* and *relaxation* of muscle cells.
2. Explain what is meant by *antagonistic muscle system*.
3. Explain how segmentation, longitudinal muscles, circular muscles, and a hydrostatic skeleton are functionally interrelated in earthworms.
4. Describe the general design of the human skeletal system. Name six bones (or groups of bones) that compose an arm. Name the parts of the axial skeleton.
5. Explain the various roles of osteoblasts, osteoclasts, cartilage models, long bones, and epiphyseal plates in the development of human bones.

Self-Quiz Questions

Fill-in-the-Blanks

All motor systems are based on effector cells that are able to (1) _____ and (2) _____, and on the presence of a medium against which the (3) _____ force may be applied. Longitudinal and circular muscle layers work as an (4) _____ muscle system, in which the action of one motor element opposes the action of another. A membrane filled with fluid resists compression and can act as a (5) _____ skeleton. Arthropods have antagonistic muscles attached to an (6) [choose one] ☐ endoskeleton ☐ exoskeleton.

Labeling   Identify each indicated part of the accompanying illustration.

(7) ————   (10) ————   (13) ————

(8) ————   (11) ————   (14) ————

(9) ————   (12) ————   (15) ————

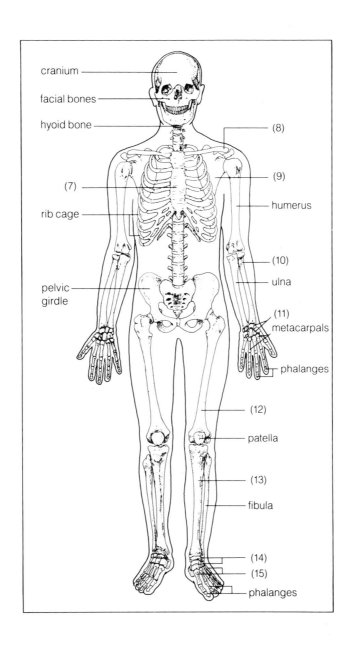

cranium

facial bones

hyoid bone

(8)

(9)

(7)

humerus

rib cage

(10)

ulna

pelvic girdle

(11) metacarpals

phalanges

(12)

patella

(13)

fibula

(14)

(15)

phalanges

MUSCLE STRUCTURE AND FUNCTION
Mechanisms of Muscle Contraction
Control of Muscle Contraction

Summary

A skeletal muscle is made up of hundreds of muscle fibers that are bound by connective tissue into muscle bundles, which are attached to bone. A motor neuron and all the muscle fibers it controls are called a motor unit. Each muscle fiber consists of fine, threadlike *myofibrils*, which in turn are composed of parallel actin and myosin filaments arranged in alternating bands. Z lines define the limits of each *sarcomere*, the fundamental unit of muscle contraction.

Combined decreases in length of individual sarcomeres account for contraction of the whole muscle. Contraction of a skeletal muscle cell is initiated by action potential(s) from neuron(s), causing an action potential along the muscle membrane, which in turn causes calcium ions to be released from the sarcoplasmic reticulum that surrounds the myofibrils. Each myofibril contains actin and myosin (protein filaments that lie parallel to each other). Calcium ions clear the binding sites on the actin filaments, freeing them to interact with crossbridges extending outward from the myosin filaments.

The hydrolysis of ATP releases free energy that is used to slide myosin past the actin in an enzyme-controlled mechanism that telescopes the protein filaments and thus shortens the muscle. Calcium pumps eventually transport calcium ions back to the sarcoplasmic reticulum, actin no longer interacts with myosin, and the sarcomere returns to the relaxed state. Because the nervous system controls the release of calcium from the sarcoplasmic reticulum, it exerts control over contraction itself.

Key Terms

skeletal muscle	I band	rigor
smooth muscle	deltoid	rigor mortis
cardiac muscle	pectoralis major	glycogen
muscle	triceps	creatine phosphate
muscle bundle	biceps	sarcolemma
muscle fiber (= muscle cell)	external oblique	sarcoplasmic reticulum
	rectus abdominis	transverse tubule system
myofibril	sartorius	neuromuscular junction
myofilaments	rectus femoris	acetylcholine
actin filament	gastrocnemius	troponin
myosin filament	tibialis anterior	tropomyosin
sarcomere	sliding-filament model	attachment site
Z line	cross-bridges	
A band	myosin head	

Objectives

1. Describe the fine structure of a muscle fiber; use terms such as *myofibril*, *sarcomere, motor unit, actin,* and *myosin.*
2. List, in sequence, the biochemical and fine structural events that occur during the contraction of a skeletal muscle fiber. Then explain how the fiber relaxes.
3. Refer to Figure 28.8 of your main text, and indicate (a) a muscle used in sit-ups, (b) another used in dorsally flexing and inverting the foot, and (c) another used in flexing the elbow joint.

Self-Quiz Questions

Sequence

Arrange in order of decreasing size.

(1) _____   A. Muscle fiber
(2) _____   B. Myofilament
                  C. Muscle bundle
(3) _____   D. Muscle
(4) _____   E. Myofibril
                  F. Myosin head
(5) _____
(6) _____

Fill-in-the-Blanks

Each myofibril contains (7) _____ filaments, which have cross-bridges, and (8) _____ filaments, which are thin and lack cross-bridges. The repetitive fundamental unit of muscle contraction is the (9) _____. The sarcoplasmic reticulum releases (10) _____ ions as a consequence of charge reversal.

Labeling

Identify each indicated part of the accompanying illustration (p. 250).

(11) _____   (14) _____   (17) _____

(12) _____ _____   (15) _____ _____   (18) _____ _____

(13) _____   (16) _____ _____   (19) _____ _____

28-III
(pp. 395–396)

SKELETAL-MUSCULAR INTERACTIONS
SUMMARY

Summary

In skeletal muscle systems, *reciprocal innervation* of reflexes between antagonistic muscle pairs is the usual basis of coordinated contractions. Motor neurons signal groups of muscles how to contract and move bones at joints. Muscle insertions move close to their joints to move some body part. A limb can be extended and rotated around a joint because of the way pairs or groups of muscles are arranged in relation to their associated joints. *Muscle spindles* provide the nervous system with information about how a muscle is being stretched. Muscles that are being stimulated repeatedly do not have time to relax; they are maintained in a state of contraction called *tetanus*. Tetanic contractions, rather than twitches, are the usual mode of contraction.

Key Terms

lever	stretch-sensitive receptors	twitch
origin	muscle spindle	tetanus
insertion	motor unit	muscle tone
reciprocal innervation		

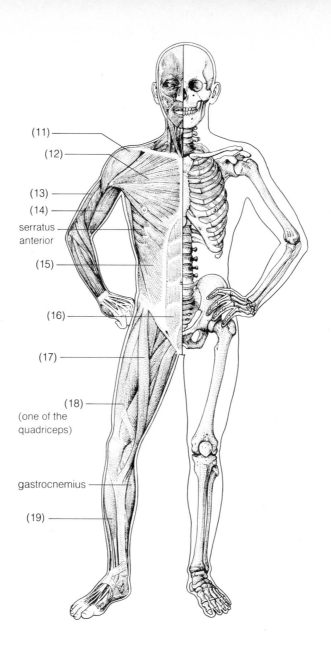

(11)

(12)

(13)

(14)

serratus
anterior

(15)

(16)

(17)

(18)
(one of the
quadriceps)

gastrocnemius

(19)

Objectives

1. Explain how an endoskeleton and an antagonistic muscle system are functionally related in the human action of moving the thumb away from the index finger and in moving the two together. Draw the general shapes of the bones involved and then try to draw in muscles showing how shortening of the muscles causes the action mentioned.
2. Define *reciprocal innervation* and explain how it helps coordinate motor elements.
3. Distinguish *twitch* contractions from *tetanic* contractions.

**Self-Quiz
Questions**

Fill-in-the-Blanks

(1) _____ ordinarily do not contribute to normal, sustained muscle

contractions, but (2) _____ contractions are the normal mode of

contraction. Low levels of tetanic contractions are collectively referred to as

(3) _____ _____. In skeletal muscle, (4) _____ _____ of reflexes between antagonistic muscle pairs is the basis for coordinated contractions.

CHAPTER TEST

UNDERSTANDING AND INTERPRETING KEY CONCEPTS

For Questions 1–5, choose from these answers:
   (a) clavicle
   (b) femur
   (c) humerus
   (d) phalanx
   (e) scapula

___ (1) The bone in the upper arm is the _____.

___ (2) Any bone in a toe or finger is called a _____.

___ (3) The collar bone is the _____.

___ (4) Another name for the shoulder blade is the _____.

___ (5) The thigh bone is the _____.

___ (6) Reciprocal innervation of reflexes between antagonistic muscle pairs _____.
   (a) is the usual basis of coordinated contractions
   (b) explains the mechanism for the operation of the calcium pump
   (c) refers to the adjustments in lens width that focus an image precisely on the retina
   (d) causes rods to interfere with the stimulation of cones
   (e) All of the above

___ (7) Muscle tone refers to _____.
   (a) the length of time it takes for a muscle to contract
   (b) the speed at which a motor unit recovers after contraction
   (c) the fact that some motor units remain contracted even when the total muscle is considered to be relaxed
   (d) the state in which a muscle is left after reciprocal innervation of antagonistic muscle pairs has occurred
   (e) the state in which a muscle bundle is left after a twitch contraction

For Questions 8–11, choose from these answers:
   (a) cartilaginous
   (b) fibrous
   (c) hinge
   (d) synovial
   (e) None of the above

___ (8) Vertebral discs with small amounts of movement are examples of _____ joints.

___ (9) Nonmoving joints between skull bones are examples of _____ joints.

___ (10) Freely movable bones that are separated by a fluid-filled cavity compose a _____ joint.

___ (11) Knees and elbows are _____ joints.

___ (12) Which of the following statements does *not* describe normal activity in a motor system?
  (a) All motor systems require the presence of some medium or structural element against which force can be applied.
  (b) In a resting muscle, energy is stored in the form of tropomyosin.
  (c) In a skeletal muscle system, coordinated contraction depends on reciprocal innervation of motor neurons to antagonistic muscle pairs.
  (d) Calcium ions are required for contraction in all muscle cells.
  (e) In vertebrates, only skeletal muscle actually moves the body through the environment.

CHAPTER TEST

INTEGRATING AND APPLYING KEY CONCEPTS

If humans had an exoskeleton rather than an endoskeleton, would they move differently from the way they do now? Name any advantages or disadvantages that having an exoskeleton instead of an endoskeleton would present in human locomotion.

29

CIRCULATION

General Objectives

1. Describe the composition and functions of blood.
2. Explain how the cardiovascular systems of vertebrates differ from those of invertebrates.
3. Explain the factors that cause blood to exist under different pressures.
4. Describe the composition and function of the lymphatic system.

29-I
(pp. 397–401)

CIRCULATION SYSTEMS: AN OVERVIEW
CHARACTERISTICS OF BLOOD
    Functions of Blood
    Blood Volume and Composition

Summary

A *circulatory system* and the blood it transports through the animal body carries vital material to cells, carries products and wastes from them, and helps maintain an internal environment that is favorable for cell activities. Flatworms and less complex animals lack networks of blood transport vessels because none of the body cells is so far from the external world as to be unable to obtain the necessary oxygen and nutrients or to eliminate wastes directly. In more complex animals, various circulatory systems have evolved for exchanging materials between the internal and external environments. The *closed circulatory systems* of some invertebrates and all vertebrates contain blood enclosed within the walls of one or more hearts and blood vessels. The *open circulation systems* of many invertebrates (including insects) pump blood from the heart into one or more blood vessels that open directly into tissue spaces. The blood bathes nearby cells and then seeps sluggishly back to the heart. Animals with closed circulatory systems often have a supplementary *lymph vascular system*, which recovers and purifies interstitial fluid and returns it to the major blood vessels.

Vertebrate blood is composed of *plasma* (55 percent) and formed elements (45 percent). Plasma is mostly water (91.5 percent) and plasma proteins (7 percent) such as albumin, the globulins, and fibrinogen; however, nutrients, dissolved gases, hormones, and wastes are also dissolved or suspended in it. *Red blood cells* transport O_2; *leukocytes* defend against tissue damage and infections; and *platelets* are involved in forming blood clots.

Key Terms

blood	fibrinogen	white blood cells
heart	lipoproteins	lymphocytes
blood vessels	erythrocytes	neutrophils
closed circulation system	red blood cells	monocytes
open circulation system	oxyhemoglobin	eosinophils
lymph vascular system	carbaminohemoglobin	basophils
plasma	cell count	B cells
plasma proteins	erythropoietin	T cells
albumin	leukocytes	megakaryocytes
alpha and beta globulins	stem cells	platelets
gamma globulin		

Objectives

1. Explain why some organisms are able to manage without a system of pumps and vessels to distribute body fluids.
2. Distinguish between *open* and *closed* circulation systems.
3. State what the lymph vascular system does.
4. Describe, using percentages of volume, the composition of human blood.
5. Name the plasma components and their functions.
6. Explain how red blood cells manage to carry both O_2 and CO_2 at different times.
7. State where erythrocytes, leukocytes, and platelets are produced.
8. Distinguish each of the five types of leukocytes from each other in terms of structure and functions.

Self-Quiz Questions

Fill-in-the-Blanks

Blood is a highly specialized fluid (1) _____ tissue that helps stabilize internal (2) _____ and equalize internal temperature throughout an animal's body. Organisms with (3) _____ circulation systems generally also have a supplementary (4) _____ _____ _____ that recovers and purifies interstitial fluid and returns it to the major blood vessels. Oxygen binds with the (5) _____ atom in a hemoglobin molecule and forms the substance (6) _____. When oxygen levels in tissues are low (as they would be if you moved from Phoenix, Arizona, to Santa Fe, New Mexico), the kidneys secrete a substance that combines with a plasma protein, converting it into the hormone known as (7) _____, which stimulates the production of erythrocytes by (8) _____ _____. (9) _____ _____ are immature cells not yet fully differentiated. (10) _____ and monocytes are highly mobile and phagocytic; they chemically detect, ingest, and destroy bacteria, foreign matter, and dead cells. (11) _____ (thrombocytes) are cell fragments that aid in forming blood clots.

29-II
(pp. 401–404)

CARDIOVASCULAR SYSTEM OF VERTEBRATES
Blood Circulation Routes
The Human Heart

Summary

Arteries, which branch into smaller tubes called *arterioles*, carry blood away from the heart. Through controls over their musculature, *arterioles* can vary the resistance to blood flow along different routes; hence they control the distribution of blood flow throughout the body. Many thin-walled *capillaries* diverge from the arterioles and represent an immense cross-sectional exchange area between blood and interstitial fluid. Capillaries converge into *venules*, which converge into *veins*. Because of their great elasticity, veins and venules can function as blood volume reservoirs during times of low metabolic output. Terrestrial vertebrates have two major blood transport routes: *pulmonary* circulation to and from the lungs and *systemic* circulation to and from the rest of the body.

Zones of the vertebrate heart that receive blood being returned from the body are called *atria*; other zones called *ventricles* pump the blood out. Fishes have two-chambered hearts, with one atrium and one ventricle; amphibians have two atria and one ventricle; and birds and mammals have two atria and two ventricles. The *pacemaker* of the heart has the fastest inherent rate of firing. Each contraction period (systole) is followed by a rest period (*diastole*), in which the heart muscle cells relax and the heart fills; each systole and its diastole constitute one cardiac cycle.

Key Terms

arteries
arterioles
capillaries
venules
veins
cardiovascular system
capillary bed
pulmonary circuit
systemic circuit
myocardium
pericardium

endocardium
endothelium
lumen
atrium, atria
ventricle
atrioventricular valve
semilunar valve
superior vena cava
inferior vena cava
coronary circulation
aorta

cardiac cycle
systole
diastole
pulmonary artery
cardiac conduction system
sinoatrial node
cardiac pacemaker
atrioventricular node
  (= AV node)
intercalated disk

Objectives

1. Describe how the vertebrate heart has evolved from that of fishes to that of mammals.
2. Trace the path of blood in the human body. Begin with the aorta and name all major components of the circulatory system that the blood passes through before it returns to the aorta.

3. Explain what causes a heart to beat. Then describe how the rate of heartbeat can be slowed down or speeded up.
4. List the factors that cause blood to leave the heart and the factors that cooperate to return blood to the heart.

Self-Quiz Questions

Fill-in-the-Blanks   A receiving zone of a vertebrate heart is called a(n) (1) _____; a departure zone is called a(n) (2) _____. Each contraction period is called (3) _____; each relaxation period is (4) _____.

The heart is a pumping station for two major blood transport routes: the (5) _____ circulation to and from the lungs and the (6) _____ circulation to and from the rest of the body.

Labeling   Identify each indicated part of the accompanying illustration.

(7) _____

(8) _____ _____ _____

(9) _____ _____

(10) _____ _____

(11) _____ _____ _____

(12) _____ _____

(13) _____ _____ _____

(14) _____ _____ _____

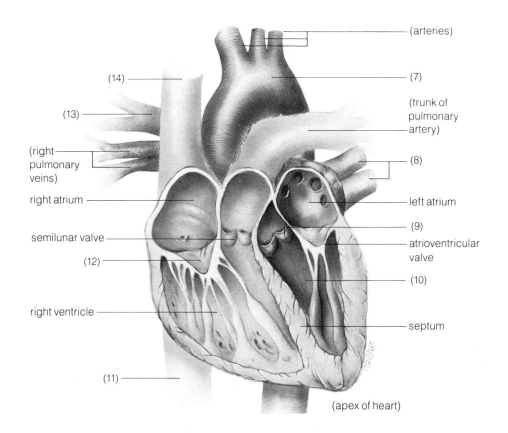

(arteries)

(14)

(13)

(7)

(trunk of pulmonary artery)

(right pulmonary veins)

(8)

right atrium

left atrium

semilunar valve

(9)

atrioventricular valve

(12)

(10)

right ventricle

septum

(11)

(apex of heart)

CARDIOVASCULAR SYSTEM OF VERTEBRATES (cont.)
Blood Pressure in the Vascular System

Summary

The pulse pressure is the difference between the systolic and the diastolic readings. Blood pressure is not the same at the start and end of the circuit. The pressure gradient that heart contractions establish is aided by pressure generated by muscle pumps and by respiration; these forces together work to return venous blood to the heart.

Arteries are pressure reservoirs that keep blood flowing while the ventricles are relaxing. Arteries have large diameters that offer little resistance to flow so that blood pressure drops little in the arteries. *Arterioles* can increase or decrease in diameter and thereby offer variations in resistance to flow. Arterioles can offer much resistance to flow so that a major drop in pressure can occur in arterioles. *Capillary beds* are diffusion zones for exchanges between blood and interstitial fluid; they present less total resistance to flow because, collectively, capillaries have a greater cross-sectional area than do the arterioles leading to the beds; there is some drop in pressure here. *Venules* overlap capillaries to some extent in function; they control capillary pressure somewhat. *Veins* are more than low-resistance transport tubes; they also are blood volume reservoirs. Veins are highly distensible and are important in adjusting flow volume back to the heart. Blood pressure is very low in veins.

Key Terms

blood pressure
arteries
endothelium
arteriosclerosis
systolic pressure
pulse pressure

sphygmomanometer
arterioles
vasodilation
vasoconstriction
capillary

interstitial fluid
filtration
absorption
venules
veins

Objectives

1. Explain what causes high pressure and low pressure in any animal circulatory system. Then relate this to the human circulatory system by explaining where major drops in blood pressure occur.
2. Understand the events happening in capillary beds and explain how you think those events change the composition of the blood.
3. Describe how the structures of arteries, capillaries, and veins differ.
4. Explain how veins and venules can act as reservoirs of blood volume.

Self-Quiz Questions

Fill-in-the-Blanks

A(n) (1) _____ carries blood away from the heart. A(n) (2) _____ is a blood vessel with such a small diameter that red blood cells must flow through it single-file; its wall consists of no more than a single layer of (3) _____ cells. In each (4) _____ _____, small molecules move between the bloodstream and the (5) _____ fluid. (6) _____ are in the walls of veins and prevent backwashing. Both (7) _____ and (8) _____ serve as temporary reservoirs for blood volume.

If false, explain why.

_____ (9) The pulse rate is the difference between the systolic and the diastolic pressure readings.

_____ (10) Because the total volume of blood remains constant in the human body, blood pressure must also remain constant throughout the circuit.

29-IV
(pp. 408–414)

CARDIOVASCULAR SYSTEM OF VERTEBRATES (cont.)
   Commentary: **On Cardiovascular Disorders**
   Regulation of Blood Flow
   Blood Typing
   Hemostasis

Summary

Various cardiovascular disorders occur in humans. Two disorders of the arterial walls, *arteriosclerosis* and *atherosclerosis*, promote high blood pressure, which can lead to the rupture of small blood vessels. The rupture of such vessels in the brain can cause a stroke. A *coronary occlusion* is a blockage of the heart's own blood supply. *Hypertension*—a gradual increase in arterial blood pressure—is one of the most prevalent killers in the United States.

Blood is typed according to the identification markers present on red blood cell surfaces. In ABO blood typing, type A blood has A markers, type B has type B markers, type AB has both, and type O has neither. Blood can also be typed according to whether the Rh marker is present (Rh+) or absent (Rh−) on red blood cell surfaces.

Blood pressure is raised and lowered through the entire cardiovascular system as follows: When blood pressure rises too much, the *medulla oblongata* signals the heart to slow down and the arterioles to dilate; blood pressure soon falls. If it falls too much, the medulla signals the heart rate to increase and the arterioles to constrict. The hormone *angiotensin* causes vasoconstriction; *epinephrine* can cause vasoconstriction or vasodilation depending on which receptors are on the arterioles that are affected.

The blood vessels of many invertebrates and of all vertebrates constrict in a rapid reflex response to hemorrhage. In vertebrates, this muscular reflex is followed by platelets adhering and clumping, plugging the rupture and releasing vasoconstrictors that prolong the muscle spasm. A clot forms as the blood is converted into a solid gel; the clot then retracts into a compact mass that draws ruptured walls back together. The formation of a blood clot involves a series of at least thirteen reactions; if one important substance is missing, as in hemophilia, a clot will not form.

Key Terms

hypertension	angiography	bradycardia
coronary artery disease	coronary bypass surgery	tachycardia
heart attack	laser angioplasty	coronary occlusion
stroke	high-density lipoproteins, HDL	ventricular fibrillation
atherosclerosis		medulla oblongata
arteriosclerosis	low-density lipoproteins, LDL	epinephrine
atherosclerotic plaque		angiotensin
thrombus	omega-3 fatty acids	antibodies
embolus	arrhythmia	foreign cells
angina pectoris	electrocardiogram (EKG)	

ABO blood typing
agglutination
  (= clumping)
Rh blood typing
erythroblastosis fetalis

platelet plug formation
coagulation (= clotting)
collagen
fibrin

intrinsic clotting
  mechanism
extrinsic clotting
  mechanism

Objectives

1. Distinguish a *stroke* from *atherosclerosis* and a *stroke* from a *coronary occlusion*.
2. Describe how hypertension develops, how it is detected, and whether it can be corrected.
3. State the significance of high- and low-density lipoproteins to cardiovascular disorders.
4. Explain how the medulla oblongata and hormones influence blood pressure.
5. Describe how blood is typed for the ABO blood group and for the Rh factor.
6. List the homeostatic events that occur when small blood vessels are damaged.
7. List in sequence the events that occur in the formation of a blood clot.

Self-Quiz Questions

Fill-in-the Blanks

One cause of (1) _____ is the rupture of one or more blood vessels in the brain. A (2) _____ _____ blocks a coronary artery. (3) _____ is a term for a formation that can include cholesterol, calcium salts, and fibrous tissue. It is not healthful to have a high concentration of (4) _____-density lipoproteins in the bloodstream. The (5) _____ _____ is the principal organ that controls blood pressure in the entire cardiovascular system. Bleeding is stopped by several mechanisms that are referred to as (6) _____; the mechanisms include blood vessel spasm, (7) _____ _____ _____, and blood (8) _____. Once the platelets reach a damaged vessel, through chemical recognition they adhere to exposed (9) _____ fibers in damaged vessel walls. When attached, the platelets release ADP (which attracts still more platelets) and (10) _____ ions, which promote clumping into a plug. Cascade reactions act on a soluble plasma protein, fibrinogen, which is then assembled into long, insoluble (11) _____ fibers. These trap blood cells and components of plasma. Under normal conditions, a clot eventually forms at the damaged site.

29-V
(pp. 415–417)

LYMPHATIC SYSTEM
Lymph Vascular System
Lymphoid Organs
SUMMARY

Summary

The *lymph vascular system* accumulates, purifies, and returns excess fluid and a small amount of proteins to the bloodstream from which they came. It also transports fats absorbed from the digestive tract to the bloodstream and carries foreign particles and cellular debris to disposal centers (lymph nodes). *Lymph* is the tissue fluid in the vessels of the lymphatic system. Cells of the immune system are strategically dispersed throughout the body, especially in tissues of the lymphoid system: bone marrow, thymus, lymph nodes, spleen, appendix, tonsils, adenoids, and tissues associated with the small intestine.

Key Terms

lymphatic system	right lymphatic duct	lymphocytes
lymph	thoracic duct	plasma cells
lymph vascular system	lymphoid organs	macrophages
lymph capillaries	lymph nodes	spleen
lymph vessels	tonsils	thymus

Objectives

1. List three functions of the lymph vascular system.
2. List as many tissues and organs of the lymphoid system as you can.

Self-Quiz Questions

Labeling

Identify each indicated part of the accompanying illustration (p. 261).

(1) _____

(3) _____

(5) _____

(2) _____ _____ _____

(4) _____ _____

(6) _____ _____

Fill-in-the-Blanks

(7) _____ vessels reclaim fluid lost from the bloodstream, purify the blood of microorganisms, and transport (8) _____ from the (9) _____ _____ to the bloodstream.

CHAPTER TEST

UNDERSTANDING AND INTERPRETING KEY CONCEPTS

____ (1) Most of the oxygen in human blood is transported by _____.
(a) plasma
(b) serum
(c) platelets
(d) hemoglobin
(e) leukocytes

____ (2) When the oxygen level in human tissues is low, erythropoietin production is stimulated by enzymes secreted by the _____; the red blood cell count then increases.
(a) spleen
(b) lungs
(c) liver
(d) pancreas
(e) kidneys

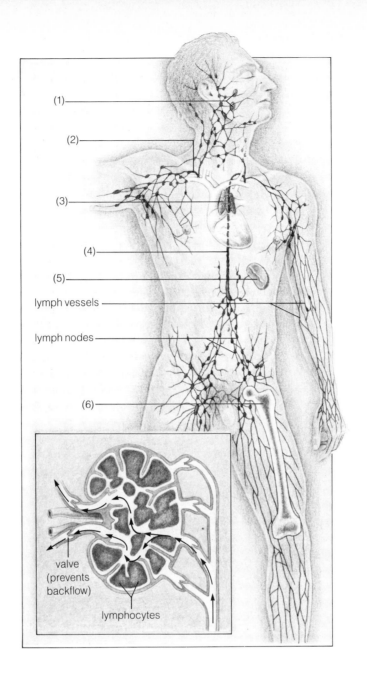

(1)———

(2)———

(3)———

(4)———

(5)———

lymph vessels———

lymph nodes———

(6)———

valve
(prevents
backflow)

lymphocytes

——— (3) Open circulatory systems generally lack ———.
 (a) a heart
 (b) arterioles
 (c) capillaries
 (d) veins
 (e) arteries

——— (4) Red blood cells originate in the ———.
 (a) liver
 (b) spleen
 (c) yellow bone marrow
 (d) thymus gland
 (e) red bone marrow

___ (5) Hemoglobin contains _____.
  (a) copper
  (b) magnesium
  (c) sodium
  (d) calcium
  (e) iron

___ (6) The pacemaker of the human heart is the _____.
  (a) sinoatrial node
  (b) semilunar valve
  (c) inferior vena cava
  (d) superior vena cava
  (e) atrioventricular node

___ (7) During systole, _____.
  (a) oxygen-rich blood is pumped to the lungs
  (b) the heart muscle tissues contract
  (c) the atrioventricular valves suddenly open
  (d) oxygen-poor blood from all parts of the human body, excepting the lungs, flows toward the right atrium
  (e) None of the above

___ (8) _____ are reservoirs of blood pressure in which resistance to flow is low.
  (a) Arteries
  (b) Arterioles
  (c) Capillaries
  (d) Venules
  (e) Veins

___ (9) Which is the proper sequence in clotting?
  (a) Prothrombin, fibrinogen, fibrin, thrombin, clot
  (b) Thrombin, prothrombin, fibrin, fibrinogen, clot
  (c) Prothrombin, fibrinogen, thrombin, fibrin, clot
  (d) Prothrombin, thrombin, fibrinogen, fibrin, clot
  (e) Prothrombin, fibrin, fibrinogen, thrombin, clot

___ (10) The lymphatic system is the principal avenue in the human body for transporting _____.
  (a) fats
  (b) wastes
  (c) carbon dioxide
  (d) amino acids
  (e) interstitial fluids

CHAPTER TEST

INTEGRATING AND APPLYING KEY CONCEPTS

You observe that some people appear as though fluid had accumulated in their lower legs and feet. Their lower extremities resemble those of elephants. You inquire about what is wrong and are told that the condition is caused by the bite of a mosquito that is active at night. Construct a testable hypothesis that would explain (1) why the fluid was not being returned to the torso, as normal, and (2) what the mosquito did to its victims.

Crossword
Number Four

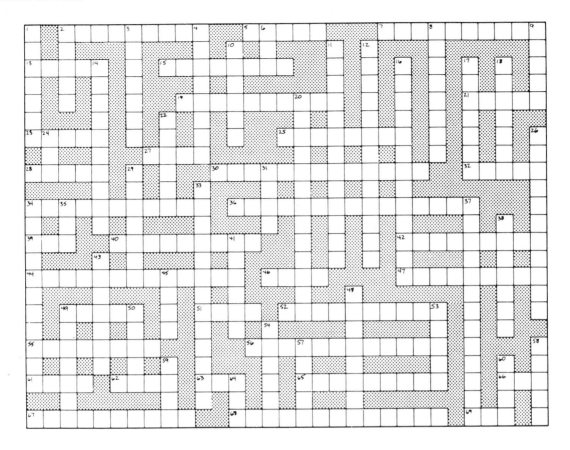

This puzzle covers material from Unit 3, as well as from Unit 5 (main text). You may want to refer back to those chapters for some of the answers. The terms in this puzzle are not necessarily biology terms.

Across

2. one-half of a replicated chromosome
5. material surgically attached to or inserted into a bodily part
7. transmission of disease from an original site to one or more sites elsewhere in the body, as in tuberculosis or cancer
13. a branch modified in the form of a sharp woody spine
15. a pattern used as a guide in making something accurately
19. when DNA is in the form of chromatin
21. a speech form that is peculiar to itself within the usage of a particular language
23. a haploid cell that develops directly into a new haploid body
25. gives rise to digestive glands and the gut lining
27. an individual entity regarded as an elementary constituent of a whole
28. specific DNA regions that tell how to build different kinds of RNA
30. groups of similar cells produce structures of predefined shapes
32. physical characteristic
34. characteristics of two phenotypes appear
36. cells undergo _____ as they become more specialized
39. a kind of nucleic acid
40. an arrangement and grouping of chromosomes on the basis of size, number, and shape

42. embryonic _____: one body part differentiates according to signals it receives from an adjacent body part
44. many-celled
46. _____ cross, in which a dominant-appearing individual of unknown genotype is crossed with a known homozygous recessive individual
47. organelle that makes the precursors to ribosomes
49. process of predictable cellular deterioration
51. _____ Grey, 1875–1939; American author
52. independent _____: the random distribution of chromosomes
55. any heritable change in the genes or chromosomes of an organism
56. at the end of this process, three layers of embryonic tissues will exist
61. an herb; also a wise person
62. a long-lived population of cancerous human cells kept in culture
63. eggs
65. designating an abnormal growth that tends to metastasize
66. male sheep
67. the structure that holds two chromatids together
68. dictated in advance of some other event
69. Middle English "no"

Down

1. the separation of duplicated DNA into two equivalent parcels
2. an organism that is genetically identical to its parent
3. a process that reduces the parental diploid number of chromosomes to the haploid state
4. a _____ allele masks expression of the other allele
6. a successful rodent
8. produced in response to an antigen
9. male gamete
10. one of several alternative forms of a gene
11. the production of sex cells
12. said of partners that, together, make up a whole
14. a bead on a nucleoprotein fiber
16. happens as a result of fertilization
17. a sequence of three nucleotide bases that specifies a particular amino acid
18. a region of apparent contact between two homologous chromosomes
20. a method of sampling uterine fluid and fetal cells
22. not malignant
24. American short-story writer, poet, and journalist, 1809–1849
26. cytoplasmic division that generally follows nuclear division
29. all genes located on a given chromosome compose a _____ group
31. a region of a polytene chromosome with open loops from homologous bands extending outward
33. the fusion and mingling of the hereditary material of two gametes
34. distinct threadlike structures that bear hereditary information in eukaryotes
35. a type of nucleic acid
37. a substance made of amino acids and nucleotides
38. having two complete sets of chromosomes
41. the larger of the two kinds of nucleotide bases
43. a protistan with hairlike organelles
45. discovered that, in mammalian females, only the genes of one X chromosome are expressed
48. more ornamental

49. stimulates antibody production
50. _____, composer of the *Grand Canyon Suite*
53. belief
54. haploid sex cell
57. a circumscribed, noninflammatory growth that serves no useful physiological function
58. a clear watery fluid that contains white blood cells and acts to remove bacteria and certain proteins from the tissues
59. an entire X chromosome that remains compact during interphase is a _____ body
60. that which a predator consumes
64. an energy-storage molecule

30

IMMUNITY

General Objectives

1. Describe typical external barriers that organisms present to invading organisms.
2. Understand the process involved in the nonspecific inflammatory response.
3. Understand how vertebrates (especially mammals) recognize and discriminate between self and nonself tissues.
4. Distinguish between antibody-mediated and cell-mediated patterns of warfare.
5. Describe some examples of immune failures and identify as specifically as you can which weapons in the immunity arsenal failed in each case.

30-I
(pp. 419–422)

NONSPECIFIC DEFENSE RESPONSES
    Barriers to Invasion
    Phagocytes
    Complement System
    Inflammation
SPECIFIC DEFENSE RESPONSES: THE IMMUNE SYSTEM
    The Defenders: An Overview
    Recognition of Self and Nonself

Summary

Most animals have some sort of surface barrier that prevents invasions by viruses, bacteria, and foreign substances; examples are skin, mucous membranes, scales, and a protective exoskeleton. In addition to these external barriers, many produce and excrete substances and cells that destroy would-be invaders.

The *inflammatory response* involves localized warmth, redness, swelling, phagocytosis, and repair to the damaged tissues. The *vertebrate immune system* includes a variety of cells (helper T cells, B cells, macrophages, killer T cells, and suppressor T cells) and cell products (complement, antibodies, antitoxins). This system is concerned with recognizing microbial invaders, as well as cells or substances that do not belong in a given tissue, and selectively eliminating or neutralizing them.

The vertebrate body becomes able to distinguish self from nonself cells during embryonic development. In all mammals, a group of genes called the *major histocompatibility complex* controls the production of surface recognition factors present on all nucleated cells. Except in the case of identical twins, the cells of different individuals have different patterns of surface recognition factors. T cells recognize the differences and destroy the nonself invaders.

Key Terms

Jenner
vaccination
exocrine glands
lysozyme
stem cells
complement system
nonspecific defense
   response
specific defense response
lyse, lytic, lysis
inflammatory response
mast cells
histamine

T lymphocytes
B lymphocytes
vertebrate immune
   system
specificity
memory
macrophage
helper T cells
B cells
killer T cells
natural killer (NK) cells
suppressor T cells
memory cells

antibodies
lymphokines
interleukins
perforin
"cell-mediated" response
"antibody-mediated"
   response
nonself cells
self cells
major histocompatibility
   complex, MHC markers
antigens

Objectives

1. List and discuss four nonspecific defense responses that serve to exclude microbes from the body.
2. Explain how the complement system is related to an inflammatory response.
3. List the three general types of cells that form the basis of the vertebrate immune system.
4. Describe the physical basis that establishes self cells and nonself cells.

Self-Quiz Questions

Fill-in-the-Blanks

Among single-celled organisms, (1) _phagocytosis_ was a means of ingesting food that probably proved adaptive in defense. (2) _Antibodies_ are Y-shaped proteins that bind specific foreign targets and thereby tag them for destruction by phagocytes or by activating the complement system. The (3) _Major histocompatibility complex_ controls the production of surface markers that identify specific nucleated cells. The nonspecific defense system known as (4) _complement_ stimulates phagocytes to follow the gradient of these small proteins to the offending invaders.

SPECIFIC DEFENSE RESPONSES: THE IMMUNE SYSTEM (cont.)
    Primary Immune Responses
    Commentary: **Cancer and the Immune System**
    Control of Immune Responses
    Antibody Diversity and the Clonal Selection Theory
    Secondary Immune Response
IMMUNIZATION

Summary

Specific immune responses amplify the nonspecific ones by recognizing and disposing of invaders in a highly discriminatory way. *Lymphocytes* are divided into two groups: *T cells*, which spend some time in the thymus where they acquire specific substances on their cellular surfaces that enable them to battle with viruses, fungi, bacteria, parasites, cancer cells, and graft cells; and *B cells*, which differentiate into the plasma cells that secrete the *antibodies* that set up invaders not yet infecting individual cells for subsequent destruction by macrophages.

In the *antibody-mediated* immune response, B cells produce antibodies (after first contacting a specific antigen) that circulate freely in the body's tissues. Antibodies that meet an antigen-bearing target bind to it and mark it for disposal by other immune agents such as macrophages and complement proteins. Bacteria, the *extracellular* phases of viruses, and certain parasitic fungi and protozoans are the main targets of antibodies.

In the *cell-mediated* immune response, T cells that have MHC markers on their surfaces and that are activated by a specific antigen undergo clonal multiplication. Some of these cells are held as memory lymphocytes, but others differentiate into three different kinds of T cells that attack infected cells whose recognition marker pattern is disturbed. Turned-on killer T cells destroy virally infected, mutant, or cancerous cells by punching holes in them with perforin and other secreted substances.

Current theories about the immune system suggest that cancer cells may arise in your body every day as a result of mutations induced by events such as viral attack, chemical bombardment, or irradiation. As fast as cancer cells arise, circulating sensitized lymphocytes become activated and begin cloning themselves rapidly and secreting lymphokines, which stimulate voracious macrophages to destroy the mutant cells. Surgery, immune therapy, interferons, and monoclonal antibodies each provide ways of combating the various forms of cancer.

According to the *clonal selection theory*, a lymphocyte is activated when an antigen binds with one of its recognition proteins. The activated lymphocyte multiplies quickly, and its descendants make up a clone of cells that are armed to fight in any future clashes with the antigen that selected them. The theory explains how a person has "immunological memory" and can mobilize a more rapid response to a subsequent invasion by the same antigen.

Key Terms

primary immune response	Ig A antibody	targeted drug therapy
"virgin" B cells	Ig D antibody	secondary immune response
plasma cells	clonal selection theory	memory lymphocytes
immunoglobulins	clone	immunization
Ig M antibody	immune therapy	vaccine
Ig G antibody	interferons	booster shot
Ig E antibody	monoclonal antibodies	passive immunity
	Milstein and Kohler	

Objectives

1. Explain what the clonal selection theory is and tell what it helps to explain.
2. Describe how recognition proteins and antibodies are made. State how they are used in immunity.
3. Distinguish between the antibody-mediated warfare pattern and the cell-mediated warfare pattern.
4. Distinguish the roles of T cells from the roles of B cells.
5. Explain what is meant by *primary immune pathway*, as contrasted with *secondary immune pathway*.
6. Describe two ways that people can be immunized against specific diseases.
7. Explain what immunosurveillance theory has to say about the development of cancer. Discuss the measures available when cancer cells slip through the immunosurveillance network.
8. Explain what monoclonal antibodies are and tell how they are currently being used in passive immunization and cancer treatment.

Self-Quiz Questions

Fill-in-the-Blanks

The (1) _____ _____ _____ explains how an individual has immunological memory, which is the basis of a secondary immune response; the theory also explains in part how *self* cells are distinguished from (2) _____ cells in the vertebrate immune response. Two distinct (3) _____ populations carry out specific immune responses: (4) _____ _____ and (5) _____ _____. Both kinds arise from (6) _____ _____ in bone marrow. Some of these progeny move to the thymus where they acquire specific (7) _____ on their cell surfaces; in doing so, they become (8) _____ _____. (9) _____ _____ clones indirectly attack their targets through the production of antibodies. The (10) _____ _____ _____ is the route taken during a first-time contact with an antigen. Deliberately provoking the production of memory lymphocytes is known as (11) _____. (12) _____ refers to cells that have lost control over cell division. Milstein and Kohler developed a means of producing large amounts of (13) _____ _____.

(14) _____ drug therapy hooks up anticancer drug molecules with monoclonal antibodies, which locate precisely where the cancer cells are and deliver the drug to them only.

ABNORMAL OR DEFICIENT IMMUNE RESPONSES
  Allergies
  Autoimmune Disorders
  Deficient Immune Responses
  Case Study: The Silent, Unseen Struggles
  Commentary: **Acquired Immune Deficiency Syndrome (AIDS)**
SUMMARY

Summary

Allergies and autoimmune disease are examples of the immune system damaging the body instead of protecting it. Rheumatoid arthritis and some cases of Type 1 diabetes are examples of autoimmune disorders.

AIDS is a virally caused immune disorder that changes the growth and behavior of T4 lymphocytes. T4 cells perform several vital functions in both the antibody- and cell-mediated immune responses. The Atlanta Centers for Disease Control has predicted that the current 1988 estimate of AIDS cases (more than 73,000) is likely to increase to 300,000 cases by 1991 unless effective measures are used to control the spread of this incurable illness. Five million or more people may be killed by AIDS within the next decade.

Key Terms

allergy	autoimmune response	human immunodeficiency
Ig E antibodies	rheumatoid arthritis	virus, HIV
histamine	Ig G antibodies	retrovirus
prostaglandins	acquired immune	helper T cells
asthma	deficiency syndrome,	T4 cells
hay fever	AIDS	
"blocking antibodies"		

Objectives

1. Distinguish *allergy* from *autoimmune disease.*
2. Describe how AIDS specifically interferes with the human immune system.

Self-Quiz Questions

Fill-in-the-Blanks

(1) _Allergy_ is an altered secondary response to a normally harmless substance that may actually cause injury to tissues. (2) _Autoimmune disease_ is a disorder in which the body mobilizes its forces against certain of its own tissues. AIDS is a constellation of disorders that follow infection by the (3) _human immunodeficiency virus_. In the United States, transmission has occurred most often among intravenous drug abusers who share needles and (4) _male homosexuals_. Mutant cells sensitize and activate lymphocytes; some become (5) _memory cells_, and others release (6) _lymphokines_—substances that arouse macrophages. Aroused macrophages war against any invaders, including cancer cells. (7) _Interferons_ are a group of small proteins that perform antiviral activity and induce resistance to a wide range of viruses.

Matching

Match each letter with its best mate.

(8) ___ allergy

(9) ___ antibody

(10) ___ antigen

(11) ___ macrophage

(12) ___ clone

(13) ___ complement

(14) ___ histamine

(15) ___ MHC marker

(16) ___ plasma cell

(17) ___ T cell

A. Begins its development in bone marrow, but matures in the thymus gland
B. Cell that has directly or indirectly descended from the same parent cell
C. A potent chemical that causes blood vessels to dilate and let protein pass through the vessel walls
D. Y-shaped immunoglobulin
E. A nonself marker
F. The progeny of turned-on B cells
G. A group of about fifteen proteins that participate in the inflammatory response
H. An altered secondary immune response to a substance that is normally harmless to other people
I. The basis for self-recognition at the cell surface
J. Principal perpetrator of phagocytosis

CHAPTER TEST

UNDERSTANDING AND INTERPRETING KEY CONCEPTS

___ (1) All the body's phagocytes are derived from stem cells in the _____.
(a) spleen
(b) liver
(c) thymus
(d) bone marrow
(e) thyroid

___ (2) The plasma proteins that are activated when they contact a bacterial cell are collectively known as the _____ system.
(a) shield
(b) complement
(c) Ig G
(d) MHC
(e) HIV

___ (3) _____ are divided into two groups: T-cells and B-cells.
(a) Macrophages
(b) Lymphocytes
(c) Platelets
(d) Complement cells
(e) Cancer cells

___ (4) _____ produce and secrete antibodies that set up bacterial invaders for subsequent destruction by macrophages.
(a) B cells
(b) Phagocytes
(c) T cells
(d) Bacteriophages
(e) Thymus cells

_a_

(5) Antibodies are shaped like the letter _____.
   (a) Y
   (b) W
   (c) Z
   (d) H
   (e) E

_e_

(6) The markers for every cell in the human body are referred to by the letters _____.
   (a) HIV
   (b) MBC
   (c) RNA
   (d) DNA
   (e) MHC

_e a_

(7) Plasma cells _____.
   (a) die within a week after they are produced
   (b) develop from B cells
   (c) manufacture and secrete antibodies
   (d) do not divide and form clones
   (e) All of the above

_C x c_

(8) Clones of B or T cells are _____.
   (a) being produced continually
   (b) known as memory cells
   (c) only produced when their surface proteins recognize other specific proteins
   (d) interchangeable
   (e) produced and mature in the bone marrow

_a_

(9) Whenever the body is reexposed to a specific sensitizing agent, Ig E antibodies cause _____.
   (a) prostaglandins and histamine to be produced
   (b) clonal cells to be produced
   (c) histamine to be released
   (d) the immune response to be suppressed
   (e) None of the above

_b_

(10) The leading cause of death among transplant patients is _____.
   (a) failure of the MHC in plasma cells to do its work
   (b) pneumonia
   (c) loss of a vital organ when the transplant fails
   (d) an excessive number of antigens being released into the bloodstream
   (e) a transplant reaction similar to that which causes blood transfusion deaths

---

CHAPTER TEST

INTEGRATING AND APPLYING KEY CONCEPTS

Suppose that you wish to get rid of forty-seven warts that you have on your hands, by treating them with monoclonal antibodies. Outline the steps that would have to be taken.

31

RESPIRATION

General Objectives

1. Understand the behavior of gases and the types of respiratory surfaces that participate in gas exchange.
2. Understand how the human respiratory system is related to the circulatory system, to cellular respiration, and to the nervous system.
3. List some of the things that go awry with the respiratory system and describe the characteristics of the breakdown.

31-I
(pp. 436–440)

SOME PROPERTIES OF GASES

RESPIRATORY SURFACES
    Integumentary Exchange
    Gills
    Tracheas
    Lungs

OVERVIEW OF VERTEBRATE RESPIRATION

Summary

The overall exchange of gases among cells, the blood, and the environment is known as respiration. Respiratory systems are based on the diffusive properties of oxygen and carbon dioxide, which tend to move from a region of higher partial pressure to a region of lower partial pressure. Any respiratory membrane must be kept moist on both surfaces in order for gas exchange to occur. When moisture retention is not a problem (as it is not for water-dwelling animals), gas exchange is enhanced by the folding outward of a membrane, such as an external gill and by *countercurrent flow mechanisms*, in which water flows past the bloodstream in the opposite direction. When moisture retention is a problem (as it can be for land-dwelling animals), gas exchange is enhanced by the formation of an inward-directed pocket in a membrane, as occurs in air sacs and lungs.

In most terrestrial vertebrates, air moves by bulk flow into and out of the lungs. The gases in the new air diffuse across the respiratory surface of the lungs. Bulk flow of blood to and from the lungs (pulmonary circulation) helps dissolved gases diffuse into and out of lung capillaries. Gases then diffuse from blood to the interstitial fluid and thence to individual body cells.

Key Terms

respiration	external gills	spiracles
rate of diffusion	internal gills	lung
partial pressure	countercurrent flow	airways
integumentary exchange	lamellae	blood vessels
gill	tracheas	pulmonary ventilation

Objectives

1. Describe the ways that respiration differs in aquatic and terrestrial animals.
2. Define *countercurrent flow mechanism* and explain how it works. State where such a mechanism is found.
3. Describe how incoming oxygen is distributed to the tissues of insects, and contrast this with the process that occurs in mammals.

Self-Quiz Questions

Fill-in-the-Blanks

A (1) _____ is an outfolded, thin, moist membrane endowed with blood vessels. Gas transfer is enhanced by a (2) _____ _____ _____, in which water flows past the bloodstream in the opposite direction. Insects have (3) _____: chitin-lined air tubes leading from the body surface to the interior. (4) _____ are openings at the body surface of land arthropods. At sea level, atmospheric pressure is about (5) _____ mm Hg, and oxygen represents about (6) _____ percent of the total volume. Lungs originated as pockets of the front of the (7) _____. In some fishes, lung sacs became modified into (8) _____ _____: buoyancy devices that help keep the fish from sinking.

HUMAN RESPIRATORY SYSTEM
Air-Conducting Portion
Gas Exchange Portion
Lungs and the Pleural Sac
VENTILATION
Inhalation and Exhalation
Lung Volumes

Summary

In the human respiratory system, air normally enters the body via the nose and the nasal cavity, where it is cleaned, warmed, and moistened. The air then moves through the *pharynx* and down through the cartilage-reinforced tubes called the *larynx* (voice-box) and *trachea* (windpipe). The trachea diverges into two *bronchi*, which diverge further into bronchioles. Air finally ends up in tiny pouches called *alveoli*. It is across the epithelium of these alveoli that gas exchange occurs.

In the human respiratory system, oxygen moves inward and carbon dioxide moves outward through the same branched tubes. Gas exchange in the same tubes is possible because of rhythmic changes in pressure gradients between the lungs and the atmosphere. The *diaphragm*, a muscular partition that separates the chest and abdominal cavities, moves downward and flattens during normal *inhalation*; simultaneously, the rib cage moves outward and upward, the volume of the chest cavity increases, and its internal pressure drops. Because atmospheric pressure remains the same, air rushes into the respiratory tract. During *exhalation*, the diaphragm and rib cage return to their resting position, the volume of the chest cavity decreases, and the resultant rise in pressure on the lungs forces air outward. The maximum volume of air that can move into and out of your lungs in a single breath is the *vital capacity*. At rest, about 500 ml of air compose a normal breath. When all of the air that you can exhale is exhaled, a *residual volume* of about 1,000 ml still remains.

Key Terms

nasal cavities	respiratory bronchioles	pleural sac
pharynx	intercostal muscles	intrapleural space
larynx	alveolus, -oli	pleurisy
epiglottis	alveolar ducts	inhalation
true vocal cords	alveolar sac	exhalation
glottis	diaphragm	"tidal volume"
trachea	thoracic cavity	"vital capacity"
bronchus, -chi	abdominal cavity	"residual volume"

Objectives

1. List all the principal parts of the human respiratory system and explain how each structure contributes to transporting oxygen from the external world to the bloodstream.
2. Describe the relationship of the human lungs to the pleural sac and to the thoracic cavity.

Self-Quiz Questions

Fill-in-the-Blanks

During inhalation, the (1) _____ moves downward and flattens, and the (2) _____ _____ moves outward and upward; when these things happen, the chest cavity volume (3) [choose one] ☐ increases ☐ decreases, and the internal pressure (4) [choose one] ☐ rises ☐ drops ☐ stays the same. The rate of breathing is governed by a respiratory center in the (5) _____ _____, which monitors signals coming in from arterial walls, from blood vessels, and from other brain regions. The (6) _____ _____ surrounds each lung. In succession, air passes through the nasal cavities, pharynx, and (7) _____, past the epiglottis into the (8) _____ (the space between the true vocal cords), into the trachea, and then to the (9) _____, (10) _____, and alveolar ducts. Exchange of gases occurs across the epithelium of the (11) _____.

Labeling

Identify each indicated part of the accompanying illustration (p. 277).

(12) _____ _____   (16) _____ _____   (19) _____

(13) _____   (17) _____   (20) _____ _____

(14) _____   (18) _____   (21) _____ _____

(15) _____

31-III
(pp. 443–449)

GAS EXCHANGE AND TRANSPORT
    Gas Exchange in Alveoli
    Gas Transport Between Lungs and Tissues
MATCHING AIR FLOW AND BLOOD FLOW DURING VENTILATION
    Neural Control Mechanisms
    Local Control Mechanisms
    Hypoxia
HOUSEKEEPING AND DEFENSE IN THE RESPIRATORY TRACT
    When Defenses Break Down
    The Heimlich Maneuver
SUMMARY

Summary

Driven by its partial pressure gradient, oxygen diffuses from alveolar air spaces through interstitial fluid and into the blood capillaries. Carbon dioxide, driven by its partial pressure gradient, diffuses in the reverse direction. As oxygen moves into the blood plasma and then into the red blood cells, four oxygen molecules form weak bonds with hemoglobin, forming *oxyhemoglobin*; this enables

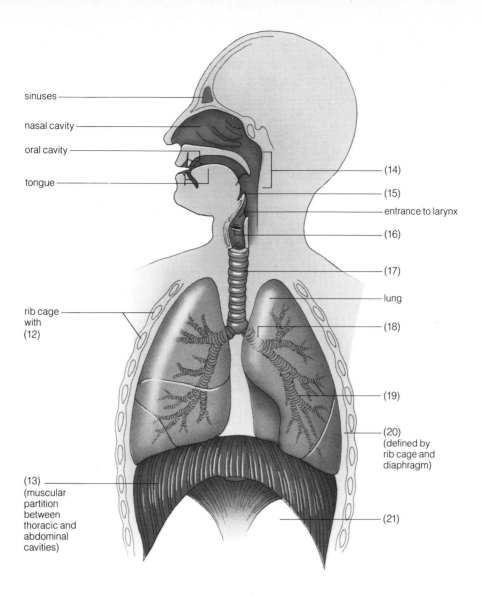

sinuses

nasal cavity

oral cavity

tongue

(14)

(15)

entrance to larynx

(16)

(17)

lung

(18)

(19)

(20)
(defined by
rib cage and
diaphragm)

(21)

rib cage
with
(12)

(13)
(muscular
partition
between
thoracic and
abdominal
cavities)

the blood to carry about seventy times as much oxygen as plasma alone could. Oxyhemoglobin gives up oxygen when the O_2 concentration is low, CO_2 concentrations are high, temperature is elevated, and pH values are low. These conditions are characteristic of tissues showing increased metabolic activity.

The waste product, carbon dioxide, diffuses into red blood cells and is converted into *bicarbonate ions* (which diffuse back into the plasma); about 23 percent of the carbon dioxide is converted into *carbaminohemoglobin*, and in this form it is transported to the lung capillaries, where it is again converted to CO_2 and then exhaled.

Gas exchange is most efficient when the rate of air flow is matched with the rate of blood flow. Neural controls that regulate respiration are located in the *reticular formation*, in the *medulla oblongata*, and in the walls of arteries near the heart.

Bronchitis is characterized by degradation of bronchial tissue and the formation of scar tissue. If excess CO_2 becomes trapped by scar tissue in the alveoli, the normally pink, elastic lung sacs become dry and perforated; the outcome is *emphysema*, with a severe reduction in the efficiency of gas exchange. Lung cancer is believed to be caused by compounds found in coal tar and cigarette smoke, which are chemically modified in the body into carcinogens that interfere with normal gene expression.

Key Terms

oxyhemoglobin	aortic bodies	emphysema
HbCO$_2$,	hypoxia	"smoker's cough"
carbaminohemoglobin	carbon monoxide	lung cancer
HCO$_3^{-1}$, bicarbonate ion	poisoning	methylcholanthrene
carbonic anhydrase	ciliated epithelium	coal tar
carotid bodies	chronic bronchitis	Hemlich maneuver

Objectives

1. Explain why oxygen diffuses from alveolar air spaces, through interstitial fluid, and across capillary epithelium. Then explain why carbon dioxide diffuses in the reverse direction.
2. Explain why oxygen diffuses from the bloodstream into the tissues far from the lungs. Then explain why carbon dioxide diffuses into the bloodstream from the same tissues.
3. Explain the roles of oxyhemoglobin and carbaminohemoglobin.
4. Describe what happens to carbon dioxide when it dissolves in water under conditions normally present in the human body.
5. State the role played by carbonic anhydrase in carbon dioxide transport.
6. List the structures that are involved in detecting carbon dioxide levels in the blood and in regulating the rate of breathing. Name the location of each structure.
7. Distinguish *bronchitis* from *emphysema*. Then explain how lung cancer differs from emphysema.

Self-Quiz Questions

Fill-in-the-Blanks

Oxygen is said to exert a (1) _____ _____ of 760/21 or 160 mm Hg. (2) _____ _____ alone moves oxygen from the alveoli into the bloodstream, and it is enough to move (3) _____ _____ in the reverse direction. (4) _____ is the medical name for oxygen deficiency; it is characterized by faster breathing, faster heart rate, and anxiety at altitudes of 8,000 feet above sea level. About 70 percent of the carbon dioxide in the blood is transported as (5) _____. Without (6) _____, the plasma would be able to carry only about 2 percent of the oxygen that whole blood carries. When oxygen-rich blood reaches a (7) _____ tissue capillary bed, oxygen diffuses outward, and carbon dioxide moves from tissues into the capillaries. With the assistance of (8) _____ _____, carbonic acid dissociates to form water and carbon dioxide. Clusters of cells in the (9) _____ _____ and elsewhere in the (10) _____ _____ regulate contractions of the diaphragm and intercostal muscles associated with inhalation and exhalation.

(11) _____ is the distension of lungs and the loss of gas exchange efficiency such that running, walking, and even exhaling are painful experiences. At least 90 percent of all (12) _____ _____ deaths are the result of cigarette smoking; only about 10 percent of afflicted individuals will survive.

UNDERSTANDING AND INTERPRETING KEY CONCEPTS

—— (1) Most forms of life depend on _____ to obtain oxygen and
eliminate carbon dioxide.
(a) active transport
(b) bulk flow
(c) diffusion
(d) osmosis
(e) muscular contractions

—— (2) _____ is the most abundant gas in Earth's atmosphere.
(a) Water vapor
(b) Oxygen
(c) Carbon dioxide
(d) Hydrogen
(e) Nitrogen

—— (3) With respect to respiratory systems, countercurrent flow is a
mechanism that explains how _____.
(a) oxygen uptake by blood capillaries in the lamellae of fish gills occurs
(b) ventilation occurs
(c) intrapleural pressure is established
(d) sounds originating in the vocal cords of the larynx are formed
(e) All of the above

—— (4) _____ have the most efficient respiratory system.
(a) Amphibians
(b) Reptiles
(c) Birds
(d) Mammals
(e) Humans

—— (5) Immediately before air reaches the alveoli, it passes through
the _____.
(a) bronchioles
(b) glottis
(c) larynx
(d) pharynx
(e) trachea

—— (6) During inhalation, _____.
(a) the pressure in the thoracic cavity is less than the pressure within
the lungs
(b) the pressure in the chest cavity is greater than within the lungs
(c) the diaphragm moves upward and becomes more curved
(d) the thoracic cavity volume decreases
(e) All of the above

—— (7) Hemoglobin _____.
(a) releases oxygen more readily in active tissues
(b) tends to release oxygen in places where the temperature is lower
(c) tends to hold on to oxygen when the pH of the blood drops
(d) tends to give up oxygen in regions where partial pressure of oxygen
exceeds that in the lungs
(e) All of the above

___ (8) Oxygen moves from alveoli to the blood stream _____.
  (a) whenever the concentration of oxygen is greater in alveoli than in the blood
  (b) by means of active transport
  (c) by using the assistance of carbaminohemoglobin
  (d) principally due to the activity of carbonic anhydrase in the red blood cells
  (e) All of the above

___ (9) Oxyhemoglobin releases O_2 when _____.
  (a) carbon dioxide concentrations are high
  (b) body temperature is lowered
  (c) pH values are high
  (d) CO_2 concentrations are low
  (e) All of the above occur

___ (10) Nonsmokers live an average of _____ longer than those in their mid-twenties who smoke two packs of cigarettes each day.
  (a) 6 months
  (b) 1–2 years
  (c) 3–5 years
  (d) 7–9 years
  (e) over 12 years

CHAPTER TEST

INTEGRATING AND APPLYING KEY CONCEPTS

Consider the amphibians—animals that generally have aquatic larval forms (tadpoles) and terrestrial adults. Outline the respiratory changes that you think might occur as an aquatic tadpole metamorphoses into a land-going juvenile.

32

DIGESTION AND ORGANIC METABOLISM

General Objectives

1. Know the various ways in which a digestive system can be structured. Realize the behavioral limitations of organisms with incomplete digestive systems.
2. Understand the structure and function of the human digestive system.
3. Summarize the daily nutritional requirement of a 25-year-old man who works at a desk job and exercises very little. State what he needs in energy, carbohydrates, proteins, and lipids, and name at least six vitamins and six minerals that he needs to include in his diet every day.
4. Explain how the human body manages to meet the energy and nutritional needs of the various body parts even though the person may be feasting sometimes and fasting at other times.

TYPES OF DIGESTIVE SYSTEMS AND THEIR FUNCTIONS

Summary

Nutrition is a function that includes ingesting, digesting, absorbing, and assimilating nutrients. Nutrition depends on digestion, circulation, and respiration, as well as on obtaining nutrients in the first place. An *incomplete digestive system* has only one opening, which serves both as an entrance for newly obtained food and as an exit for undigested residues. Some food is necessarily wasted in the two-way (incoming and outgoing) traffic in this unspecialized system. The more highly evolved animals have a *complete digestive system*, with one-way traffic from mouth to anus; the food passes through different regions specialized for particular digestive functions along the way. *Ingestion* involves simply taking in food. *Digestion* mechanically and chemically reduces foods into substances small enough to diffuse into the circulatory system. *Absorption* passes digested nutrients from the gut lumen into the blood or lymph, which distributes them throughout the body. *Assimilation* moves the nutrients into the cytoplasm of cells. *Elimination* expels indigestible residues from the body.

Key Terms

nutrition	complete digestive system	ruminants
systems integration	lumen	motility
digestive system	crop	secretion
extracellular fluid	gizzard	digestion
incomplete digestive system	discontinuous feeding habit	absorption
pharynx		

Objectives

1. Define *nutrition* and tell why it is not synonymous with *digestion*.
2. List the items that leave the digestive system and enter the circulatory system during the process of absorption.
3. Explain how the respiratory system is involved in nutrition.
4. Distinguish between *incomplete* and *complete* digestive systems and tell which is characterized by (a) specialized regions, (b) two-way traffic, and (c) discontinuous feeding.
5. Define and distinguish among *digestion, absorption,* and *assimilation*.

Self-Quiz Questions

Fill-in-the-Blanks

(1) _____ includes taking food into the digestive tract, breaking it down, and transporting the broken-down food into the circulatory system and thence into cells. (2) _____ is the term for taking food into the digestive tract, but (3) _____ means breaking down the food into its simplest components. The passage of food components into the circulatory system is (4) _____, but the components are not (5) _____ until they participate in reactions inside cells.

Summary

The human digestive system is divided into mouth, oral cavity, pharynx, esophagus, stomach, small and large intestines, and anus. Secretory cells in glandular organs such as the salivary glands, the liver, and the pancreas add a variety of digestive substances to the food being broken down. In the oral cavity, incoming food is moistened and mechanically degraded by the teeth and tongue; polysaccharide digestion is initiated by enzymes in the saliva. *Peristalsis* propels the food ball down the esophagus into the *stomach*, which initiates protein digestion by means of hydrochloric acid and enzymes. The stomach regulates the rate of food movement into the *small intestine*, where the digestion and absorption of most nutrients occur. Bile secreted by the *liver* and stored in the *gallbladder* emulsifies fats in the small intestine, and the resulting particles are subsequently digested by lipase there. Carbohydrate and protein digestion and absorption continue in the small intestine with the aid of secretions from the *pancreas* and *intestinal wall*. Absorption of water and minerals occurs in both the small and large intestines. Undigested food residues, together with bacteria, are coated with mucus and move on to the *anus*.

Key Terms

gastrointestinal tract
mucosa
submucosa
muscle layer
serosa
pyloric sphincter
peristalsis
segmentation
sphincters
mouth, oral cavity
tooth
molars
incisors
cuspids
saliva
salivary glands
salivary amylase

mucin
pharynx
esophagus
stomach
hydrochloric acid, HCl
chyme
pepsinogen
pepsins
peptic ulcer
small intestine
villus, villi
microvillus, -villi
pancreas
trypsin
chymotrypsin
carboxypeptidase

aminopeptidase
islets of Langerhans
insulin
glucagon
liver
bile
gallbladder
bile salt
micelle
colon, large intestine
feces
rectum
anus
appendix
appendicitis

Objectives

1. List all parts (in order) of the human digestive system that food actually passes through. Then list the auxiliary organs that contribute one or more substances to the digestive process.
2. Explain how, during digestion, food is mechanically broken down. Then explain how it is chemically broken down.
3. Describe the mechanism that prevents food from entering the lung.
4. Explain how the rate at which food is passed from the stomach to the small intestine is regulated.
5. Tell which foods undergo digestion in each of the following parts of the digestive system and state what the food is broken into: oral cavity, stomach, small intestine, large intestine.
6. List the enzyme(s) that act in (a) the oral cavity, (b) the stomach, and (c) the small intestine. Then tell where each enzyme was originally made.
7. Describe the cross-sectional structure of the small intestine and explain how its structure is related to its function.
8. Describe how the digestion and absorption of fats differs from the digestion and absorption of carbohydrates and proteins.
9. Explain what processes would be disrupted if you took antibiotics that killed all the bacteria in your colon.
10. State which processes occur in the colon (large intestine).

Self-Quiz Questions

Fill-in-the-Blanks

Saliva contains an enzyme (1) _____ that hydrolyzes starch. Contractions force the larynx against a cartilaginous flap called the (2) _____, which closes off the trachea. The (3) _____ is a muscular tube that propels food to the stomach. Any alternating progression of contracting and relaxing muscle movements along the length of a tube is known as (4) _____. The (5) _____'s most important function is to regulate the rate at which food reaches the intestine. Most digestion and absorption of nutrients occurs in the (6) _____ _____. (7) _____ is an enzyme that works in the stomach. (8) _____ is made by the liver, is stored in the gallbladder, and works in the (9) _____ _____. (10) _____ is an example of an enzyme that is made by the pancreas but works in the small intestine.

Labeling

Identify each of the numbered structures of the accompanying illustration (p. 285).

(11) _____ _____   (15) _____   (19) _____

(12) _____ _____   (16) _____ _____   (20) _____

(13) _____   (17) _____ _____   (21) _____

(14) _____   (18) _____

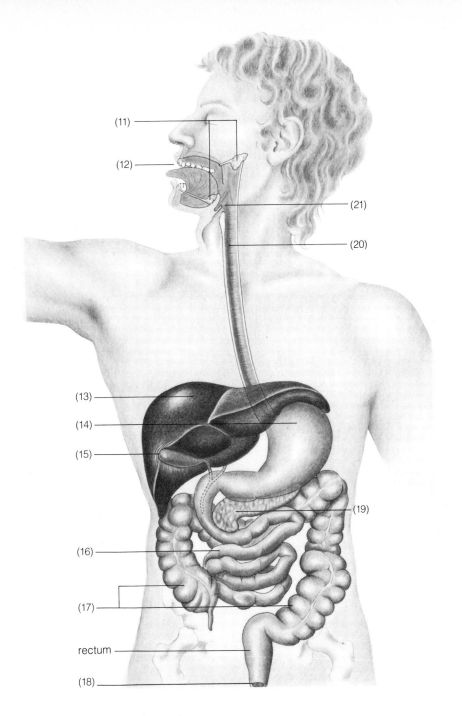

(11)

(12)

(21)

(20)

(13)

(14)

(15)

(19)

(16)

(17)

rectum

(18)

True-False

If false, explain why.

___ (22) Amylase digests starch, lipase digests lipids, and proteases break peptide bonds.

___ (23) ATP is the end product of digestion.

___ (24) Vitamin K is synthesized in the small intestine and is important in the formation of a blood clot.

___ (25) Water and minerals are absorbed into the bloodstream from the lumen of the large intestine.

___ (26) Gallstones are formed in the kidney when something changes the concentrations of bile salts, lecithin, and cholesterol in the bile.

HUMAN NUTRITIONAL NEEDS
   Energy Needs
   Carbohydrates
   Fats
   Proteins
   Vitamins and Minerals
   Objective and Subjective Views of Obesity

Summary

Carbohydrates, fats, and proteins are the fundamental sources of energy and raw materials for growth and bodily maintenance. *Vitamins* are accessory substances that are required in small amounts for the normal functioning of key enzymes in metabolism and for the synthesis of substances important in the structure of specific cells. The continual intake of at least thirteen different vitamins is essential in maintaining human health. *Minerals* are inorganic materials (such as calcium, magnesium, phosphorus, and iron) that are also important in normal enzyme functioning, in phosphorylation reactions, in maintaining normal osmotic balances, and as components of important substances used in the structure (for example, in bones and teeth) as well as the function (for example, in cytochromes and hemoglobin) of healthy organisms. On average, about 32–42 grams of protein, 250–500 grams of carbohydrates, and 66–83 grams of fat per day taken in a well-balanced diet supply the essential vitamins and minerals. Residents of the United States generally eat too much refined sugar, cholesterol, red meat, and salt, and too little fiber. As a result, high blood pressure, heart disorders, diabetes, cancer of the colon, and bad teeth have replaced diseases caused by microorganisms as the main afflictions that hasten death and disability today.

Key Terms

bulk	vitamin B$_1$, thiamine	calcium
calorie	vitamin B$_2$, riboflavin	phosphorus
kilocalorie	niacin	magnesium
polyunsaturated fat	vitamin B$_{12}$	potassium
saturated fat	vitamin C, ascorbic acid	sodium
essential amino acids	vitamin A, retinol	iron
net protein utilization	vitamin D	obesity
vitamins	vitamin E, tocopherol	anorexia nervosa
minerals	vitamin K, phylloquinone	bulimia
trace elements		

Objectives

1. Compare the contributions of carbohydrates, proteins, and fats to human nutrition with the contributions of vitamins and minerals.
2. Distinguish *vitamins* from *minerals* and state what is meant by *net protein utilization*.
3. Name four minerals that are important in human nutrition and state the specific role of each.
4. Explain how vitamins A, D, E, and K differ from other vitamins; tell the possible outcomes of (a) deficiency and (b) excess.
5. State the contribution of vitamin C to the human diet, and tell what happens if there is not enough vitamin C in the diet. Then name two of the B vitamins, state their contributions, and describe the symptoms of a deficiency of each.
6. Summarize the 1979 U.S. Surgeon General's report that presented ideas for promoting health by eating properly.

Self-Quiz Questions

Fill-in-the-Blanks

A deficiency of vitamin (1) _____ produces rickets in children and osteomalcia in adults; a deficiency of the mineral (2) _____ produces similar outcomes. It is possible to accumulate too much vitamin (3) _____, which is a constituent of rhodopsin (visual pigment), and vitamin (4) _____, which promotes growth and mineralization of bones. Vitamin (5) _____ functions in forming a blood clot. A deficiency of (6) _____ promotes pellagra, a disease characterized by sores on the skin and gut wall and nervous and mental disorders. A deficiency of vitamin (7) _____ gives rise to scurvy. The mineral (8) _____ is a necessary participant in phosphorylation reactions. (9) _____ is needed to make thyroid hormones. (10) _____ is needed to make hemoglobin; a deficiency of it leads to a form of anemia.

32-IV
(pp. 465–469)

ORGANIC METABOLISM
The Vertebrate Liver
Absorptive and Post-Absorptive States
Controls Over Organic Metabolism
Case Study: Feasting, Fasting, and Systems Integration
SUMMARY

Summary

The *liver* is the largest glandular organ in vertebrates; it detoxifies blood; participates in the metabolism of carbohydrates, fats, and proteins; and regulates the organic components of blood. Lipids reach the liver (where they are assembled, stored, or broken down into compounds such as Acetyl-CoA) by way of the general systemic circulation. Excess glucose is stored as glycogen by the liver. Excess amino acids are converted to forms that can be fed into the Krebs cycle as an alternative source of energy.

Bile, which emulsifies fats in the small intestine, is made in the liver. The liver, hypothalamus, pituitary, pancreas, adrenal glands, and muscles are all involved in regulating the level of glucose in the blood. During the absorptive state, glucose moves into cells, where it can be used for energy and where the excess can be stored. During the *post*-absorptive state, most cells use fat as the main energy source. Stored fats are broken down and some are converted to glucose, the principal energy source of brain cells. *Insulin*, a pancreatic hormone, enhances the uptake of glucose into body cells; *glucagon*, another pancreatic hormone, stimulates the breakdown of glycogen into its glucose subunits.

Key Terms

liver	post-absorptive state	epinephrine
hepatic portal vein	beta cells	norepinephrine
glycogen	insulin	adrenocorticotropic
detoxification	alpha cells	hormone, ACTH
absorptive state	glucagon	glucocorticoid hormones

1. List four functions of the liver.
2. Describe how your body protects itself from excess glucose in the bloodstream.
3. List the sequence of events that occurs as your blood sugar level drops in the course of sustained exertion coupled with no intake of foods.

Self-Quiz Questions

Fill-in-the-Blanks

Amino acid conversions in the liver form (1) _____, which is potentially toxic to cells; the liver immediately converts this substance to (2) _____, a much less toxic waste product that is expelled by the urinary system from the body. Beta cells of the pancreas secrete (3) _____, and alpha cells secrete (4) _____, a hormone that commands liver cells to convert (5) _____ (a storage starch) into glucose subunits. Under hypothalamic commands, the adrenal medulla begins secreting (6) _____ and (7) _____, which stop (8) _____ synthesis in the liver, stop (9) _____ uptake in muscles, and promote the shift from "burning" (10) _____ during cellular respiration to "burning" fats.

CHAPTER TEST

UNDERSTANDING AND INTERPRETING KEY CONCEPTS

___ (1) The process that moves nutrients into the blood or lymph is _____.
  (a) ingestion
  (b) absorption
  (c) assimilation
  (d) digestion
  (e) none of the above

___ (2) The enzymatic digestion of proteins begins in the _____.
  (a) mouth
  (b) stomach
  (c) liver
  (d) pancreas
  (e) small intestine

___ (3) The enzymatic digestion of starches begins in the _____.
  (a) mouth
  (b) stomach
  (c) liver
  (d) pancreas
  (e) small intestine

___ (4) The greatest amount of absorption of digested nutrients occurs in
the _____.
  (a) stomach
  (b) pancreas
  (c) liver
  (d) colon
  (e) duodenum

___ (5) Glucose moves through the membranes of the small intestine mainly
by _____.
  (a) peristalsis
  (b) osmosis
  (c) diffusion
  (d) active transport
  (e) bulk flow

___ (6) Which of the following is *not* found in bile?
  (a) Lecithin
  (b) Salts
  (c) Digestive enzymes
  (d) Cholesterol
  (e) Pigments

___ (7) The average American consumes approximately _____ pounds of
sugar per year.
  (a) 25
  (b) 50
  (c) 75
  (d) 100
  (e) 125

___ (8) Of the following, _____ has (have) the highest net protein
utilization.
  (a) milk
  (b) eggs
  (c) fish
  (d) meat
  (e) bread

___ (9) Obesity is defined as being _____ percent above the ideal weight
as developed by insurance companies.
  (a) 5
  (b) 10
  (c) 15
  (d) 20
  (e) 25

___ (10) A deficiency of vitamin _____ causes rickets in children and
osteomalcia in adults.
  (a) A
  (b) B
  (c) C
  (d) D
  (e) E

___ (11) The element needed by humans for blood clotting, nerve impulse transmission, and bone and tooth formation is _____.
(a) magnesium
(b) iron
(c) calcium
(d) iodine
(e) zinc

CHAPTER TEST

INTEGRATING AND APPLYING KEY CONCEPTS

Suppose that you could not eat solid food for two weeks and that you had only water to drink. List in correct sequential order the measures that your body would take to try to preserve your life. Mention the command signals that are given as one after another critical point is reached, and tell which parts of the body are the first and the last to make up for the deficit.

33

TEMPERATURE CONTROL AND FLUID REGULATION

General Objectives

1. Understand the degrees to which ectotherms, endotherms, and heterotherms can control their body temperatures. Be able to explain how heat gain and loss occur in birds and mammals and how these animals maintain a steady body temperature.
2. Explain how the chemical composition of extracellular fluid is maintained by mammals.

33-I
(pp. 472–476)

CONTROL OF BODY TEMPERATURE
    Temperatures Suitable for Life
    Heat Gains and Heat Losses
    Classification of Animals Based on Temperature
    Temperature Regulation in Mammals
    Commentary: Falling Overboard and the Odds for Survival

Summary

All animals make controlled as well as obligatory exchanges with the external environment to maintain a hospitable internal environment. Animals gain heat by way of metabolism, and they exchange heat with the environment by processes called radiation, conduction, convection, and evaporation.

Almost all animals must maintain their body temperature within the narrow limits of 0°–40° C. The body temperatures of *ectotherms* rise and fall with environmental changes, although they can influence them to a minor extent through metabolic and behavioral responses to temperature change. Although some birds and mammals continuously maintain a relatively constant body temperature and are considered to be endotherms, most are *heterothermic* and do not maintain precisely the same body temperature at all times. In mammals, the *hypothalamus* is the seat of *temperature control*. Skin thermoreceptors monitor shifts in environmental temperature, and core thermoreceptors within the body signal internal temperatures to the regulatory centers. Two principal types of responses to cold temperatures occur: heat production and heat retention. In response to excessive heat, blood is shunted to the skin surface and heat is radiated, sweat gland secretion is enhanced, epinephrine secretion is inhibited, and exothermic reactions in the body are generally suppressed. *Hypothermia*—a drop in body temperature below tolerance levels—is a common cause of death and disability among humans in boating accidents and on cold weather expeditions.

Key Terms

radiation
conduction
convection
evaporation
ectotherms

behavioral temperature
    regulation
endotherms
heterotherms

shivering
hypothermic
hyperthermic
fever

Objectives

1. Distinguish between *ectotherms* and *endotherms*; give two examples of each group.
2. Explain how endotherms maintain their body temperature when environmental temperatures fall.
3. Explain how endotherms maintain their body temperature when environmental temperatures rise 3 to 4 Fahrenheit degrees above standard body temperature.
4. List the ways in which ectotherms are disadvantaged by not being able to maintain a particular body temperature. Describe the things that ectotherms can do to lessen their vulnerability.
5. Define *hypothermia* and state the situations in which a human might experience the disorder.

**Self-Quiz
Questions**

Fill-in-the-Blanks

In mammals, the (1) _____ is the seat of temperature control.

Thermoreceptors located deep in the body are called (2) _____

thermoreceptors. The adrenal medulla secretes (3) _____ when heat

production is called for. A drop in body temperature below tolerance levels is

referred to as (4) _____.

True-False

If false, explain why.

___ (5) Some modern reptiles are ectotherms.

___ (6) Jackrabbits are endotherms.

CONTROL OF EXTRACELLULAR FLUID
Water Gains and Losses
Solute Gains and Losses
Urinary System of Mammals
Overview of Urine Formation

Summary

Animals must have mechanisms for maintaining within some fairly narrow range whatever body fluid concentration and composition are necessary for cell functioning. All animals must regulate individual ion concentrations in the body, and all animals must eliminate all waste products that are potentially toxic.

The mammalian *kidney*, an intricately structured organ that regulates water and ion content, is part of a tubular network called the *urinary system*. The discharged fluid—*urine*—contains water, salts, nitrogen-containing wastes such as urea, and substances that the body cannot metabolize. The composition and volume of urine are adjusted as internal conditions change. *Nephrons* are tubular components of the kidney that filter and purify a large volume of blood each day. *Filtration* is an entirely passive process in which essentially protein-free plasma is driven by a relatively high-pressure difference from the knot of capillaries that make up each *glomerulus* across two layers of cells into the cavity of the *Bowman's capsule*. With the plasma go substances both useful and not useful to the human body. As the plasma passes along the length of the nephron tubule, the useful substances (water and most solutes) are *reabsorbed* by both active and passive means into the surrounding *peritubular capillaries*. Substances not useful (excess water, solutes) are *secreted* from peritubular capillaries into the nephron by both active and passive transport. *Filtration*, *reabsorption* of water and solutes, and *secretion* of excess and toxic substances by the nephrons (which are surrounded by capillaries in the kidney) all influence the ultimate composition and volume of urine; thus, they influence how much water and solute the body conserves. The longer the loop of Henle, the greater the animal's capacity to conserve water and to concentrate solutes for excretion in urine.

Key Terms

thirst behavior
excretion
nutrients
mineral ions
waste products
ammonia (NH_3)
deamination reaction
urea (NH_2CONH_2)
uric acid
urine
kidneys

cortex, cortical region
medulla
nephrons
collecting ducts
renal pelvis
ureter
urinary bladder
urethra
urinary system
urination
kidney stones

glomerulus
glomerular capillaries
Bowman's capsule
proximal tubule
loop of Henle
distal tubule
peritubular capillaries
filtration
bulk flow
reabsorption
secretion

Objectives

1. List successively the parts of the human urinary system that constitute the path of urine formation and excretion.
2. Locate the processes of filtration, reabsorption, and tubular secretion along a nephron, and tell what makes each process happen.
3. List three soluble by-products of animal metabolism that are potentially toxic.

Self-Quiz Questions

True-False

If false, explain why.

___ (1) All animals must regulate individual ion concentrations in the body.

___ (2) Water reabsorption into capillaries is achieved by diffusion and active transport.

___ (3) When the body rids itself of excess water, urine becomes more dilute.

___ (4) The subunits in the human kidney that process urine are called *nephridia*.

___ (5) Of the three principal waste products excreted from the human urinary bladder, uric acid is generally most toxic.

Fill-in-the-Blanks

(6) _____ is an entirely passive event that begins in the Bowman's capsule. Membrane (7) _____ and the (8) _____ difference across the membrane are the reasons kidneys can (9) _____ an astonishing volume of fluid each day. During (10) _____, all but about 1 percent of the water and most of the (11) _____ that entered the nephron are returned to the bloodstream. Some substances that appear in urine in greater amounts than were carried into the nephron to begin with are actively transported from the (12) _____ _____ into the nephron tubules, an event known as tubular secretion. (13) _____ carry urine away from the kidney to the (14) _____ _____, where it is stored until it is released via a tube called the (15) _____, which carries urine to the outside.

Labeling

Identify each indicated part of the accompanying illustration (p. 295).

(16) _____ _____   (20) _____ _____

(17) _____ _____   (21) _____ _____

(18) _____ _____   (22) _____ _____ _____

(19) _____ _____

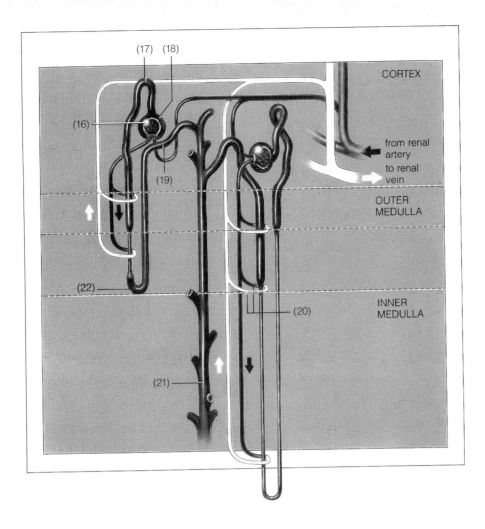

33-III
(pp. 480–487)

CONTROL OF EXTRACELLULAR FLUID (cont.)
    A Closer Look at Filtration
    A Closer Look at Reabsorption
    Acid-Base Balance
    Commentary: **Kidney Failure, Bypass Measures, and Transplants**
    Case Study: On Fish, Frogs, and Kangaroo Rats

SUMMARY

Summary

Changes in diet, muscular activity, and metabolic activity alter body fluid volume and composition. Urine composition is regulated by two homeostatic controls, both governed by the *hypothalamus*: a hormonal action, which affects the amount of water and solutes excreted in the urine; and a *thirst mechanism*, which affects water intake. Although *antidiuretic hormone* (ADH) is the principal hormone that regulates fluid movement through both vascular and urinary tubes, other hormones such as *aldosterone* also influence solute concentrations. A sodium concentration gradient is established along the loops of Henle through a process called *countercurrent multiplication*. If sodium concentrations and extracellular fluid volume drop, an auxiliary *renin-angiotensin system* will come into play and cause aldosterone to be secreted from the adrenal cortex. *Aldosterone* enhances the reabsorption of sodium by the distal tubules and collecting ducts, which in turn causes the blood pressure in the glomeruli to rise.

The respiratory system and other organ systems work together with the kidneys in maintaining acid-base balance; only the urinary system can eliminate excess amounts of H^+ and restore the body's buffers.

When kidneys malfunction, solute levels get out of balance. Substances may accumulate in the bloodstream until they reach toxic levels that can lead to nausea, fatigue, loss of memory, and (in advanced cases) death. A kidney dialysis machine can be used to purify blood and restore the proper solute balance through a process called hemodialysis. An alternative approach is a kidney transplant.

Key Terms

afferent arterioles
hydrostatic pressure
efferent arteriole
glomerulonephritis
kidney dialysis machine
hemodialysis
peritoneal dialysis
obligatory water loss

osmotic gradient
countercurrent
    multiplication
isotonic
hypertonic
antidiuretic hormone,
    ADH

thirst center
juxtaglomerular apparatus
renin
angiotensin I
angiotensin II
aldosterone
hypertension

Objectives

1. List some of the factors that can change the composition and volume of body fluids.
2. Describe the process of countercurrent multiplication and explain its specific role in excretion.
3. Name and describe two homeostatic mechanisms that regulate the composition of urine.
4. State explicitly how the hypothalamus, posterior pituitary, and distal tubules of the nephrons are interrelated in regulating water and solute levels in body fluids.
5. List two kidney disorders and explain what can be done if kidneys become too diseased to work properly.
6. Describe the role of the kidney in maintaining the pH of the extracellular fluids between 7.35 and 7.45.

Self-Quiz Questions

Fill-in-the-Blanks

(1) _____ _____ is the principal hormone that regulates fluid movement through vascular and urinary tubes. (2) _____ _____ establishes a solute concentration gradient between the nephron and the interstitial fluid that surrounds the loop of Henle. The (3) _____ contains a nerve cell cluster that is sensitive to concentrations of sodium and some other solutes in blood; it also contains a (4) _____ center that detects a rise in salt levels (or a drop in water volume). A kidney (5) _____ machine can be used to restore the proper solute balance when kidneys are too diseased to function properly.

True-False   If false, explain why.

___ (6) Countercurrent multiplication establishes a sodium concentration gradient in the loop of Henle.

CHAPTER TEST

UNDERSTANDING AND INTERPRETING KEY CONCEPTS

___ (1) The most toxic waste product of metabolism is _____.
   (a) water
   (b) uric acid
   (c) urea
   (d) ammonia
   (e) carbon dioxide

___ (2) An entire subunit of a kidney that purifies blood and restores solute and water balance is called a _____.
   (a) glomerulus
   (b) loop of Henle
   (c) nephron
   (d) ureter
   (e) None of the above

___ (3) In humans, the thirst center is located in the _____.
   (a) adrenal cortex
   (b) thymus
   (c) heart
   (d) adrenal medulla
   (e) hypothalamus

___ (4) The longer the _____, the greater is an animal's capacity to conserve water and to concentrate solutes to be excreted in the urine.
   (a) loop of Henle
   (b) proximal tubule
   (c) ureter
   (d) Bowman's capsule
   (e) collecting tubule

___ (5) During reabsorption, sodium ions cross the proximal tubule walls into the interstitial fluid principally by means of _____.
   (a) phagocytosis
   (b) countercurrent multiplication
   (c) bulk flow
   (d) active transport
   (e) All of the above

___ (6) Filtration of the blood in the kidney takes place in the _____.
   (a) loop of Henle
   (b) proximal tubule
   (c) distal tubule
   (d) Bowman's capsule
   (e) All of the above

—— (7) _____ controls the concentration of solutes in urine.
  (a) Insulin
  (b) Glucagon
  (c) Antidiuretic hormone
  (d) Thyroxin
  (e) Epinephrine

—— (8) Hormonal control over excretion primarily affects _____.
  (a) Bowman's capsules
  (b) distal tubules
  (c) proximal tubules
  (d) the urinary bladder
  (e) loops of Henle

—— (9) The last portion of the excretory system passed by urine before it is eliminated from the body is the _____.
  (a) renal pelvis
  (b) bladder
  (c) ureter
  (d) collecting ducts
  (e) urethra

—— (10) Desert animals excrete _____ as their principal nitrogenous waste in a highly concentrated urine.
  (a) urea
  (b) uric acid
  (c) ammonia
  (d) amino acids
  (e) ADH

CHAPTER TEST

INTEGRATING AND APPLYING KEY CONCEPTS

The hemodialysis machine used in hospitals is expensive and time-consuming. So far, artificial kidneys capable of allowing people who have nonfunctional kidneys to purify their blood by themselves, without having to go to a hospital or clinic, have not been developed. Which aspects of the hemodialysis procedure do you think have presented the most problems in developing a method of home self-care? If *you* had an unlimited budget and were appointed head of a team to develop such a procedure and its instrumentation, what strategy would you pursue?

34

PRINCIPLES OF REPRODUCTION AND DEVELOPMENT

THE BEGINNING: REPRODUCTIVE MODES
  Asexual Reproduction
  Sexual Reproduction
  Some Strategic Problems in Having
  Separate Sexes
STAGES OF DEVELOPMENT
  Gametogenesis
  Fertilization
  Cleavage
  Gastrulation
  Organogenesis
  Post-Embryonic Pathways of Development

MECHANISMS OF DEVELOPMENT
  Developmental Information in the Egg
  Cell Differentiation
  Mechanisms Underlying Morphogenesis
  Pattern Formation
AGING AND DEATH
  Commentary: Death in the Open
SUMMARY

General Objectives

1. Understand how asexual reproduction differs from sexual reproduction. Know the advantages and problems associated with having separate sexes.
2. Describe early embryonic development and distinguish each: oogenesis, fertilization, cleavage, gastrulation, and organ formation.
3. Explain how a spherical zygote becomes a multicellular adult with arms and legs.

34-I
(pp. 489–492)

THE BEGINNING: REPRODUCTIVE MODES
  Asexual Reproduction
  Sexual Reproduction
  Some Strategic Problems in Having Separate Sexes

Summary

Animal reproductive strategies range from asexual processes, through development of unfertilized eggs into adults, through self-fertilization, to sexual reproduction between separate male and female forms. *Asexual* processes include regeneration and budding; neither case promotes genetic variability because the offspring cells are identical copies of the parents. Fission and mitosis are considered asexual forms of reproduction because reproductive cells are not produced. Asexual reproduction is advantageous to organisms that are highly adapted to their surroundings as long as the surroundings remain stable. *Sexual* processes include parthenogenesis, hermaphroditism with self- or cross-fertilization, and separate male forms releasing sperms either outside or inside separate female forms. Sexual strategies are considered advantageous to populations of organisms that inhabit unstable surroundings.

Key Terms

larva
sexual reproduction
asexual reproduction
fission
budding
gemmules
clones

parthenogenesis
hermaphrodites, -ditism
viviparous
ovoviviparous
oviparous
synchronous
external fertilization

internal fertilization
testes
penis
vagina
ovary
yolk

Objectives

1. Distinguish between regeneration and budding. State which organisms can accomplish each process.
2. Define *parthenogenesis* and state the organisms in which it occurs. List one advantage and one disadvantage of parthenogenesis.
3. Describe the conditions in which hermaphroditism with cross-fertilization is a more successful strategy than separate sexes using internal fertilization. Identify an organism that uses hermaphroditism.
4. Describe the conditions in which hermaphroditism with self-fertilization would be a more successful strategy than separate internal fertilization. Identify an organism that uses hermaphroditism with self-fertilization.
5. Describe the conditions in which separate sexes using external fertilization would be a more successful strategy than separate sexes using internal fertilization.
6. Explain why evolutionary trends in many groups of organisms tend toward developing more complex sexual strategies rather than retaining simpler, asexual strategies.
7. Define *oviparous*, *viviparous*, and *ovoviparous*. Cite an example of an animal that goes through each of the three developmental strategies.

Self-Quiz Questions

Fill-in-the-Blanks

At the end of fertilization a (1) _____ is formed. (2) _____ is a form of cell division in which the offspring cells do not increase in size. In many animals the embryo develops into a motile, independent (3) _____, which extends the food supply and range of the population. Asexual processes of reproduction include (4) _____, which is common in prey animals, and (5) _____, which is common in animals such as *Hydra* and other cnidarians. (6) _____ is the cleavage and subsequent differentiation of an unfertilized egg into an adult. Earthworms and parasitic tapeworms are (7) _____; each organism has both male and female reproductive organs and produces both eggs and sperms. Eggs are produced in (8) _____, and sperms are produced in (9) _____.

STAGES OF DEVELOPMENT
    Gametogenesis
    Fertilization

Summary

Gametogenesis follows meiosis and enlarges and polarizes the eggs of most complex eukaryotes; it also includes in the egg many *maternal messages*, which not only control development until gastrulation begins and the embryonic genotype is activated but also lay the groundwork for all development that follows. *Polarization* is the unequal distribution of structures and substances in the oocyte (and mature egg), including RNA transcripts of the maternal DNA. This unequal distribution is the initial determinant in the structural patterning of the embryo. Sperm penetration into the egg activates a series of events within the egg cytoplasm, and the mingling of DNA from both parents establishes the diploid genotype; all of these activities constitute *fertilization*.

    Gene expression also varies through time in animals. During gametogenesis, sperms acquire different structures and function differently from eggs. Fertilization is completed when sperm and egg nuclei fuse, restoring the diploid number of chromosomes.

Key Terms

gametogenesis	growth	animal pole
fertilization	tissue specialization	vegetal pole
cleavage	oocyte	cortex
gastrulation	mRNA transcripts	gray crescent
organogenesis	polarity	

Objectives

1. Characterize gametogenesis, fertilization, and cleavage according to the principal events that occur in each.
2. List sequentially all the major phases in animal development and define the activities that make each phase unique.
3. Explain what causes polarity to occur during oocyte maturation in the mother and state how polarity influences later development.

**Self-Quiz
Questions**

True-False

If false, explain why.

___ (1) In complex eukaryotes, development until gastrulation is governed by DNA in the nucleus of the zygote.

___ (2) Sperm penetration into the cytoplasm of the egg brings about specific structural changes and chemical reactions.

Fill-in-the-Blanks

In animals, gene expression also varies through time. (3) _____, the formation and maturation of eggs and sperm, is considered the first stage of animal development. Rich stores of substances become assembled in localized regions of the (4) _____ cytoplasm. When sperm and egg unite and their

DNA mingles and is reorganized, the process is referred to as (5) _____.
(6) _____ includes the repeated mitotic divisons of a zygote that segregate the egg cytoplasm into a cluster of cells; the entire cluster is known as a (7) _____. (8) _____ is the process that regroups cells and arranges them into embryonic tissue layers. Copperheads show (9) _____: Fertilization is internal; the fertilized eggs develop inside the mother's body, without additional nourishment; and the young are born live. Birds show (10) _____: Eggs with large yolk reserves are released from and develop outside the mother's body. (11) _____ _____ activated at fertilization direct the initial stages of development until gastrulation occurs. In amphibian eggs, sperm penetration on one side of an egg causes pigment granules on the opposite side of the egg to flow toward the (12) _____ _____. A lightly pigmented area called the (13) _____ _____ results. It is a visible marker of the site where the (14) _____ _____ will be established and where gastrulation will begin.

34-III
(pp. 493–497)

STAGES OF DEVELOPMENT (cont.)
    Cleavage
    Gastrulation
    Organogenesis

Following fertilization, animal development proceeds through stages of *cleavage* (the subdivision of the zygote into cellular compartments), *gastrulation* (the formation of embryonic tissue layers), *organogenesis*, and growth. Initially, development is guided by regionally positioned messages in the egg cytoplasm. The destiny of cell lineages is in part established according to which sector of the egg cytoplasm is inherited by the first embryonic cells formed during cleavage.

Cleavage is the compartmentalization of the zygote into many cells that differ due to their differing assemblages of cytoplasmic constituents. Even at this time, there are structural and functional differences between cells that result from the translation of RNA that was transcribed from maternal DNA and from the variable amounts of yolk. During cleavage, the zygote is divided into many smaller cells by mitosis to form a *blastula*, but no growth or expression of the embryonic genotype occurs yet.

Gastrulation involves the formation of two or more layers of embryonic tissues: generally *ectoderm*, which forms the body covering and nervous system; *endoderm*, which forms the inner lining of the gut and the gut derivatives; and (in all but sponges) *mesoderm*, which forms the muscles, the skeleton, and the circulatory system. With gastrulation, the embryonic genotype is transcribed and translated into its own unique assemblage of proteins, and the influences of the maternal system wane. Gastrulation also is characterized by the movement of masses of cells; endoderm ends up on the interior of the embryo and ectoderm on the outside. When the embryonic tissues have been established, organogenesis occurs by growth, tissue movements, and specialization, forming the rudimentary organs.

Key Terms

dorsal lip of blastopore
yolk plug
cleavage
blastula
blastocoel
blastodisk
extraembryonic
  membranes

inner cell mass
blastocyst
gastrulation
germ layers
endoderm
mesoderm
ectoderm

gastrula
archenteron
neural tube
incubate
somites
primitive streak

Objectives

1. List and describe the events that occur during cleavage.
2. Explain how the amount of yolk in an ovum can influence an animal's cleavage pattern.
3. Define gastrulation and state what process begins at this stage that did not happen during cleavage.
4. Name each of the three embryonic tissue layers and the organs formed from each.
5. Compare the early stages of frog and chick development (Figures 34.6 and 34.8) with respect to (a) egg sizes and (b) type of cleavage pattern (incomplete and complete).

**Self-Quiz
Questions**

True-False

If false, explain why.

____ (1) During gastrulation, maternal controls over gene activity are activated and begin the process of differentiation in each cell's nucleus.

Fill-in-the-Blanks

The third stage of animal development, (2) _____, is characterized by the subdividing and compartmentalization of the zygote; no growth occurs at this stage, and usually a hollow ball of cells, the (3) _____, is formed. The fourth stage, (4) _____, is concerned with the formation of ectoderm, mesoderm, and endoderm, the (5) _____ layers of the embryo; at the end of this stage, the (6) _____ is formed. Ectoderm eventually will give rise to skin epidermis and the (7) _____ system; endoderm forms the inner lining of the (8) _____ and associated digestive glands. Mesoderm forms the circulatory system, the (9) _____, and the muscles. Cleavage of a frog's zygote is total because there is little enough yolk that cleavage membranes can subdivide the entire cytoplasmic mass; in the chick, however, there is so much yolk that cleavage membranes cannot. Cleavage is therefore said to be incomplete, and the chick grows from a primitive streak on the surface of the (10) _____ mass into a chick embryo complete with wing and leg buds and beating heart during the first (11) _____ days.

STAGES OF DEVELOPMENT (cont.)
    Post-Embryonic Pathways of Development
MECHANISMS OF DEVELOPMENT
    Developmental Information in the Egg
    Cell Differentiation

Summary

Developmental strategies in multicellular organisms vary from *indirect development*, which includes a larval stage to *direct development*, in which the embryo is nourished either externally or internally by the mother.

The intricate life cycles of multicelled eukaryotes are typified by changes in form and function that depend not only on cell growth and division but also on *differentiation*. In this process, cells having identical sets of DNA become structurally and functionally different from one another over time. Different portions of the total set of instructions are used in different cells at different times. Controls over this gene expression lead to variations in the way that materials are patterned and arranged as the life cycle unfolds. *Which* genes are expressed in a given cell depends on the nature of cytoplasmic changes induced by the environment and by interactions with other cells. In *regulative* patterns of development, genes that were shut down in differentiated cells can become activated again when missing body parts must be replaced.

Finally, by further growth and tissue specialization, the various body parts acquire the structural and physiological properties necessary for performing specialized tasks.

Key Terms

adult	regeneration	stem cells
larva	genetically equivalent	John Gurdon
metamorphosis	cell differentiation	*Xenopus laevis*
direct development	morphogenesis	identical twins
indirect development	"uncommitted"	

Objectives

1. Distinguish direct from indirect development.
2. Define what is meant by *larva*. Distinguish metamorphosis from morphogenesis.
3. Define *differentiation* and give two examples of cells in a multicellular organism that have undergone differentiation.
4. Explain why the differentiation of cells in a multicellular organism goes hand in hand with division of labor and the integration of life processes.

Self-Quiz Questions

Fill-in-the-Blanks

A larva necessarily must undergo (1) _____ in order to become a juvenile.

Any development program that includes an adaptive larval stage is called

(2) _____ development. If a zygote is split, either naturally or

experimentally, into two equal parts, the result is (3) _____ _____.

Differentiation requires (4) _____ gene expression in certain cells at

certain times.

If false, explain why.

___ (5) Muscle cells that contract and nerve cells that relay messages are examples of cells that have undergone differentiation from the initial ball of embryonic cells.

34-V
(pp. 500–508)

MECHANISMS OF DEVELOPMENT (cont.)
    Mechanisms Underlying Morphogenesis
    Pattern Formation
AGING AND DEATH
    Commentary: **Death in the Open**
SUMMARY

Summary

Morphogenesis (the growth, shaping, and spatial coordination of tissues and organs according to predefined patterns) depends on controls over expression of genes, which affect the extent, direction, and rate of growth. Morphogenesis also depends on chemical communication links, physical attachments, and physical restraints imposed by neighboring cells in the developing embryo.

During morphogenesis, three types of *morphogenetic* movements cause embryonic parts to move from one site to another. Individual cells may follow specific paths in response to chemical gradients or may respond to adhesive cues. Epithelial sheets fold inward or outward, creating tissue layers, body cavities, and evaginations. In some cases, whole organs move. In morphogenesis, *controlled cell death* causes the folding of parts, the development of slits, the hollowing-out of tubes, the shaping of bones, and the opening of mouth, nostrils, and ears. Sometimes a mutation blocks the capacity of specific cells to respond to the signal from elsewhere to die, so those cells delay dying until they are no longer supported by other parts of the body.

During the formation of organs, one body part differentiates according to signals it receives from an adjacent body part; this process is known as *embryonic induction*. In all animals, the coordinated development of body parts in specific regions depends on some combination of (1) the influence of cytoplasmic substances within the unfertilized egg and (2) inductive interactions between cells at later stages of development.

Normal cells of complex eukaryotes have limited life spans characteristic of their species. Following growth and division, cells begin to deteriorate in a predictable manner, a process called *aging*. Aging is built into the life cycle of all organisms in which differentiated cells show considerable specialization. Aging is not simply the result of an accumulation of environmental damage. Aging and death may be coded largely in DNA; perhaps signals from the cytoplasm of cells nearby or some distance away in the differentiated body activate those messages, informing the cell that it is time to die. Every cell that lives must eventually die or at least be incorporated into its offspring cells.

Key Terms

active cell migrations	ooplasmic localization	homeotic mutation
chemotaxis	embryonic induction	controlled cell death
adhesive cues	Spemann	brain neurohormone
neural plate	inducer substances	juvenile hormone
neural tube	*Drosophila*	ecdysone
pattern formation	limited division potential	aging
Moorhead and Hayflick	imaginal disks	collagen

Objectives

1. Define *morphogenesis*. If it differs from organogenesis, indicate how.
2. List three of the developmental factors that control and guide morphogenesis.
3. Explain how substances released from cells can induce changes in nearby cells.
4. Describe the relationship that develops between the retina and the lens as a result of embryonic induction.
5. State what Spemann's transplant of the optic cup demonstrated and indicate how it advanced our understanding of the ways that parts are fit together during development.
6. Give three examples of structures in you that were formed by controlled cell death. Describe how those structures would have appeared if controlled cell death had not occurred.
7. State any evidence that suggests that either DNA or the reception of external signals may be involved in aging and death.
8. Define *aging* and state why the death of cells is believed not to be caused solely by an accumulation of environmental insults.
9. Contrast the degree of finality in the death of prokaryotes with that of complex multicellular eukaryotes.
10. Describe the behavior of elephants and humans with regard to the disposal of the dead bodies of their own kind. Then state which populations, human or elephant, do a more efficient job of recycling their atoms through the ecosystem of which they are part.

Self-Quiz Questions

True-False

If false, explain why.

___ (1) The aging and death of a cell may be coded in large part in its DNA; external signals activate those DNA messages and tell the cell that it is time to die.

___ (2) A process of predictable cellular deterioration is built into the life cycle of all organisms that consist of differentiated cells that show considerable specialization.

___ (3) Body parts become folded, tubes become hollowed-out, and eyelids, lips, noses, and ears all become slit or perforated by controlled cell death.

Fill-in-the-Blanks

(4) _____ involves the growth, shaping, and spatial coordination necessary to form functional body units. Embryonic cells move in response to adhesive cues and (5) _____ _____. Sometimes entire organs (such as testes in human males) change position in the developing organism but the inward or outward folding of (6) _____ _____ is seen more often. Spemann demonstrated that the process known as (7) _____ _____ occurs in salamander embryos, where one body part differentiates because of signals it receives from an adjacent body part. The influence of cytoplasmic substances in particular locations of the unfertilized egg on development after

fertilization is called (8) _____ _____. A kitten's eyes opening after

birth is an example of (9) _____ _____ _____ in action.

UNDERSTANDING AND INTERPRETING KEY CONCEPTS

___ (1) Animals such as birds lay eggs with large amounts of yolk; embryonic
development happens within the egg covering outside the mother's
body. This developmental strategy is called _____.
(a) ovoviviparity
(b) viviparity
(c) oviparity
(d) parthenogenesis
(e) None of the above

___ (2) The process of cleavage most commonly produces a(n) _____.
(a) zygote
(b) blastula
(c) gastrula
(d) third germ layer
(e) organ

___ (3) Imaginal disks are characteristic of the embryonic development
of _____.
(a) frogs
(b) fruit flies
(c) chickens
(d) sea urchins
(e) humans

___ (4) Which of the following forms undergoes indirect development?
(a) Frogs
(b) Snakes
(c) Whales
(d) Horses
(e) Hawks

___ (5) The differentiation of a body part in response to signals from an
adjacent body part is _____.
(a) contact inhibition
(b) ooplasmic localization
(c) embryonic induction
(d) pattern formation
(e) None of the above

___ (6) A homeotic mutation _____.
(a) may cause a leg to develop on the head where an antenna should
grow
(b) affects the expression of imaginal disks
(c) affects morphogenesis
(d) may alter the path of development
(e) All of the above

Principles of Reproduction and Development *307*

___ (7) Shortly after fertilization, the zygote is subdivided into a multicelled embryo during a process known as _____.
   (a) meiosis
   (b) parthenogenesis
   (c) embryonic induction
   (d) cleavage
   (e) invagination

___ (8) Muscles differentiate from _____ tissue.
   (a) ectoderm
   (b) mesoderm
   (c) endoderm
   (d) parthenogenetic
   (e) yolky

___ (9) The gray crescent is _____.
   (a) formed where the sperm penetrates the egg
   (b) next to the dorsal lip of the blastopore
   (c) the yolky region of the egg
   (d) where the first mitotic division begins
   (e) formed opposite from where the sperm enters the egg

CHAPTER TEST

INTEGRATING AND APPLYING KEY CONCEPTS

If embryonic induction did not occur in a human embryo, how would the eye region appear? What would happen to the forebrain and epidermis? If controlled cell death did not happen in a human embryo, how would its hands appear? Its face?

35

HUMAN REPRODUCTION AND DEVELOPMENT

General Objectives

1. Describe the structure and function of the male and female reproductive tracts.
2. Outline the principal events of prenatal development.
3. Know the principal means of controlling human fertility.
4. Know how breast and testicular cancer can be detected by self-examination.

35-I
(pp. 509–512)

PRIMARY REPRODUCTIVE ORGANS
MALE REPRODUCTIVE SYSTEM
   Spermatogenesis
   Sperm Movement Through the Reproductive Tract
   Hormonal Control of Male Reproductive Functions

Summary

In human males and females, as in other separately sexed animals, the primary reproductive organs are *testes* and *ovaries*; the remaining reproductive structures are accessory reproductive organs. Ovaries and testes not only produce gametes by meiosis, they also secrete important sex hormones that maintain gamete production, normal sexuality, and secondary sex characteristics.

Sperm that have matured in the *seminiferous tubules* of the testes during gametogenesis move into the *epididymis* (plural: epididymides), finish maturation, and become motile. Sexual activity propels them along the *vas deferens* (plural: vasa deferentia) upward and posteriorly to the junction behind the urinary bladder where the two vasa deferentia merge with the *ejaculatory duct*. From here, the sperm are propelled along the urethra, receiving secretions from the *seminal vesicles, prostate,* and *bulbourethral glands*. Collectively, the sperm and secretions (mucus, nutrients, water, and prostaglandins) are called *semen*.

Testosterone, produced by the *interstitial cells* of the testes, stimulates *spermatogenesis*, governs the form and function of the male reproductive tract, develops and maintains male sexual behavior, and maintains male *secondary sexual characteristics* (deep voice and growth of beard and pubic hair).

Key Terms

gonads	primary spermatocytes	semen
testes (sing.: testis)	secondary spermatocytes	prostaglandins
ovaries	spermatids	testosterone
sperm	Sertoli cells	penis
secondary oocytes	acrosome	follicle-stimulating
secondary sexual trait	epididymis	hormone, FSH
ejaculatory duct	vas deferens	luteinizing hormone, LH
scrotum	seminal vesicle	gonadotropin-releasing
seminiferous tubules	prostate gland	hormone, GnRH
interstitial cells	bulbourethral glands	accessory glands

Objectives

1. Distinguish between primary and secondary sexual traits and between gonads and accessory reproductive organs.
2. List in order the stages that compose spermatogenesis.
3. Follow the path of a mature sperm from the seminiferous tubules to the urethral exit. List every structure encountered along the path and state the contribution to the nurture of the sperm.
4. Diagram the structure of a sperm, label its components, and state the function of each.
5. Name the four hormones that directly or indirectly control male reproductive function. Diagram the negative feedback mechanisms that link the hypothalamus, anterior pituitary, and the testes in controlling gonadal function.

Self-Quiz Questions

Fill-in-the-Blanks

Spermatogenesis occurs in the (1) _____ _____, but sperms become somewhat motile in the (2) _____ of the male. The interstitial cells of the testis produce (3) _____. Accessory reproductive organs of the human male include the (4) _____ _____, which either stores most of the sperm or, during sexual activity, moves them along by peristaltic action. The seminal vesicles and (5) _____ contribute secretions that make up most of

the seminal fluid. A cap over most of the head of each sperm contains
(6) _____ _____ that function in egg penetration. The (7) _____
of each sperm contains mitochondria, which provide the energy necessary for
motility. (8) _____ secreted by the prostate gland into semen stimulate
uterine contractions in the female.

Labeling   Label the structures in the illustration.

(9) _____ _____   (11) _____ _____   (13) _____

(10) _____ _____   (12) _____ _____   (14) _____ _____

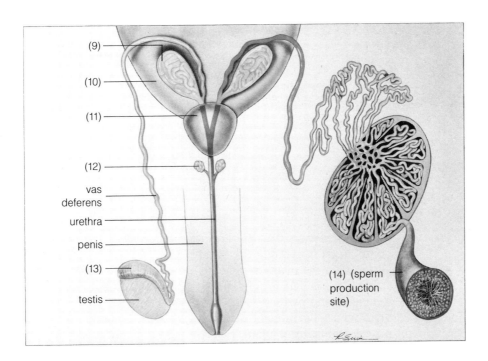

35-II
(pp. 513–519)

FEMALE REPRODUCTIVE SYSTEM
    Oogenesis
    Menstrual Cycle: An Overview
    Control of Ovarian Function
    Control of Uterine Function
SEXUAL UNION AND FERTILIZATION

Summary

The *ovarian cycle* consists of oocyte maturation in a fluid-filled *follicle within the ovary*, the transformation of that cavity into the *corpus luteum* (endocrine tissue) when the follicle ruptures and the mature ovum escapes from it, and a concurrent production of ovarian hormones. The *menstrual cycle* consists of profound changes in the *endometrium* (epithelial lining) and muscles of the *uterus*, as well as glandular secretions, all of which prepare the uterus for implantation. In the absence of implantation, the ovarian and menstrual cycles are repeated monthly.

Some evidence suggests that the *midcycle surge of luteinizing hormone* stimulates production of enzymes that can dissolve the thin membranes between the follicle and the surroundings; thus, the follicle ruptures and ovulation occurs. If fertilization does not occur, the corpus luteum degenerates during the last days of the menstrual cycle and the levels of estrogen and progesterone fall rapidly. Deprived of its hormonal support, the highly developed endometrium disintegrates and the menstrual period begins.

Coitus (sexual intercourse) generally involves the stimulation of the glans penis and vulva by friction, collection of blood in the spongy vascular tissue of both sexes, and rhythmic muscular contractions that force seminal fluid from the vesicles and prostate into the male urethra and thence into the vagina during ejaculation. *Orgasm,* with its involuntary muscular contractions and associated sensations of warmth and release of tension may or may not be reached by either sex during coitus, but if the male ejaculates, the female can become pregnant whether or not she experiences orgasm.

Key Terms

estrogen	labia minora	corpus luteum
progesterone	clitoris	follicular phase
oviduct	granulosa cells	antrum
uterus	primary follicle	midcycle surge of LH
myometrium	zona pellucida	luteal phase
endometrium	secondary oocyte	endometriosis
cervix	ovulation	coitus
vagina	ovum	orgasm
vulva	estrous cycle	mature ovum
labia majora	menstrual cycle	capacitation

Objectives

1. Indicate the ways that the human menstrual cycle differs from estrus as it occurs in most mammals.
2. Distinguish the ovarian cycle from the menstrual cycle and explain how the two cycles are synchronized by hormones from the anterior pituitary, hypothalamus, and ovaries.
3. State which hormonal event brings about ovulation and which other hormonal events bring about the onset and finish of menstruation.
4. Trace the path of a sperm from the seminiferous tubules to where fertilization normally occurs. Mention in correct sequence all major structures of the male and female reproductive tracts that are passed along the way and state the principal function of each structure.
5. Compare the function of the interstitial cells of the testis with the function of the ovarian follicle and corpus luteum.
6. List the physiological factors that bring about erection of the penis during sexual stimulation and bring about ejaculation.
7. List the similar events that occur in both male and female orgasm.

Self-Quiz Questions

Fill-in-the-Blanks

Ovaries produce the important sex hormones (1) _____ and (2) _____.

In the female, (3) _____ are passageways that channel ova from the ovary into the (4) _____, which houses the embryo during pregnancy. In most mammals, the (5) _____ cycle is a predictably recurring time when the female becomes sexually receptive to the male. During ejaculation, a (6) _____ closes off the bladder so that urine cannot mix with semen. Each oocyte is contained in a spherical chamber called a (7) _____. At (8) _____, the follicle ruptures, and the ovum escapes from the ovary and is swept by ciliary action into the (9) _____, the road to the uterus. The ruptured follicle now changes into secretory tissue called the (10) _____ _____. Estrogen causes the (11) _____ (epithelial uterine lining) to thicken. (12) _____ stimulates glands in the thickened tissues to secrete various substances. The midcycle peak of (13) _____, the level of which overshoots (14) _____, triggers ovulation. The decrease of all of the hormones that regulate the monthly cycle brings about (15) _____.

Labeling   Identify each indicated part of the accompanying illustration.

(16) _____   (20) _____   (23) _____ _____

(17) _____   (21) _____   (24) _____

(18) _____   (22) _____ _____   (25) _____

(19) _____ _____

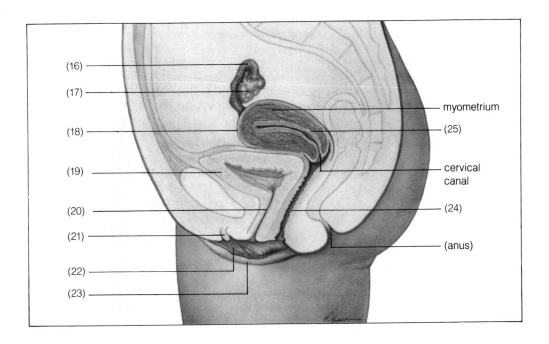

PRENATAL DEVELOPMENT
  First Week of Development
  Extraembryonic Membranes
  The Placenta
  Embryonic and Fetal Development
  Birth
  Lactation
  Case Study: Mother as Protector, Provider, Potential Threat

Summary

During the first month of human development, cleavage, gastrulation, and organ formation occur. By the end of the first month, the embryo has begun to take on recognizably human characteristics. Two months of a relatively slow development for the main organs follow. At the end of the first three-month period (trimester) the term *embryo* is replaced by *fetus* because the developing individual now resembles a miniature adult. By the end of the seventh month, fetal development appears to be relatively complete, but fewer than 10 percent of infants born at this stage survive; by the ninth month, survival chances increase about 95 percent. The embryo is particularly vulnerable to damaging substances from the maternal bloodstream during the first six months after fertilization because that is the critical period of organ formation. Poor nutrition during the last trimester affects all fetal organs but is most damaging to the brain; in the weeks just before and just after delivery, the human brain undergoes its greatest growth. Mothers who smoke cigarettes every day throughout pregnancy tend to produce smaller infants who risk heart disease and decreased reading abilities.

Key Terms

blastocyst	allantois	second trimester
trophoblast	amnion	third trimester
chorionic gonadotrophin	chorion	lactation
embryonic disk	umbilical cord	colostrum
implantation	placenta	thalidomide
extraembryonic	chorionic villi	fetal alcohol syndrome,
membranes	first trimester	FAS
yolk sac	fetus	

Objectives

1. Describe the events that occur during the first month of human development. State how long cleavage and gastrulation require, when organogenesis begins, and what is involved in implantation and placenta formation.
2. State when the embryo begins to be referred to as a fetus and at what point at least ten percent of births result in survival.
3. Explain why the mother must be particularly careful of her diet, health habits, and life style during the first trimester after fertilization (especially the first six weeks) and during the last trimester.

**Self-Quiz
Questions**

Fill-in-the-Blanks

Fertilization generally takes place in the (1) _____; five or six days after conception, (2) _____ begins as the blastocyst burrows inside the endometrium. Extensions from the chorion fuse with the endometrium of the uterus to form a (3) _____, the organ of interchange between mother and fetus. By the beginning of the (4) _____ trimester, all major organs have formed; the offspring is now referred to as a(n) (5) _____.

35-IV
(pp. 526–536)

Commentary: **Cancer in the Human Reproductive System**
CONTROL OF HUMAN FERTILITY
    Some Ethical Considerations
    Possible Means of Birth Control
    In Vitro Fertilization
    Commentary: **Sexually Transmitted Diseases**
SUMMARY

Summary

Breast and *testicular cancers* often can be detected through routine self-examination; these techniques are discussed and illustrated.

Each week the human population grows by about 1,700,000 people and outstrips the resources of many countries. A variety of birth control methods are used to reduce this astounding birth rate. *Extremely effective birth control* methods include abstinence, vasectomy, and tubal ligation. *Highly effective* methods include the Pill, IUDs, and condoms plus spermicide. *Effective* methods include vaginal foams, diaphragms with spermicide, good brands of condoms, and vaginal sponge impregnated with spermicide. Douches are unreliable. Cheap brands of condoms, withdrawal, and the rhythm method based on temperature typically are only 70–76 percent effective. At present, sterilization is generally irreversible.

Abortion terminates a pregnancy by dislodging the implanted embryo and removing it from the uterus; control of conception in the first place is a preferable route to birth control.

Sometimes *in vitro* fertilization is recommended for women who are unable to become pregnant due to blockages of their oviducts. Preovulatory eggs are detected by laparoscopy and removed by suction. Freshly obtained diluted sperm are combined with the ova, and a few hours later fertilization occurs. The cleavage stages are stored for two to four days in a saline solution that supports development; then one is drawn up in a fine tube and transferred to the uterus, where implantation may occur and development proceeds normally.

Sexually transmitted diseases have reached epidemic proportions and have very serious social consequences. Among the STDs are AIDS, gonorrhea, genital herpes, chlamydial infections, syphilis, and pelvic inflammatory disease. There is *no* cure for AIDS; during the next decade, as many as 50–100 million could be infected worldwide.

Key Terms

mammography	spermicidal foam	abortion
mammogram	spermicidal jelly	miscarriages,
biopsy	diaphragm	spontaneous abortion
radical mastectomy	condoms	in vitro fertilization
lumpectomy	IUD, intrauterine device	gonorrhea
metastasis	the Pill	syphilis
abstinence	sexually transmitted	AIDS
rhythm method	diseases, STDs	chlamydia
withdrawal	vasectomy	systemic herpes
douching	tubal ligation	

Objectives

1. Describe how a woman examines herself for breast cancer and how a man examines himself for testicular cancer.
2. Identify the factors that encourage and discourage methods of human birth control.
3. Identify the three most effective birth control methods used in the United States and the four least effective birth control methods.
4. State which birth control methods help to prevent venereal disease.
5. Describe two different types of sterilization.
6. State the physiological circumstances that would prompt a couple to try in vitro fertilization.
7. For each STD described in the Commentary, know (a) the causative organism and (b) the symptoms of the disease.

Self-Quiz Questions

Fill-in-the-Blanks

On Earth each week, (1) _____ more babies are born than people die. Each year in the United States, we still have about (2) _____ unwed teenage mothers and (3) _____ abortions. The most effective method of preventing conception is complete (4) _____. (5) _____ are about 85–93 percent reliable and help prevent venereal disease. A (6) _____ is a flexible, dome-shaped disk, used with a spermicidal foam or jelly, that is placed over the cervix. The most widely used contraceptive is the Pill—an oral contraceptive of synthetic (7) _____ and (8) _____ that suppress the release of (9) _____ from the pituitary and thereby prevent the cyclic maturation and release of eggs. Two forms of surgical sterilization are vasectomy and (10) _____ _____.

CHAPTER TEST

UNDERSTANDING AND INTERPRETING KEY CONCEPTS

For Questions 1–5, choose from these answers:
(a) AIDS
(b) Chlamydial infections
(c) Genital herpes
(d) Gonorrhea
(e) Syphilis

____ (1) _____ is a disease caused by a spherical bacterium (*Neisseria*) with pili; it is curable by prompt diagnosis and treatment.

____ (2) _____ is a disease caused by a spiral bacterium (*Treponema*) that produces a localized ulcer (a chancre).

____ (3) _____ is an incurable disease caused by a retrovirus (an RNA-based virus).

___ (4) _____ is a disease caused by an obligate, intracellular parasite that migrates to regional lymph nodes, which swell and become tender.

___ (5) _____ is an extremely contagious viral infection (DNA-based) that causes sores on the facial area and reproductive tract; it is also incurable.

For Questions 6–8, choose from these answers:
   (a) blastocyst
   (b) allantois
   (c) yolk sac
   (d) oviduct
   (e) cervix

___ (6) The _____ lies between the uterus and the vagina.

___ (7) The _____ is a pathway from the ovary to the uterus.

___ (8) The _____ results from the process known as cleavage.

For Questions 9–12, choose from these answers:
   (a) interstitial cells
   (b) seminiferous tubules
   (c) vas deferens
   (d) epididymis
   (e) prostate

___ (9) The _____ connects a structure on the surface of the testis with the ejaculatory duct.

___ (10) Testosterone is produced by the _____.

___ (11) Meiosis occurs in the _____.

___ (12) Sperms mature and become motile in the _____.

CHAPTER TEST

INTEGRATING AND APPLYING KEY CONCEPTS

What rewards do you think a society should give a woman who has at most two children during her lifetime? In the absence of rewards or punishments, how else can a society encourage women not to have abortions and yet ensure that the human birth rate does not continue to increase?

36

POPULATION GENETICS, NATURAL SELECTION, AND SPECIATION

General Objectives

1. Understand how variation occurs in populations and how changes in allele frequencies can be measured.
2. Know how mutations, gene flow, and population size can influence the rate and direction of population change.
3. Describe four kinds of selection mechanisms that help shape populations.
4. Describe three modes of speciation.

POPULATION GENETICS
    Sources of Variation
    The Hardy-Weinberg Baseline for Measuring Change
    Factors Bringing About Change
MUTATION
FACTORS RELATED TO POPULATION SIZE
    Genetic Drift
    Founder Effect
    Bottlenecks
GENE FLOW

Summary

When Carl von Linné studied the diversity of organisms, he believed that *species* didn't change. In general, a species encompasses all of those actually or potentially interbreeding populations that are reproductively isolated from other such groups. Linné's approach was *typological*: An individual was selected as being the perfect type for the species, on the basis of a somewhat arbitrary choice of what "perfect" physical features were. Now the *population concept*, which holds that variation in populations not only encourages evolutionary change but also is the product of evolutionary change, is followed. In a sexually reproducing population, all the genotypes taken together compose a *pool of alleles*. For any one locus on the chromosome, some alleles occur more often than others; thus, variation can be thought of in terms of *allele frequencies*.

When allele frequencies for a given gene locus remain constant through succeeding generations, the population is said to be at *Hardy-Weinberg equilibrium*, which is maintained only when all of a set of specific conditions occur. Rarely are all these conditions met, but many populations are close enough to the ideal population that equilibrium equations are applicable. If all individuals in a population have an equal chance of surviving and reproducing, then the frequency of each allele in the population should remain constant from generation to generation (that is, be in *genetic equilibrium*). In nature, allele frequencies may change all the time. Some disturbances that cause changes are (1) mutation, (2) genetic drift, (3) gene flow, and (4) the selection pressures that bring about differential survival and reproduction of genotypes within a population.

Key Terms

Carl von Linné	$2pq$	Hardy-Weinberg
typological	alleles	equilibrium
genotype	population	allele frequencies
phenotype	$p + q = 1$	genetic drift
Hardy-Weinberg principle	p	gene flow
genetic equilibrium	q	natural selection
mutation	genotypic frequencies	founder effect
p^2	q^2	bottlenecks

Objectives

1. Distinguish the typological approach to diversity from the currently used population concept.
2. Distinguish the terms *population* and *species*.
3. Explain how frequency distributions of traits can be useful in studies of evolution.
4. Show how the Hardy-Weinberg equations allow geneticists to measure changes in allele frequencies in population. In Figure 36.2, determine what would happen to *AA*, *Aa*, and *aa* if the value of p changed to 0.8 and the value of q changed to 0.2.

5. Give an example illustrating that genotype frequencies can change even though allele frequencies have remained the same. Then relate the stability of allele ratios to genetic equilibrium.
6. List four conditions that must be met before Hardy-Weinberg equilibrium can occur. Then explain why, if these conditions are seldom met in natural populations, geneticists use the Hardy-Weinberg equations.
7. Define *mutation* and explain how mutations cause allele frequencies to change in natural populations.
8. Define *genetic drift* and explain how mutations cause allele frequencies to change in natural populations.
9. List two of the causes of gene flow and explain how gene flow can cause allele frequencies to change in natural populations.

Self-Quiz Questions

Fill-in-the-Blanks

In a(n) (1) _____ approach to studying diversity in populations, an individual is selected as the perfect representative type for the species, on the basis of a somewhat arbitrary choice of what "perfect" representative features are. A(n) (2) _____ is a group of individuals of the same species that occupy a given area at a specific time. The (3) _____ _____ _____ allows researchers to establish a theoretical reference point (baseline) against which changes in allele frequency can be measured. Variation can be expressed in terms of (4) _____ _____: the relative abundance of different alleles carried by the individuals in that population. The stability of allele ratios that would occur if all individuals had equal probability of surviving and reproducing is called (5) _____ _____. Over time, allele frequencies tend to change through infrequent, but inevitable (6) _____, which are the original source of genetic variation. Random fluctuation in allele frequencies over time due to chance occurrence alone is called (7) _____ _____; it is more pronounced in small populations than in large ones. (8) _____ _____ associated with immigration and/ or emigration also changes allele frequencies.

Problems

(9) In a population, 81 percent of the organisms are homozygous dominant, and 1 percent are homozygous recessive. Find (a) the percentage of heterozygotes, (b) the frequency of the dominant allele, and (c) the frequency of the recessive allele.

(10) In a population of 200 individuals, determine how many individuals are (a) homozygous dominant, (b) homozygous recessive, and (c) heterozygous for a particular locus if $p = 0.8$.

NATURAL SELECTION
   Modes of Natural Selection
   Stabilizing Selection
   Directional Selection
   Disruptive Selection
   Sexual Selection
   Selection and Balanced Polymorphism

Summary

Natural selection is no longer in the realm of pure theory; examples of natural selection have been demonstrated in different populations. The transient polymorphism observed among peppered moths appears to be a case of *directional selection*: because of a specific change in the environment, a heritable trait occurs with increasing frequency, and the whole population tends to shift in a parallel direction. The development of pesticide-resistant pests is another example of directional selection.

Disruptive selection, in which two or more distinct polymorphic varieties are favored and become increasingly represented in a population, tends to split up a population. Batesian mimicry among females of the African swallowtail butterfly and their foul-tasting models provides an example of disruptive selection that is probably caused by predation.

A process in which a form already well adapted to a given environment is selected for and maintained, even as extreme variants are selected against, is called *stabilizing selection*. For more than 400 million years, the form of horseshoe crabs has remained essentially the same. Horseshoe crabs probably represent a balanced system of the best working combination of traits in an environment that has remained relatively stable. Presumably, extremely variant individuals were continually eliminated, so little structural change has occurred through evolutionary time.

Selection pressures generally favor one allele such that other alleles for the same locus are suppressed. However, two or more alleles for the same locus may persist simultaneously in a population at a frequency greater than can be accounted for by newly arising mutations alone; such a situation is called *polymorphism*.

The survival value of variant alleles must be weighed in the context of the environment in which they are being expressed; they are not advantageous or disadvantageous in themselves. The HbA (normal hemoglobin) and HbS (sickle-cell hemoglobin) form a *balanced polymorphism* that has been maintained in populations throughout West and Central Africa for as long as malaria has been prevalent.

Key Terms

adaptive
differential reproduction
stabilizing selection
directional selection
disruptive selection
polymorphism

mark-release-recapture
mimicry
mimic
model
differential fertility

balanced polymorphism
HbS allele
Plasmodium falciparum
differential mortality
sexual dimorphism

Objectives

1. Define *directional selection* and give two examples.
2. Define *disruptive selection* and give an example.
3. Explain how horseshoe crabs have managed to remain essentially the same for 400 million years.

4. Define *balanced polymorphism*. Explain why inheriting a gene for HbS (sickle-cell hemoglobin) might be advantageous for some humans living under certain situations.

5. Contrast transient (i.e., one that does not persist) polymorphism and balanced polymorphism.

Self-Quiz
Questions

Fill-in-the-Blanks

(1) _____ _____ occurs when a specific change in the environment causes a heritable trait to occur with increasing frequency and the whole population tends to shift in a parallel direction. (2) _____ _____ favors the development of two or more distinct polymorphic varieties such that they become increasingly represented in a population and the population is split up into different phenotypic variations. (3) _____ _____ provide an excellent example of stabilizing selection because they have existed essentially unchanged for 400 million years. Because deaths of HbA homozygotes due to malaria were balanced by deaths of HbS homozygotes brought about by sickle-cell anemia, (4) _____ _____ at the sickle-cell locus was maintained in populations in regions where malaria was prevalent. Populations in which two or more forms of a trait persist are said to show (5) _____ at that gene locus. When individuals of different phenotypes in a population differ in their ability to survive and reproduce, their alleles are subject to (6) _____ _____.

36-III
(pp. 550–554)

EVOLUTION OF SPECIES
    Divergence
    When Does Speciation Occur?
    Reproductive Isolating Mechanisms
    Modes of Speciation
SUMMARY

Summary

Evolution may be thought of as successive changes in allele frequencies brought about by such occurrences as mutation, genetic drift, gene flow, and selection pressure. Usually studies of evolution focus on some local breeding unit: a population of localized extent within a larger population system. The usual barriers to allele exchange between local breeding units of a large population are those that create *geographic isolation*: severe storms, major floods, earthquakes, the uplift of mountain ranges, and the separation of continents. Following geographic isolation, mutation, genetic drift, and selection pressures may operate on the different local breeding units. Over time, these forces may lead to *divergence*: a buildup of differences in allele frequencies between isolated populations of the same species.

Speciation has occurred when some restriction on interbreeding between populations is followed by enough genetic differentiation that, even if individuals from those populations do not make contact, they cannot or will not interbreed. *Isolating mechanisms* are aspects of structure or function that prevent interbreeding between populations that are undergoing or have undergone speciation. Differences in reproductive structure or physiology, in timing of reproduction, in behavior, and in ecological factors may serve as isolating mechanisms. Local populations that are geographically separated are called allopatric. *Allopatric speciation*, in which geographic separation and gradual divergence lead to the reproductive isolation of two populations, is the most common pattern in nature; it is generally caused by a disturbance in the environment. When distinct local breeding units coexist in the same geographic range, they are called sympatric. In *sympatric speciation*, two or more populations occupying the same distribution range are thought to undergo reproductive isolation before genetic differentiation transforms them into separate species. Especially among plants, *polyploidy* and/ or hybridization are two other speciation routes; about 40 percent of all flowering plant species are polyploid. Wheat is a successful polyploid, and Kentucky bluegrass hybridizes with a number of related species.

Key Terms

speciation	mechanical isolation	parapatric speciation
species	behavioral isolation	sympatric speciation
divergence	hybrid inviability	polyploidy
reproductive isolating mechanism	allopatric speciation	hybridization

Objectives

1. Describe the usual way that divergence occurs in animal populations and species evolve.
2. Define *local breeding unit* and explain its place in studies of evolution.
3. List three potential causes of geographic isolation and indicate how each might promote divergence.
4. Contrast partial reproductive isolation with complete reproductive isolation.
5. List four examples of isolating mechanisms.
6. Contrast allopatric and sympatric speciation.
7. Define *polyploidy* and indicate how human technology has utilized polyploidy.

Self-Quiz Questions

Fill-in-the-Blanks

The usual barriers to allele exchange between local breeding units of a larger population are those that create (1) _____ _____, which occurs during severe storms, earthquakes, floods, geologic uplift and subsidence, and the long-term separation of continents. Subsequently, mutation, (2) _____ _____, and selection pressures may operate on the different local breeding units. Over time, these forces may lead to (3) _____: a buildup of differences in allele frequencies (or genetic differentiation) between isolated populations of the same species. (4) _____ is said to have occurred when some restriction on interbreeding between populations is followed by enough genetic differentiation that, even if individuals from those populations do make contact, they cannot or will not interbreed. When related plant species come into flower at nonoverlapping times, their (5) _____ isolation is complete. (6) _____ speciation, the most common pattern in nature, occurs when geographic separation of two populations, accompanied by gradual divergent evolution between them, leads to reproductive isolation. Especially among plants, (7) _____ and/or hybridization are two other speciation routes; about 40 percent of all flowering plants are (8) _____.

CHAPTER TEST

UNDERSTANDING AND INTERPRETING KEY CONCEPTS

For Questions 1–3, choose from these answers:
  (a) p
  (b) q
  (c) p^2
  (d) $2pq$
  (e) q^2

___ (1) _____ represents the genotype frequency of the homozygous dominant individuals in a population.

___ (2) _____ represents the allele frequency of the dominant trait.

___ (3) _____ represents the frequency of the heterozygotes in a population.

For Questions 4–6, choose from these answers:
    (a) mimicry
    (b) disruptive selection
    (c) genetic drift
    (d) directional selection
    (e) stabilizing selection

___ (4) Cockroaches becoming increasingly resistant to pesticides is an example of _____.

___ (5) The founder effect is a special case of _____.

___ (6) Different alleles that code for different forms of hemoglobin have led to _____ in the midst of the malarial belt that extends through West and Central Africa.

For Questions 7–10, choose from these answers:
    (a) sexual dimorphism
    (b) isolating mechanisms
    (c) polyploidy
    (d) allopatric speciation
    (e) genetic equilibrium

___ (7) Cardinals, mallard ducks, gorillas, and walruses are all classic examples of _____.

___ (8) Hardy and Weinberg invented an ideal population that was in _____ and used it as a baseline against which to measure evolution in real populations.

___ (9) Plants that blossom at different times and chromosome sets that no longer can match up effectively during fertilization are examples of _____.

___ (10) Many domestic varieties of fruits, vegetables, and grains have evolved larger vegetative and reproductive structures through _____.

CHAPTER TEST

INTEGRATING AND APPLYING KEY CONCEPTS

Can you imagine any way in which directional selection may have occurred or may be occurring in humans? Which factor(s) do you suppose are the driving force(s) that sustain(s) the trend? Do you think the trend could be reversed? If so, by what factor(s)?

37

PHYLOGENY AND MACROEVOLUTION

General Objectives

1. Understand the various classification schemes and realize the difficulty in determining from fossil evidence where the limits of one species end and the limits of another begin.
2. Be able to cite what biologists generally accept as evidence that supports their belief in evolution. Explain how observations from comparative morphology and comparative biochemistry are used to reconstruct the past.
3. Know the time boundaries of the five geologic eras and identify the principal organisms associated with each era.
4. Understand the factors that encourage increased rates of speciation and the formation of larger taxonomic groups. Know also the factors that bring about extinction and replacement of species.

37-I
(pp. 555–563)

Summary

Change may occur within an established lineage (as, for example, in large, long-legged birds). Gradual changes in traits such as horn size occur as early genera evolve into advanced ones. According to the *gradualistic model*, this evolutionary pattern is the main trend in the history of life. Not all transitions in the fossil record are gradual; some major groups of organisms appeared relatively abruptly and were already highly developed and diverse when they did appear. Several explanations have been proposed to account for such abrupt appearances. The rapid crossing of adaptive thresholds, the *punctuational model* states, occurs in times of intense pressure in small, isolated populations. This model also holds that rapid speciation has been the principal mode and that most morphological change has occurred then. Sometimes important gaps occur in the fossil record.

Not all organisms become fossilized when they die. The quality and completeness of the fossil record varies as a function of the types of organisms (for example, shelled versus soft-bodied), the places where they died (eroding soils versus natural traps favoring fossilization), and the geologic stability of the region (tectonically active zones versus undisturbed sedimentary plains). Abrupt climatic or geographic changes encourage some populations to expand and diversify rapidly.

The geologic time scale initially was defined as a progression of five broad eras: *Archean*, in which the earliest bacteria-like forms originated, *Proterozoic* ("first life"), *Paleozoic* ("ancient"), *Mesozoic* ("middle life"), and *Cenozoic* ("modern life"). Within the past three decades, fairly firm boundaries have been assigned to these geologic intervals by using radioactive dating methods on an enormous number of rock samples taken from all over Earth. These methods are based on comparing the amounts of radioactive isotopes (such as of uranium, thorium, potassium, and strontium) with the amounts of their stable decay products in different kinds of rocks. Each radioactive element has its own characteristic decay rate, and this rate cannot be modified by physical or chemical changes in the environment.

As we move from the most ancient fossilized cells that resemble bacteria, the fossil organisms preserved within younger and younger rocks become progressively more abundant, more diverse, and more like modern forms. As environments change, selection pressures change—and the character of the population changes. If one population somehow expands into two different settings, different traits will be selected for in each setting. With time, the two populations will accumulate more and more structural and behavioral differences, until members of the two populations can no longer interbreed when brought back together; at this point, two different species have evolved from one ancestral species.

Key Terms

species	gradualism	molecular clock
category of relationship	punctuation	DNA hybridization studies
higher taxa	fossilization	
macroevolution	comparative morphology	Archean
phenotype	morphological	Proterozoic
phylogeny	divergence	Paleozoic
evolutionary taxonomy	convergence	Mesozoic
lineage	comparative biochemistry	Cenozoic
cladistics	neutral mutation	half-life

Objectives

1. Contrast the gradualistic model with the punctuational model.
2. Explain why there are gaps in the fossil record.
3. Explain why transitions in the fossil record do not always occur gradually.
4. Define *selection pressure*, give two examples of selection pressures, and indicate how selection pressures can promote differential survival and reproduction of genotypes within a population.
5. Identify the five major eras of the geologic time scale, their time spans, and principal organisms that lived during each era.

6. Explain how radioactive dating is used to estimate the age of rocks. Use terms such as *isotope*, *decay product*, and *half-life* in your explanation.
7. Explain how speciation can occur.

Self-Quiz Questions

Fill-in-the-Blanks

Evolution can be viewed either by the (1) _____ model, which envisions speciation as accounting for only a small amount of large-scale change, or by the (2) _____ model, which sees higher taxa originating from the rapid crossing of adaptive thresholds. The (3) _____ era extended from 2.5 billion years ago to about 570 million years ago; initially heterotrophic and autotrophic (4) _____ existed, but between 900 and 750 million years ago (5) _____ had appeared. The (6) _____ era was characterized by "ancient life," with invertebrates, fishes, and amphibians coexisting principally with aquatic plants. As environments change, (7) _____ pressures also change, and so populations change in character. If subgroups of one original population that have been living separated for many generations can no longer interbreed when brought back together, (8) _____ is said to have occurred. The classification scheme known as (9) _____ depicts relatedness only in terms of branch points in the lines of descent. Molecular methods (including immunological comparisons and (10) _____ _____ studies) can be used to construct a branching sequence for a lineage.

37-II
(pp. 563–569)

MACROEVOLUTIONARY PATTERNS
 Evolutionary Trends
 Adaptive Radiation
 Origin of Higher Taxa
 Extinction and Replacement
SUMMARY

Summary

Differences in the relative rates of morphological change, speciation, and extinction can lead to highly divergent patterns in the development of lineages, and evolutionary trends become evident at the level of higher taxa (classes, orders, families).

When populations of the same species become adapted for exploiting different resources in the environment in specialized ways, they are said to have undergone *adaptive radiation*; the dispersal of Darwin's finches is an example of adaptive radiation.

Extinction, the disappearance of groups of organisms, is one of the major patterns of evolution. Some factors such as the absence of genetic diversity or having a narrow distribution range or small population size may place a group on the brink of extinction. In recent times, human-induced disturbances are pushing an alarming number of species over the edge at an accelerated rate that far exceeds anything that has occurred in the past.

Key Terms

microevolutionary
macroevolution
units of evolution
speciation
adaptive radiation
adaptive zone

Darwin's finches
key innovation
marsupials
monotremes
extinction

replacement
background extinction
mass extinction
Permian
Cretaceous

Objectives

1. Understand how classes, orders, and families come to thrive, become extinct, or undergo sustained trends in their general pattern of characteristics.
2. Describe some of the factors that have brought about extinction and replacement during the Ordovician, Devonian, Permian, Triassic, and Cretaceous periods.
3. Explain why marsupials and monotremes did not participate in the adaptive radiation of mammals into new niches during the first 10–12 million years of the Cenozoic era. Consult Figure 37.11 in your main text.

Self-Quiz Questions

Fill-in-the-Blanks

(1) _____ is the disappearance of an entire taxon. Bursts of evolutionary activity among populations of related species that fill environmental niches with many new species are called (2) _____ _____. Throughout the Mesozoic era (when dinosaurs thrived), low diversity persisted but was followed by rapid appearance and evolution of many different (3) _____ mammals from 65–53 million years ago. Continental movement and substantial climatic change may have caused the late (4) _____ mass extinction about 220 million years ago. Collision with a(n) (5) _____ may have brought about dinosaur extinction during the late (6) _____ period.

CHAPTER TEST

UNDERSTANDING AND INTERPRETING KEY CONCEPTS

For Questions 1–4, choose from these answers:
  (a) Archean
  (b) Cenozoic
  (c) Mesozoic
  (d) Paleozoic
  (e) Proterozoic

___ (1) Mammals, birds and flowering plants evolved mostly during the
_____ era.

___ (2) The _____ era ended with the great Permian extinction about 240
million years ago.

___ (3) The _____ era ended with the massive Cretaceous extinction that
wiped out the dinosaurs.

___ (4) The _____ era included the first forms of life evolving into more
complex multicelled types.

For Questions 5–7, choose from these answers:
   (a) gradualistic model
   (b) punctuational model
   (c) convergence
   (d) divergence
   (e) adaptive radiation

___ (5) Penguins and porpoises serve as examples of _____.

___ (6) Kangaroos, koala bears, and opossums serve as examples of
_____.

___ (7) Darwin's finches on the Galápagos Islands serve as an example of
_____.

For Questions 8–10, choose from these answers:
   (a) Cladistics
   (b) Punctuation
   (c) Gradualism
   (d) Paleontology
   (e) Extinction

___ (8) _____ is a scheme of classification that depicts relatedness in
terms of branch points in the lines of descent.

___ (9) _____ is a model that says that most morphological change
occurs rapidly during speciation.

___ (10) _____ is the study of fossils.

CHAPTER TEST

INTEGRATING AND APPLYING KEY CONCEPTS

Imagine that in the next decade three more Chernobyl disasters happen, the
oceans acquire critical levels of carcinogenic pesticides that work their way up
the food chains, and the ozone layer shrinks dramatically in the upper atmosphere. Describe the macroevolutionary events that you believe might happen.

38

ORIGINS AND THE
EVOLUTION OF LIFE

**General
Objectives**

1. Describe how life might have spontaneously arisen on Earth approximately 3.5 billion years ago.
2. Understand how ancient prokaryotes are thought to have changed the primeval atmosphere of Earth and diverged into three primordial lineages.
3. Outline the theory that accounts for the origin and rise of eukaryotes.
4. Describe the shifty shenanigans of tectonic plates in the Paleozoic, Mesozoic, and Cenozoic eras.

38-I
(pp. 571–574)

ORIGIN OF LIFE
  The Early Earth and Its Atmosphere
  Spontaneous Assembly of Organic Compounds
  Speculations on the First Self-Replicating Systems

Summary

A cohesive theory linking the flow of earth's history and the history of life is being developed. Billions of years in the past, our solar system was born from an exploding star. Our planet took form through the gravitational compression of dust and debris that swirled around the primordial sun. Five billion years ago, Earth was a cold, homogeneous mass, but through contraction and radioactive heating it developed a core that was dense and hot. By 3.7 billion years ago, a crust that periodically erupted volcanic outpourings from the inferno below had formed, and a dense atmosphere of hydrogen, nitrogen, methane, ammonia, hydrogen sulfide, and water vapor was being retained by Earth's gravitational field as our planet settled into an orbit around the sun. Temperatures on the planet favored the development of life, and somewhere between 3.7 and 3.5 billion years ago, living organisms appeared in the mineral-rich primeval seas. A variety of experiments by Miller, Fox, and others have demonstrated that all the building blocks required for life can form under abiotic conditions, given the primitive Earth atmosphere and an abundant supply of energy. The mechanism by which independently formed organic molecules combined into a permanent reproducible living system is not yet known.

Key Terms

Stanley Miller   stellar explosions   microspheres
templates   clay crystals   liposomes
Cairns-Smith   left-handed forms

Objectives

1. List in sequence the events and processes that cooperated to make Earth a planet suitable for supporting life. Indicate how long ago these processes were occurring.
2. List the principal elements needed to support life and tell the form they took in the atmosphere and/or in the oceans.
3. Describe the basic ideas of the theory of the possible chemical evolution of the first primitive organism. Present evidence that supports or undermines the theory.
4. State any alternatives to the proposed theory of chemical evolution and the evidence that supports or undermines those alternatives.

Self-Quiz Questions

Fill-in-the-Blanks

About (1) [choose one] ☐ 10 ☐ 4.6 ☐ 3.8 ☐ 3.2 billion years ago, our planet took form as a cold homogeneous mass. By (2) [choose one] ☐ 4.6 ☐ 3.8 ☐ 3.2 ☐ 2 billion years ago, the Earth had a hot, dense core and was hurtling through space as a thin-crusted inferno. Rain fell, minerals were stripped from rocks, and the oceans were formed. According to Cairns-Smith, (3) _____ _____ probably were the absorbing agents that served to assemble (4) _____ _____ into proteins. (5) _____ are differentially permeable assemblages of polypeptides that accumulated greater concentrations of certain substances inside than were found on the outside. (6) _____ are microscopic spherical and tubular structures that are assembled from simple lipids that have single-chain tails and hydrocarbon heads.

THE AGE OF PROKARYOTES
    Prokaryotic Metabolism and a Changing Atmosphere
    Divergence Into Three Primordial Lineages
THE RISE OF EUKARYOTES
    Origin of Mitochondria
    Origin of Chloroplasts
    Beginnings of Multicellularity

Summary

Until about 3.7 billion years ago, Earth's crust may have been too unstable or too thin—and perhaps the heat flow from the interior was too great—for permanently stable land masses to form. Nevertheless, there were probably many volcanic islands rising above the primeval seas. It may have been in the shallow, near-shore waters of these islands that life appeared. Reeflike formations, perhaps formed by calcium-secreting algae, date from about 3.5 billion years ago. For about 2 billion years or so, organisms resembling modern bacteria and blue-green algae had the world much to themselves. Little or no oxygen was in the atmosphere, so anaerobic pathways must have been the style of the day. Between 2–1 billion years ago, both heterotrophic and autotrophic prokaryotes grew steadily more abundant, if not more diverse.

Between 1.4–750 million years ago, the first eukaryotes arose. Predatory forms were among them. From that time onward, fossils of both predators and prey are increasingly larger. With the rise of predatory eukaryotes, the cyanobacteria of the seas began to decline. By 570 million years ago, astonishingly diverse eukaryotes dominated the scene. The combined activity of billions of photosynthetic microorganisms generated significant levels of oxygen—oxygen that could be used to advantage by predatory heterotrophic mobile eukaryotes as they chased after their prey, and oxygen that, through aerobic pathways, could be used to store energy in ATP to build mineralized protective hard parts in both predator and prey organisms alike.

Key Terms

stromatolite	archaebacteria	Margulis
Desulfovibrio	urkaryotes	symbiosis
eubacteria	mesosomes	

Objectives

1. Describe how prokaryotic metabolism changed the primitive atmosphere of Earth.
2. Explain how early prokaryotes diverged into three primordial lineages: the true bacteria, the archaebacteria, and the now-extinct urkaryotes.
3. Describe how mitochondria and chloroplasts originated according to the symbiotic-origin-of-eukaryotes theory.

Self-Quiz Questions

Fill-in-the-Blanks

For about (1) [choose one] ☐ .5 ☐ 1 ☐ 2 ☐ 5 billion years, it seems, organisms resembling modern bacteria and blue-green algae had the world much to themselves. Little or no free (2) _____ was in the atmosphere, so

(3) _____ pathways must have been the usual metabolic pattern. Between (4) [choose one] ☐ 3.2 ☐ 1.4 ☐ 1 billion and 750 million years ago, the first eukaryotes appeared. The primitive eukaryotes were mobile, and some chased after their prey; (5) _____ respiration would clearly have been advantageous in providing energy sources for their movements. Mats of cells that resemble modern cyanobacteria and that have been found in rock formations more than 2 billion years ago are called (6) _____. *Desulfovibrio* can carry out (7) _____ _____ _____ as well as fermentation. The (8) _____ theory of eukaryote origins suggests that anaerobic amoeboid cells ingested aerobic bacteria that weren't later digested.

38-III
(pp. 578–582)

FURTHER EVOLUTION ON A SHIFTING GEOLOGIC STAGE
Life During the Paleozoic

Summary

The *Paleozoic* was characterized by several massive inundations of continents and retreats by shallow seas. By the late Cambrian, there apparently were three major land masses mostly covered by shallow seas. Two would eventually become North America and Europe, and the third (Gondwana) would be fragmented into Africa, South America, Australia, Antarctica, and parts of Asia and the eastern North American coast. All but one *invertebrate* phylum had arisen by the close of the Cambrian, and the shallow seas of the Ordovician encouraged evolutionary experimentation in the invertebrate groups. Toward the end of the Ordovician, volcanic outpourings in the mobile zone along the eastern edge of the North American land mass created the immense ancestral Appalachian Mountains as the European plate moved in. During late Silurian times, some life forms developed traits that would permit them to live on land: tough or waxy surface layers to keep from drying out, the potential for exploiting new water supplies, systems for taking in oxygen or carbon dioxide from the surrounding air, varying frameworks needed to support the body, new methods to maintain the balance of salts (ions) in body fluids, and new modes of reproduction and offspring dispersal.

As the Silurian gave way to the Devonian some 400 million years ago, the continents collided, producing a dramatic increase in dry land area; plants and then animals began their tentative adaptive forays into this vacant environment. Certain fish, left behind in drying tide pools or in evaporating freshwater ponds, were able to use their swim bladders as lunglike organs and their fins as simple limbs; with such adaptations, they could crawl from pond to pond. Thus began an evolutionary trend that gave rise to the amphibians: organisms that spend part of their lives on land but return to water to lay eggs, for their eggs have no protective shell to keep from drying out. The Carboniferous, with land masses being submerged and drained no less than fifty times, was the heyday of the amphibians. Adaptations appeared that allowed gymnosperms and reptiles to move onto higher and drier land: Gymnosperms developed seeds, and reptiles developed shelled eggs and internal fertilization; now both groups could complete their reproductive cycles without free-standing water.

As the Carboniferous gave way to the Permian, collisions of crustal plates brought all land masses together into one vast supercontinent, called Pangaea.

Lowlands and humid uplands emerged in the north, glaciation occurred in the south, and the shallow seas drained from the massive continent. Everywhere in the shrinking seas, huge numbers of marine forms became extinct.

Key Terms

plate tectonic theory
Gondwana
Laurasia
Pangaea
Tethys Sea
Cambrian
trilobites
brachiopods
Laurentia

Burgess shale
plates
mantle
oceanic ridges
Ordovician
cephalopod
Silurian
armor-plated fishes
vertebrate jaw

Devonian
rhizome
lobe-finned fishes
Carboniferous
reptiles
Permian
synapsids
therapsids

Objectives

1. Describe the principal events that occurred in each of these periods of the Paleozoic era: Cambrian, Ordovician, Silurian, Devonian, Carboniferous, Permian. State whether aquatic or terrestrial conditions prevailed and which organisms were most abundant and conspicuous during each period.
2. List the principal ideas of the plate tectonic theory and state which earthly phenomena are explained by the theory.
3. Name the changes that permitted plants and animals to shift from an aquatic environment to moist land and from moist land to dry land.
4. Name the animals and plants that were the pioneers in the transition to land.

Self-Quiz Questions

Fill-in-the-Blanks

Earth's crust has been divided into vast (1) _____, which have been moving about uneasily on top of a plastic (2) _____ and carrying the continents with them. (3) _____ was a supercontinent that became fragmented into Africa, South America, Australia, Antarctica, and parts of Asia and the eastern North American coast. By the late Cambrian, (4) _____, mud-burrowing, mud-crawling scavengers, were abundant in the shallow marine environment. All but one (5) _____ phylum had appeared by the close of the Cambrian. (6) _____ fishes appeared by the late Ordovician, and the Appalachian Mountains were being built. The Devonian period is known for giving rise to the (7) _____: organisms adapted to life both in water and on land. The (8) _____ period is characterized by flat, low coastlines where immense swamp forests became established. Collisions of crustal plates during the Permian brought all land masses together into one vast supercontinent, (9) _____; the shallow seas were drained and massive extinctions occurred.

FURTHER EVOLUTION ON A SHIFTING GEOLOGIC STAGE (cont.)
    Life During the Mesozoic
    The Cenozoic: The Past 65 Million Years
PERSPECTIVE
SUMMARY

Summary

During the ensuing *Mesozoic* era, gymnosperms and reptiles (especially the dino-saurs) had the competitive edge as Pangaea began to break up and the three major land masses began moving their separate ways, creating the basin of the Atlantic Ocean in the process. Mammals arose in the Mesozoic, but they did not become well established until the predatory dinosaurs disappeared suddenly from the Earth at the end of the Cretaceous period. The reasons for this sudden demise are not known; there were no apparent profound shifts in climate and no sudden, massive upheavals in the crust. By *Cenozoic* times major reorgani-zation was occurring among all the crustal plates that remained from the breakup of the supercontinent. North America, Europe, and Africa were moving their separate ways; brittle fragmentation of coastlines, severe vulcanism, and massive uplifting of mountain ranges promoted a variety of cooler, drier environments. Sea levels changed, as did Earth's overall temperature. Grasslands were formed; plant-eating animals and their predators evolved and occupied them. With the land essentially cleared of major reptilian predators, the *mammals* and *birds* even-tually diversified, along with the *flowering plants*. The tropical forests were frag-mented into a patchwork of new environments, and many forest inhabitants were forced into new life styles in mountain highlands, deserts, and plains. One such evicted form gave rise to the human species.

Adaptive success is ensured only as long as there is responsiveness to the environment and only as long as there is dynamic stability between the require-ments and the demands of organisms making their homes together. If the record of Earth history tells us anything at all, it is that life in one form or another has survived disruptions of the most cataclysmic sort.

Key Terms

Mesozoic Seas	*Archaeopteryx*	Cretaceous-Tertiary
thecodonts	*Protoavis*	boundary
archosaurs		Cenozoic era

Objectives

1. Describe the tectonic events that were shaping land environments during Mesozoic and Cenozoic times, and state the types of organisms that evolved to exploit the different types of environments.
2. Explain how a knowledge of Earth's geologic history helps explain why there is such a great diversity of organisms on Earth today.
3. State whether you think humans are successfully adapted to their environ-ments, and provide some evidence to support your statement.

**Self-Quiz
Questions**

Fill-in-the-Blanks

At the dawn of the (1) _____ era, the supercontinent Pangaea was established and the greatly diverse multitudes of rather stable marine communities were reduced to relatively few species. On land during this era, the (2) _____ were becoming the dominant plants and the therapsid (3) _____ dwindled to near extinction. Early in the Triassic period, there was a modest radiation of the thecodont reptiles, which included the ancestors of the (4) _____ (the ruling reptiles). Some small dinosaurs apparently were (5) _____ with an insulating coat of feathers. (6) _____ may be among the earliest transitional forms between dinosaurs and birds, which began their diversification about 100 million years ago. (7) _____ and (8) _____ had the competitive edge during Jurassic times; the climate was warm and humid with plains, mountains, and vast lagoons. Their ultimate challengers were to be the little ratlike (9) _____ scurrying through the shrubbery of the Jurassic and Cretaceous periods, whose populations exploded into the biotic vacuum established by the abrupt extinction of the dinosaurs at the end of the Cretaceous. In much of the world, the concurrent shifts in climate led to the emergence of extensive, semiarid, cooler (10) _____, into which herbivores and their predators radiated. During the Cenozoic era, (11) _____ and (12) _____ plants evolved and dominated most environments.

CHAPTER TEST

UNDERSTANDING AND INTERPRETING KEY CONCEPTS

For Questions 1–4, choose from these answers:
- (a) Archaean
- (b) Cenozoic
- (c) Mesozoic
- (d) Paleozoic
- (e) Proterozoic

____ (1) Dinosaurs and gymnosperms were the dominant forms of life during the _____ era.

____ (2) The flowering plants and insects coevolved throughout much of the _____ era.

____ (3) The composition of Earth's atmosphere changed during the _____ era from one that was anaerobic to one that was aerobic.

____ (4) Invertebrates, primitive plants, and primitive vertebrates were the principle groups of organisms on Earth during the _____ era.

For Questions 5–7, choose from these answers:
- (a) Gondwana
- (b) Laurasia
- (c) Pangaea

(d) Tethys Sea

(e) Burgess shale

___ (5) Remarkably well-preserved soft-bodied animals have been found in the fine silts that composed the _____.

___ (6) _____ was an enormous continent that began to break up during the Mesozoic era.

___ (7) _____ eventually formed most of South America, Africa, Antarctica, Australia, and parts of the Indian subcontinent.

___ (8) The primitive atmosphere of Earth did not contain _____.
(a) inert gases
(b) free oxygen
(c) water vapor
(d) free nitrogen
(e) carbon dioxide

___ (9) The most primitive forms of life were the _____.
(a) eubacteria
(b) urkaryotes
(c) archaebacteria
(d) eukaryotes
(e) protists

___ (10) According to Lynn Margulis, the organelles of the eukaryotes _____.
(a) are obligate parasites
(b) evolved separately from the nucleus
(c) are descendants of symbiotic organisms engulfed by a larger organism
(d) represent inpouchings of the plasma membrane
(e) are structures that broke off from the nucleus

CHAPTER TEST

INTEGRATING AND APPLYING KEY CONCEPTS

As Earth becomes increasingly loaded with carbon dioxide and various industrial waste products, how do you think living forms on Earth will evolve to cope with these changes?

39

VIRUSES, MONERANS, AND PROTISTANS

General Objectives

1. List five specific viruses that cause human illness and describe how the viruses do their dirty work.
2. Describe the principal moneran forms and ways of living.
3. Describe the four categories of protistans. Tell how protistans differ from monerans, viruses, and multicellular eukaryotes. Give some common names of protistans.

39-I
(pp. 590–593)

VIRUSES
General Characteristics of Viruses
Viral Infectious Cycles
Animal Viruses
Plant Viruses
Viroids and Prions

Summary

Viruses have a set of nucleic acids sheathed in a protective protein coat, but they are not capable of metabolism and cannot reproduce themselves unless they are inside a host cell. Viruses are an expression of parasitic simplification. They are the causative agents of human diseases such as smallpox, influenza, AIDS, polio, and possibly cancer. They damage livestock and crops, and they play an important role in keeping certain bacterial populations within bounds. There are presently no cures for AIDS and herpes infections in humans.

Viroids are infectious pieces of single-stranded RNA that lack protein coats and are smaller than the smallest virus. They are suspected to cause *slow-virus* diseases, which have incubation periods that may last for years. They afflict chrysanthemum and coconut crops and may be implicated in the development of cancer.

Prions are proteins that signal their host cells to replicate them; they cause slow but fatal diseases of the central nervous system in sheep and humans.

Key Terms

microbes	bacteriophage	rhinoviruses
virus	influenzaviruses	retroviruses
viral capsid	pandemic	viroids
tobacco mosaic virus,	herpesviruses	prions
TMV	Epstein-Barr virus	scrapie
"host"	mononucleosis	kuru
lytic pathway	poxviruses	Creutzfeldt-Jacob disease
lysogenic pathway		

Objectives

1. State the principal characteristics of viruses, indicate how they might have evolved, and list three diseases caused by viral agents.
2. Describe viroids and discuss how their existence might affect humans.

Self-Quiz Questions

Fill-in-the-Blanks

Individual viral particles are called (1) _____; each consists of a central (2) _____ _____ surrounded by a (3) _____ _____.

(4) _____ contain the blueprints for making more of themselves but cannot carry on metabolic activities. Influenzaviruses cause (5) _____ diseases that occur as worldwide epidemics. Chickenpox and shingles are two infections caused by DNA viruses from the (6) _____ category. Nucleic acids that lack a protein coat are called (7) _____.

39-II
(pp. 594–600)

MONERANS
Characteristics of Bacteria
Classification of Bacteria
Archaebacteria
Eubacteria
A Final Word on the "Simple" Bacteria

Summary

Some bacteria (chemosynthetic autotrophs) extract energy from inorganic molecules, some are photosynthetic, and others are autotrophs (decomposers and parasites). Two groups of bacteria have distinctly different cell walls and cell membranes, as well as different transfer RNAs and RNA polymerases. Archaebacteria such as *methanogens* and *halophiles* tolerate low-oxygen environments. Eubacterial cells all have *peptidoglycan* in their cell walls; in addition, *cyanobacteria* have the same kinds of chlorophyll molecules as do plants. Oxygen is lethal to obligate anaerobes, but *facultative anaerobes* use oxygen when it is present and switch to anaerobic pathways when it is not. Even though all bacteria have a rigid cell wall surrounding a plasma membrane, they differ somewhat in shape; *cocci* are spherical, *bacilli* are rodlike, and *spirilla* are spiral. Although each bacterium is a functionally independent unit, bacteria often remain linked together following division. Growth, reproduction, and dormancy depend on resource availability. Bacteria generally divide by asexual binary fission, although many bacteria form endospores (which resist drying out, irradiation, disinfectants, and acids).

Cholera, diphtheria, bubonic plague, and tuberculosis are diseases caused in humans by heterotrophic bacteria. Although many bacteria are agents of human diseases, the benefits derived from their recycling activities far outweigh their harmful effects.

Key Terms

monerans
photosynthetic autotrophs
chemosynthetic
  autotrophs
heterotrophs
binary fission
peptidoglycan
capsule
slime layer
pili
bacterial flagella
coccus, cocci
rod (= bacillus)
spirillum, spirilla
Gram-negative
Gram-positive
archaebacteria

methanogens
extreme halophiles
Halobacterium
bacteriorhodopsin
thermoacidophiles
Thermoplasma
eubacteria
prokaryotic
photosynthetic bacteria
cyanobacteria
Anabaena
heterocysts
green bacteria
purple bacteria
chemosynthetic bacteria
nitrifying bacteria
Azotobacter

Rhizobium
heterotrophic bacteria
rickettsias
spirochete
Lyme disease
endospores
Clostridium botulinum
obligate anaerobe
Bacillus thuringiensis
autoclaves
botulism
Escherichia coli
Lactobacillus
magnetotactic bacteria
myxobacteria
Myxococcus
fruiting bodies

Objectives

1. Distinguish *chemosynthesis* from *photosynthesis* and *obligate anaerobes* from *facultative anaerobes*.
2. Describe the diversity of "body plans" that bacteria of different groups have.
3. List three essential ways that archaebacteria differ from eubacteria and give three examples of each group.
4. Explain how endospore formation by bacteria can concern humans.
5. List three human diseases caused by bacteria.

Self-Quiz Questions

Fill-in-the-Blanks

(1) _____ autotrophs extract energy from inorganic molecules. Two types of bacterial heterotrophs are (2) _____, which break down dead organisms, and (3) _____, which obtain their nutrients from a host

organism. For (4) _____ _____, exposure to oxygen is lethal. Eubacterial cell walls are composed of (5) _____, a substance that never occurs in eukaryotes. (6) _____ include (7) _____ such as *Halobacterium* and thermoacidophiles such as *Thermoplasma*. (8) _____ have the same kind of chlorophyll pigments found in plants. Heterocysts are cells in *Anabaena* that carry out (9) _____ _____. When environmental conditions become adverse, many bacteria form (10) _____, which resist moisture loss, irradiation, disinfectants, and even acids. (11) _____ is a microbe that helps create acid mine waste. Unlike other photosynthetic bacteria, (12) _____ _____ have the same kind of chlorophyll pigments found in all true algae and plants; like bacteria, they reproduce by (13) _____.

39-III
(pp. 600–604)

PROTISTANS
    Slime Molds
    Euglenids
    Chrysophytes
    Dinoflagellates

Summary

Dinoflagellates are primitive photosynthetic eukaryotes that are motile members of the phytoplankton; many populations "bloom" suddenly in response to suddenly available nutrients. Golden algae have armor made of cellulose, pectin, and sometimes silicon dioxide. *Diatoms*, despite their glassy shells, are the major producers of marine communities. *Photosynthetic flagellates* are related to both plants and animals; although they can photosynthesize many of their own organic molecules, they need certain nutrients such as vitamins, which they absorb in a funguslike manner. The resemblances between photosynthetic and heterotrophic flagellates could signify divergence of many modern flagellates from a common ancestor. Photosynthetic protistans are probably responsible for more than half of the carbon dioxide fixation on Earth.

Key Terms

continuum of diversity	pellicle	diatoms
cellular slime molds	*Stemonitis*	xanthophylls
plasmodial slime molds	eyespot	beta-carotene
Dictyostelium discoideum	longitudinal fission	dinoflagellates
Protista	chrysophytes	*Gonyaulax*
euglenids	golden algae	red tides
Euglena		

Objectives

1. Explain how heterotrophic protistans could have acquired the capacity for photosynthesis and state the evidence to support your explanation.
2. Explain what causes red tides.
3. Tell how golden algae resemble diatoms, how the two groups differ from each other, and the ecological importance of each group.

Self-Quiz
Questions

Fill-in-the-Blanks

Euglenids reproduce by (1) _____ _____, a reproductive mode that is common among all flagellated protistans; the cell grows in circumference while all (2) _____ are being duplicated and then divides along its long axis. Some strains of *Euglena* can be converted from photosynthetic, chloroplast-containing forms to strains that are (3) _____. (4) _____ undergo explosive population growth and color the seas red or brown, causing a red tide.

39-IV
(pp. 604–608)

PROTISTANS (cont.)
Protozoans

Summary

There are four main types of animal-like protistans: the *heterotrophic flagellates* (discussed above), the *amoebas* and their relatives, the *sporozoans*, and the *ciliates*. All are unwalled single cells that live by pinocytosis and phagocytosis. Some species are nearly as small as bacteria; others can be seen with the naked eye. Amoebas move by pseudopodia and phagocytize their food by surrounding it and engulfing it; they are thought to have evolved from heterotrophic flagellates that capitalized on the ability to distort their cell surface, extending and withdrawing it. In amoebas, actin and myosin proteins are attached in an apparently random fashion to the cell membrane; when contraction occurs in one region, a pseudopodium is slowly extended. *Foraminiferans* are amoeboid relatives that secrete a hardened case of calcareous material that is peppered with tiny holes through which sticky pseudopodia extend and trap bits of phytoplankton that float by. *Sporozoans* are all parasitic, living within the bodies of animals; the host obtains the food. *Plasmodium* is the causative agent of the disease malaria; to complete its life cycle, its various developmental stages must spend time in human or bird hosts and mosquito hosts.

Ciliates are covered with thousands of cilia and have voracious appetites. They can move rapidly through their liquid environments, crawl, or simply stay put. Most ciliates have a cavity lined with cilia; these set up movements that sweep food particles into the cavity, where food vacuoles are formed. Digestion occurs in the food vacuoles, and nutrients are absorbed by the cytoplasm across the vacuole membrane. *Paramecium* and *Didinium* are examples of ciliates.

Key Terms

protozoans	amoebas	radiolarians
flagellate protozoans	*Amoeba proteus*	sporozoans
trypanosomes	*Entamoeba histolytica*	*Plasmodium*
Trypanosoma brucei	amoebic dysentery	ciliated protozoans
trichomonads	foraminiferans	*Paramecium*
Trichomonas vaginalis	heliozoans	*Didinium*
Giardia intestinalis		

Objectives

1. State the principal characteristics of the amoebas, radiolarians, and foraminiferans. Indicate how they generally move from one place to another and how they obtain food.
2. Characterize the sporozoan group, identify the group's most prominent representative, and describe the life cycle of that particular organism.
3. List the features common to most ciliates. Then describe the activities of two ciliates that have different methods of locomotion or of obtaining food.

Self-Quiz Questions

Fill-in-the-Blanks

Amoebas move by sending out (1) _____, which surround food and engulf it. (2) _____ secrete a hard exterior covering of calcareous material that is peppered with tiny holes through which sticky, food-trapping pseudopodia extend. (3) _____ is a famous (4) _____ that causes malaria. When a particular mosquito draws blood from an infected individual, (5) _____ of the parasite fuse to form zygotes, which eventually develop within the mosquito. *Paramecium* is a ciliate that lives in (6) _____ environments and depends on (7) _____ _____ for eliminating the excess water constantly flowing into the cell. *Paramecium* has a (8) _____, a cavity that opens to the external watery world. Once inside the cavity, food particles become enclosed in (9) _____ _____, where digestion takes place.

Matching

Put as many letters in each blank as are applicable.

(10) ___ *Entamoeba histolytica*

(11) ___ foraminiferans

(12) ___ *Gonyaulax*

(13) ___ *Paramecium*

(14) ___ *Plasmodium*

(15) ___ *Trichomonas vaginalis*

(16) ___ *Trypanosoma brucei*

(17) ___ *Volvox*

A. Ciliophora
B. Mastigophora
C. Rhizopoda
D. Sporozoa
E. Amoeboid protozoans
F. Flagellate protozoans
G. Multicellular photosynthetic flagellate
H. African sleeping sickness
I. Malaria
J. Traveler's diarrhea
K. Red tide
L. Primary component of many ocean sediments

ON THE ROAD TO MULTICELLULARITY
SUMMARY

Summary

A photosynthetic bacterium (*Prochloron*) that contains chlorophyll a, chlorophyll b, and carotenoids (the same pigments that are in all plants) may have given rise to the chloroplasts of green algae after symbiotic merger with a heterotrophic flagellate. Multicellular organisms are believed to have appeared when single-celled organisms—perhaps flagellates—divided and the newly formed daughter cells failed to separate. To the extent that staying together increased size, deterred smaller predators, improved mobility, or offered resistance to strong currents, the factors causing adherence were selected for and perpetuated. Colonial forms have persisted to the present. In a colony, each cell benefits from the loose association, but each acts independently and is incapable of changing its behavior according to what is happening to its neighbors. All cells feed, reproduce, and respond in the same way. When there is division of labor and interdependence, one type of cell cannot exist without the other. *Trichoplax*, a tiny blastulalike marine animal, is one of the simplest multicelled animals.

Key Terms

Plakobranchus	Chlamydomonas	multicellularity
Prochloron	*Volvox*	*Trichoplax adhaerens*

Objectives

1. Trace a sequence of events that may have transformed heterotrophic prokaryotes into photosynthetic eukaryotes.
2. Outline a possible route from unicellularity to a multicelled state with division of labor that could have been traveled by protistan ancestors.
3. Distinguish between colonial organisms and truly multicellular organisms.

Self-Quiz Questions

Fill-in-the-Blanks

(1) _____ is thought to resemble an ancestral prokaryote that may have given rise to the chloroplasts of green algae. A very simple flagellated green alga is (2) _____; cells that resemble it occur among the (3) _____— colonial organisms that straddle the fence between protistans and (4) _____. (5) _____ _____ is a tiny blastulalike marine animal that may resemble the simple multicelled animals that made their entrance during the Proterozoic. In a (6) _____, each cell benefits from a loose association with other cells, but each acts independently. In a (7) _____ organism, labor is divided up among the cells, and there is interdependence to the extent that one type of cell cannot exist without the other.

UNDERSTANDING AND INTERPRETING KEY CONCEPTS

___ (1) Which of the following diseases is *not* caused by a virus?
   (a) Smallpox
   (b) Polio
   (c) Influenza
   (d) Syphilis

___ (2) Bacteriophages are _____.
   (a) viruses that parasitize bacteria
   (b) bacteria that parasitize viruses
   (c) bacteria that phagocytize viruses
   (d) composed of a protein core surrounded by a nucleic acid coat

___ (3) Which of the following specialized structures is *not* correctly paired with a function?
   (a) Gullet–ingestion
   (b) Cilia–food gathering
   (c) Contractile vacuole–digestion
   (d) Anal pore–waste elimination

___ (4) _____ form a group of related organisms that suggests how lineages of single-celled organisms might have progressed through a colonial stage to multicellularity.
   (a) Ciliates such as *Paramecium*, *Didinium*, and *Vorticella*
   (b) Volvocales such as *Chlamydomonas* and *Volvox*
   (c) Golden algae and diatoms
   (d) Sporozoans such as *Plasmodium*, *Neisseria*, and the spirochaetes

___ (5) Population "blooms" of _____ cause "red tides" and extensive fish kills.
   (a) *Euglena*
   (b) specific dinoflagellates
   (c) diatoms
   (d) *Plasmodium*

___ (6) Exposure to free oxygen is lethal for all _____.
   (a) obligate anaerobes
   (b) bacterial heterotrophs
   (c) chemosynthetic autotrophs
   (d) facultative anaerobes

___ (7) For an organism to be considered truly multicellular, _____.
   (a) its cells must be heterotrophic
   (b) the organisms cannot be parasitic
   (c) there must be division of labor and cellular specialization
   (d) the organisms must at least be motile

__ (8) _____ all transform energy into usable forms and have complete genetic systems that maintain and reproduce themselves; hence, they are unquestionably alive.
(a) Bacteria, blue-green algae, and prions
(b) Bacteria and viroids
(c) Bacteria and rickettsias
(d) Bacteria and viruses

__ (9) When nutrients are scarce, many bacteria _____.
(a) engage in conjugation
(b) switch to photosynthesis
(c) form endospores
(d) become pathogenic

__ (10) Which of the following play an important role in the cycling of nitrogen-containing substances?
(a) Cyanobacteria
(b) Prions
(c) Viruses
(d) Photosynthetic flagellates

Matching

Match all applicable letters with the appropriate terms. A letter can be used more than once, and a blank can contain more than one letter.

(11) __ *Amoeba proteus*

(12) __ *Anabaena*

(13) __ *Clostridium botulinum*

(14) __ diatoms

(15) __ *Dictyostelium*

(16) __ foraminifera

(17) __ *Gonyaulax, Gymnodinium*

(18) __ *Herpesvirus*

(19) __ *Lactobacillus*

(20) __ *Myxococcus*

(21) __ *Paramecium*

(22) __ *Plasmodium*

(23) __ *Streptococcus*

(24) __ *Volvox*

A. Bacteria
B. Protista
C. Virus
D. Slime mold
E. Cyanobacteria
F. Photosynthetic flagellates
G. Dinoflagellates
H. Gram-positive eubacteria
I. Obtain food by using pseudopodia
J. Causes malaria
K. A sporozoan
L. A ciliate
M. Cause cold sores and a type of venereal disease
N. Live in "glass" houses
O. Live in hardened shells that have thousands of tiny holes

CHAPTER TEST

INTEGRATING AND APPLYING KEY CONCEPTS

Suppose that genetic engineers could successfully introduce chloroplasts into human zygotes and that this new combination established a symbiosis. Describe all of the changes that you can imagine would occur in the appearance and behavior of the resulting individuals.

40

FUNGI AND PLANTS

General Objectives

1. Describe the various types of fungal body plans, patterns of reproduction, and natural history.
2. Name at least one specific example of each of the five groups of true fungi.
3. Outline the evolutionary advances that converted marine algal ancestors into forms that could exist on wet land. Then state the advances that converted primitive homosporous marsh plants into dry-land flowering plants.

40-I
(pp. 611–618)

PART I. KINGDOM OF FUNGI
  The Fungal Way of Life
  Fungal Body Plans
  Overview of Reproductive Modes
  Major Groups of Fungi
  Chytrids
  Water Molds
  Zygospore-Forming Fungi
  Sac Fungi
  Club Fungi

Summary

Fungi are eukaryotes. Because their vegetative bodies are generally branched and filamentous, allowing them to come in contact with a large volume of their surroundings, they have access to raw materials that may be dilute or scarce. Like plants, many fungi have cells with large central vacuoles, and their body cells generally have walls reinforced with chitin. Also like plants, spore formation is an important part of their life cycle. However, fungi lack chloroplasts and differ from plants in other ways. Most fungi rely on enzyme secretions that promote digestion *outside* the fungal body, followed by nutrient absorption across the plasma membrane of individual cells. At least 50,000 different fungal species are recognized. Some (for example, those that cause most plant diseases and athlete's foot) are *parasitic*, but most fungal species are *saprophytic*, feeding on the remains of dead organisms or their by-products. By bringing about the decay of organic material, they help recycle such vital substances through communities of organisms.

The vegetative body of most *true fungi* is a *mycelium*—a mesh of branched, tubular filaments called *hyphae*. Most true fungi can reproduce asexually through spore formation, fission, budding, or fragmentation from a parent mycelium. Many also can reproduce sexually. *Egg fungi* (oomycetes), which have motile reproductive cells, include saprophytic water molds, parasitic downy mildews, and late blight. *Zygote fungi* (zygomycetes) are generally saprophytic and simply constructed; they include fly fungi and bread molds. *Sac fungi* (ascomycetes) bear spores in saclike structures called *asci*, which are often concentrated in a complex fruiting body called an *ascocarp*. Yeasts, morels, truffles, powdery mildews, and *Neurospora* are examples of sac fungi. *Club fungi* (basidiomycetes) are structurally the most complex of all fungi. They include edible mushrooms as well as extremely toxic ones, shelf fungi, puffballs, many rusts, and smuts. Their complex reproductive structures are called basidiocarps.

Key Terms

fungi	resting spore	morel
saprobes	zygospore	truffle
parasites	*Rhizopus stolonifer*	club fungi
mycelium, -lia	sac fungi	rusts
hypha, hyphae	ascus, asci	smuts
fungal spores	yeasts	*Agaricus bisporus*
dikaryotic stage	*Saccharomyces cerevisiae*	*Amanita muscaria*
chytrids	ascocarps	basidiospores
rhizoids	*Sarcoscypha*, the cup	basidiocarp
water molds	fungus	
Saprolegnia		

Objectives

1. Describe the general structure of fungi and its relation to their method of obtaining nutrients.
2. Distinguish between *parasitic* and *saprophytic* fungi. Mention one way in which parasitic fungi harm humans and one way in which saprophytic fungi benefit humans.
3. List the ways that fungi can reproduce.
4. List the five principal groups of fungi and give one example of an organism in each group.
5. State the fundamental contribution of fungi to ecosystems.
6. Give two examples of parasitic fungi that have played havoc with the production of crop plants.

Self-Quiz Questions

Fill-in-the-Blanks
Many fungi have cells merged lengthwise, forming tubes that have thin transparent walls reinforced with (1) _____; as in plants, (2) _____ formation is also an important part of their life cycle. Some fungal species, such as late blight, are (3) [choose one] ☐ parasitic ☐ saprophytic. The vegetative body of most true fungi is a (4) _____, which is a mesh of branched, tubular filaments called (5) _____. *Pilobilus* and *Rhizopus stolonifer* are examples of (6) _____ fungi. Sac fungi bear spores in saclike structures called (7) _____, in a complex spore-forming body called a(n) (8) _____. Morels, truffles, and yeasts are examples of (9) _____ fungi. Rusts, smuts, puffballs, and shelf fungi are examples of (10) _____ fungi.

40-II
(pp. 619–621)

PART I. KINGDOM OF FUNGI (cont.)
Imperfect Fungi
Mycorrhizae
Commentary: A Few Fungi We Would Rather Do Without
Lichens

Summary

"Imperfect fungi" (Deuteromycota) apparently lack a sexual stage. Many cause fungal infections (ringworm, athlete's foot) in humans. Some grow in damp grain and produce toxic waste products. *Phytophthora infestans* is a parasitic oomycete that causes late blight in potatoes and tomatoes. An example of its potentially devastating effect was seen in Ireland between 1845 and 1860, as the fungus got out of control in the uninterrupted fields of susceptible host plants; one-third of the population starved to death, died in the outbreak of typhoid fever that followed as a secondary effect, or fled the country. *Aspergillus* and *Penicillium* are particularly helpful to humans in producing lemon-flavored soft drinks, soy sauce, tofu, Camembert cheese, Roquefort cheese, and penicillin.

Many complex land plants live symbiotically with fungi (*mycorrhizae*) that help them absorb certain vital nutrients from the soil. *Lichens* are composite organisms comprising an alga and a fungus living interdependently: the fungus generally obtains photosynthetically derived food, and the alga enjoys improved water conservation, mechanical protection from being blown away, better gas exchange, and less overlap between individual algal cells (such as there would be in an algal crust).

Key Terms

Deuteromycetes	*Claviceps purpurea*	parasitic
"imperfect fungi"	ergotism	binary fission
Candida albicans	symbiotic relationships	budding
Aspergillus	mycorrhiza, -izae	fragmentation
Penicillium	lichen	*Arthrobotrys dactyloides*
Phytophthora infestans	saprophytic	*Trebouxia*
late blight		

1. Give two examples of fungi that participate in symbiotic relationships. Describe the separate contributions to the relationship that are made by the fungus and by the other participant.
2. Explain why the *fungi imperfecti* are viewed as "imperfect."

Self-Quiz Questions

Fill-in-the-Blanks

(1) _____ help many complex land plants absorb certain vital nutrients from the soil. (2) _____ are composite organisms that comprise an alga and fungus living interdependently. (3) _____ and athlete's foot are imperfect fungi that parasitize humans. (4) _____ is a blue mold that flavors Roquefort cheese.

40-III
(pp. 622–627)

PART II. KINGDOM OF PLANTS
Evolutionary Trends Among Plants
Classification of Algae
Red Algae
Brown Algae
Green Algae

Summary

Plants are multicellular, eukaryotic, generally photosynthetic organisms. Ancestors of modern land plants presumably evolved from green algal types. A variety of features developed over a long period of time, beginning more than 500 million years ago: (1) the development of a waxy *cuticle* that helped the plant retain water; (2) the development of *stomata* that enabled gas exchange to be regulated; (3) the development of some sort of anchoring system that extracts water and mineral ions from the soil; (4) the development of *vascular tissue* that transports substances throughout the plant and brings the photosynthetic areas up into the sunlight; (5) the evolution of gametes from motile isogametes into a *heterosporous, oogamous* habit in which pollen grains are dispersed by various agents; (6) the reduction of the haploid gametophyte generation, so that the *diploid sporophyte* (which has highly developed vascular tissues) came to occupy most of the life cycle; (7) the development of *flowers*, increasing the chances of cross-pollination and more frequent recombination of genes; and (8) the evolution of *fruits* that protect the developing embryo and serve as nutrients for the germinating plant.

The diverse body forms of the *red algae* (Rhodophyta) are adapted to a wide range of communities, but they are found especially in deep marine tropical waters. They resemble the blue-green algae with respect to their photosynthetic membranes and pigments. The *brown algae* (Phaeophyta), which are almost all marine, include kelps—some of which are anchored by holdfasts, and some of which thrive in the open sea. Most of the complexly organized kelps show alternation of generations, and one species of giant kelp has developed phloemlike tissue. The diverse *green algae* (Chlorophyta) generally dwell in freshwater communities, but there are also many species that live in salt water or in moist soil. Green algae also show alternation of generations and variation in the type of gametes produced. Chlorophytes have *chlorophyll* a and b in their plastids, store carbohydrates as starch, and have a cellulose framework within the cell wall.

Key Terms

nonvascular plants	megaspores	green algae
vascular tissues	microspores	Chlorophyta
xylem	*Porphyra*	*Chlamydomonas*
phloem	pollen grains	holdfasts
gametophyte	red algae	kelps
sporophyte	Rhodophyta	*Sargassum*
isogamy	agar	stipe
oogamy	brown algae	"floats"
plant spores	Phaeophyta	*Ulva*
homosporous	Fucus	*Acetabularia*
heterosporous	algin	*Spirogyra*

Objectives

1. Distinguish the red, brown, and green algae from one another.
2. Tell how the plant body of a green alga would have to be modified to survive in a dry-land environment.
3. Describe the basic differences between the life cycle of a green alga and the life cycle of a dry-land flowering plant.
4. List four key reproductive trends that have occurred in the evolution of plant life cycles.

Self-Quiz Questions

Fill-in-the-Blanks

In algae, simple diffusion and (1) _____ transport carry materials and wastes across external cell membranes into the surrounding water. The dominant plant body of many multicellular algae consists only of haploid cells and produces gametes by mitosis; this is the (2) _____ generation. Zygotes formed by the fusion of two gametes eventually undergo (3) _____, which leads to the formation of haploid (4) _____, each of which can then give rise to a new, conspicuous haploid body. (5) _____ algae resemble the blue-green algae with respect to their photosynthetic membranes and (6) _____, and they may have been evolutionarily derived from these. (7) _____ algae are almost all marine and include the giant (8) _____, some of which are anchored by a (9) _____, and some of which thrive in the open sea. The (10) _____ algae are found in the most diverse of environments; they have photosynthetic systems and pigment systems that are identical with those in complex (11) _____ plants.

PART II. KINGDOM OF PLANTS (cont.)
     The Land Plants
     Bryophytes
     Lycopods, Horsetails, and Ferns

Summary

Transitional land plants (*bryophytes*) developed structures for obtaining and conserving water even though *they never developed vascular tissue*, roots, or the ability to reproduce sexually in dry-land environments. The 16,000 species of bryophytes include mosses, liverworts, and hornworts, which grow close to the soil surface of moist environments. In mosses, *rhizoids* anchor the *photosynthetic gametophyte*, which supports the *dependent, stalked sporophyte*. The development of a *protected embryo sporophyte*—attached to and nourished by gametophyte tissue—must have been an important advance for life on land; it was first seen among the bryophytes.

The *primitive vascular plants* (horsetails, lycopods, and ferns) generally have underground stems (*rhizomes*) that produce aerial branches and leaves; *true roots* anchor the rhizome in the soil. Such plants survive largely because they can take in water and nutrients from the soil through roots, transport water through vascular tissue, and control water loss from the stems and leaves by means of *stomata* and waxy coverings. The development of supportive tissue enabled the primitive vascular plants to grow taller and compete more effectively for available sunlight than the nonvascular transitional plants. Although the spores of primitive vascular plants typically are dispersed by wind, the flagellated sperms require water in order to reach the female structures. And whereas the gametophyte is more conspicuous and dominant in bryophyte life cycles, *the sporophyte dominates vascular plant life cycles*. All bryophytes and most primitive vascular plants produce only one kind of spore, which grows into a gametophyte that produces either male or female sex organs (or both); this strategy is referred to as *homospory*. Some species of lycopods and ferns and all seed plants are *heterosporous*: Two kinds of spores are produced. *Megaspores* give rise to female gametophytes, and *microspores* give rise to male gametophytes.

Key Terms

bryophytes, Bryophyta	rhizoids	scalelike leaves
archegonium, -ia	lycopods, Lycophyta	ferns
antheridium, -ia	*Lycopodium*	Pterophyta
sporangium, -ia	club mosses	frond
moss	strobili, strobilus	sorus, sori
Polytrichum	horsetails, Sphenophyta	sporophyte
liverworts	*Equisetum*	heart-shaped
Marchantia	rhizomes	gametophyte
gemmae		

Objectives

1. State how the primitive vascular plants differ from bryophytes.
2. Compare the life cycles of ferns and mosses.
3. Describe *lycopods* and *horsetails*.
4. Explain what a *strobilus* is.
5. Define *homosporous* and *heterosporous* and distinguish the life cycle of one from the life cycle of the other. State which types of plants are homosporous.

Fill-in-the-Blanks

Even though bryophytes never developed complex (1) _____ tissue, roots, or the ability to reproduce (2) _____ [choose one] ☐ sexually ☐ asexually in dry-land environments, they did develop structures for obtaining water and for conserving it. The 16,000 species of bryophytes include (3) _____, liverworts, and hornworts. The leafy green plant that comes to mind when the word moss is mentioned is the (4) [choose one] ☐ sporophyte ☐ gametophyte generation. From it grows the (5) [choose one] ☐ sporophyte ☐ gametophyte generation. Bryophytes produce (6) [choose one] ☐ homospores ☐ heterospores, which germinate and grow into the (7) [choose one] ☐ sporophyte ☐ gametophyte generation; this generation produces (8) [choose one] ☐ spores ☐ gametes. When two gametes fuse, the resulting zygote divides mitotically, producing a (9) [choose one] ☐ haploid ☐ diploid (10) [choose one] ☐ sporophyte ☐ gametophyte. The development of a protected embryo (11) [choose one] ☐ sporophyte ☐ gametophyte—attached to and nourished by (12) [choose one] ☐ sporophyte ☐ gametophyte tissue— must have been an important advance for life on land. The primitive (13) _____ plants include horsetails, lycopods, and (14) _____. Instead of having upright aerial stems, most primitive vascular plants (except for tree ferns) have (15) _____, which run along or just below the surface. The (16) [choose one] ☐ sporophyte ☐ gametophyte is the conspicuous part of the fern plant life cycle.

Labeling

Identify each indicated part of the accompanying illustration (p. 355).

(17) _____   (20) _____   (23) _____

(18) _____   (21) _____   (24) _____

(19) _____   (22) _____   (25) _____

40-V
(pp. 634–640)

PART II. KINGDOM OF PLANTS (cont.)
  Existing Seed Plants
  The Gymnosperms
  Angiosperms–The Flowering Plants
SUMMARY

Summary

In gymnosperms, female gametophytes develop while still attached to the parent plant; water and nutrients required for their development are provided by the

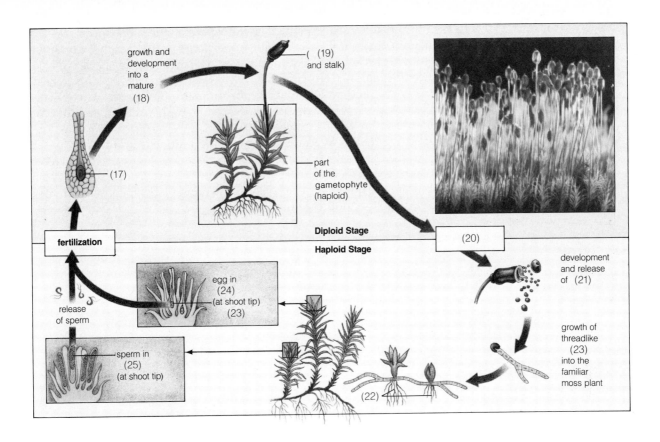

growth and
development
into a
mature
(18)

((19)
and stalk)

(17)

part
of the
gametophyte
(haploid)

Diploid Stage

Haploid Stage

fertilization

(20)

development
and release
of (21)

release
of sperm

egg in
(24)
(at shoot tip)
(23)

growth of
threadlike
(23)
into the
familiar
moss plant

sperm in
(25)
(at shoot tip)

(22)

sporophyte—the stage that is well adapted for obtaining these resources on dry land. In contrast, immature male gametophytes (pollen grains) are released from the parent plant and transported to the female gametophyte by wind, insects, and the like. Free water is not necessary for this transfer in most species. *Gymnosperms* (cycads, ginkgos, conifers, and gnetophytes) and *angiosperms* (the flowering plants) produce *seeds*. Each seed contains an *embryo sporophyte*, which generally is surrounded by internal *food reserves* that nourish it during germination and by a *seed coat* that guards against mechanical damage and extreme water loss. Seeds have taken over the dispersal function of spores.

The seeds of gymnosperms ("naked seeds") are carried on the surfaces of reproductive structures without being protected by additional tissue layers. *Cycads* (Cycadophyta) resemble squat palm trees that have cones instead of flowers and reproduce at a slow rate. *Ginkgos* (Ginkgophyta) are represented by a single modern species of tree with fan-shaped leaves that grows under protected conditions. *Conifers* (Coniferophyta), which include pine, spruce, fir, hemlock, juniper, cypress, larch, and redwood species, are the most diverse and widely distributed gymnosperms. The sporophyte is a cone-bearing woody tree or shrub with needlelike or scalelike leaves that are generally retained for several years (evergreen). Their *heterosporous*, seed-bearing life cycles vary in their details, but in general their winged pollen grains are dispersed on the wind. Conifers were dominant land plants during the Mesozoic, and their mechanisms of seed production, protection, and dispersal helped assure their success. Although their numbers are much reduced now, they are still the dominant vegetation in northern regions and at high altitudes.

In contrast to the naked seeds of gymnosperms, the angiosperm seed is contained within the ovary as it develops. *Angiosperms* (flowering plants) are grouped into two classes: monocots and dicots. The 50,000 or so species of *monocots* include palms, lilies, and orchids, as well as the principal crop plants (wheat, rice, corn, oats, rye, and barley) that support human populations. There are at least 200,000 *dicot* species (for instance, roses, beans, oaks, squash), and they are the most diverse group of flowering plants.

The vegetative bodies of plants and fungi are immobile, yet by relying on the wind, the splashing rain, and insects for dispersal, both groups have moved across the plains, to the mountains . . . to land everywhere. During the Mesozoic era, conifers were the dominant land plants; flowering plants together with their pollinating vectors have flourished in the Cenozoic era.

Key Terms

gymnosperms	conifers	megaspore mother cell
angiosperms	Coniferophyta	female gametophyte
carpel	megasporangium, -gia	tissue
cycad	megaspores	angiosperms
Cycadophyta	microspores	Anthophyta
ginkgos	pollen grain	monocot
Gingkophyta	pollen tube	dicot
gnetophytes	ovule	herbaceous
Gnetophyta	pollination	woody
Gnetum	fertilization	endosperm
Ephedra	microspore mother cell	embryo sac
Welwi tschia		

Objectives

1. Define *gymnosperm* and name the four divisions that are in the gymnosperm category.
2. Describe a representative cycad, ginkgo, and conifer.
3. Outline the principal steps of the conifer life cycle. Tell where the spores and gametes are formed. Distinguish *pollination* from *fertilization*.
4. List three parts of a seed and explain how each is produced.
5. Discuss the placement of gymnosperms in the evolutionary time scale.
6. Compare the life cycle of an angiosperm with that of a gymnosperm. See p. 180 for the comparison.

Self-Quiz Questions

Fill-in-the-Blanks

Some species of lycopods and ferns are (1) _____: Two kinds of spores are produced. In gymnosperms, megaspores give rise to (2) [choose one] □ male □ female gametophytes; these develop while still attached to the parent plant, which provides water and nutrients for their development. In contrast, immature male gametophytes (commonly known as (3) _____ _____) are released from the parent plant and are carried to the female plant by (4) _____, insects, and so on. Gymnosperms and (5) _____ (the flowering plants) produce (6) _____, each of which contains an embryo sporophyte that generally is surrounded by internal food reserves, which nourish it during germination, and by a (7) _____ _____, which guards against extreme water loss and mechanical damage. The seeds of (8) _____ are carried on the surfaces of reproductive structures without being protected by additional layers. (9) _____ resemble squat palm trees that have cones instead of flowers; they also have a slow reproductive rate. (10) _____ are represented by a single modern species of tree with fan-shaped leaves. (11) _____ have needlelike or scalelike leaves that are generally retained for several years; they were the dominant land plants of the (12) _____ era and are still dominant in northern regions and at high altitudes.

Labeling   Identify each indicated part of the accompanying illustration (p. 358).

(13) _____   (16) _____ _____   (19) _____

(14) _____ _____ _____   (17) _____   (20) _____ _____

(15) _____ _____   (18) _____

CHAPTER TEST

UNDERSTANDING AND INTERPRETING KEY CONCEPTS

____ (1) A gametophyte is _____.
  (a) a gamete-producing plant
  (b) haploid
  (c) Both a and b
  (d) The plant produced by the fusion of gametes

____ (2) "Algae" are found in the kingdom _____.
  (a) Plantae
  (b) Monera
  (c) Protista
  (d) All of the above

____ (3) Red algae _____.
  (a) are primarily marine organisms
  (b) are thought to have developed from green algae
  (c) contain xanthophyll as their main accessory pigments
  (d) All of the above

____ (4) Holdfasts, gas-filled floats, and thick leathery surfaces are found in species of _____.
  (a) red algae
  (b) brown algae
  (c) bryophytes
  (d) green algae

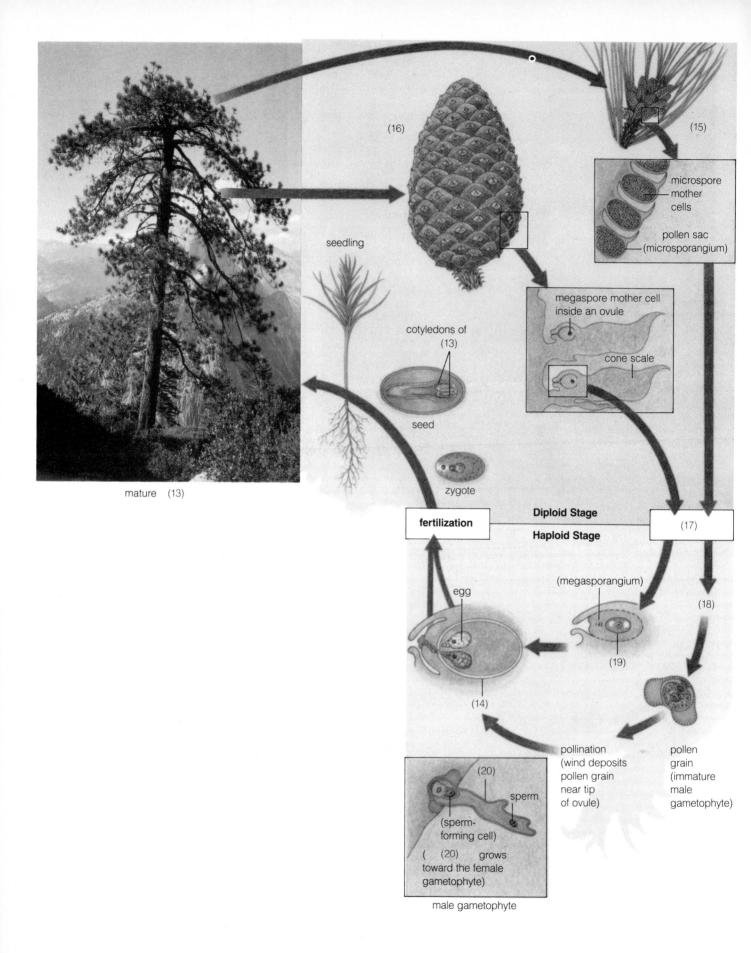

(16)

(15)

microspore
mother
cells

pollen sac
(microsporangium)

seedling

megaspore mother cell
inside an ovule

cone scale

cotyledons of
(13)

seed

zygote

mature (13)

| fertilization | **Diploid Stage** | |
| | **Haploid Stage** | (17) |

(18)

egg

(megasporangium)

(19)

(14)

pollination
(wind deposits
pollen grain
near tip
of ovule)

pollen
grain
(immature
male
gametophyte)

(20)

sperm

(sperm-
forming cell)

((20) grows
toward the female
gametophyte)

male gametophyte

___ (5) Bryophytes _____.
    (a) rely on isogamy for sexual reproduction
    (b) have vascular systems that enable them to survive on land
    (c) include lycopods, horsetails, and ferns
    (d) None of the above

___ (6) In horsetails, lycopods, and ferns, _____.
    (a) spores give rise to gametophytes
    (b) the main plant body is a gametophyte
    (c) the sporophyte bears sperm- and egg-producing organs
    (d) All of the above

___ (7) _____ are seed plants.
    (a) Cycads and ginkgos
    (b) Conifers
    (c) Angiosperms
    (d) All of the above

___ (8) _____ are widely planted in cities because of their resistance to insect predators, air pollution, and disease.
    (a) Lycopods
    (b) Ginkgos
    (c) Dutch elms
    (d) Conifers

___ (9) Anthers are _____.
    (a) attached to individual scales in a cone
    (b) female gametophytes
    (c) male reproductive organs
    (d) composed of a style and a stigma

___ (10) Most true fungi send out cellular filaments called _____.
    (a) mycelia
    (b) hyphae
    (c) mycorrhiza
    (d) asci

___ (11) Which of the following is *not* true? Chlorophytes are the likely ancestors of land plants because they _____.
    (a) all have carotenoids and xanthophylls
    (b) all have chlorophyll a and b
    (c) developed a cuticle and rhizoids that are characteristic of primitive land plants
    (d) store excess carbohydrates as starch

___ (12) The rapid expansion of angiosperms late in the Mesozoic era appears to be related to their coevolution with _____.
    (a) dinosaurs
    (b) gymnosperms
    (c) insects
    (d) mammals

___ (13) An ovary of an angiosperm develops into the _____ after fertilization.
   (a) flower
   (b) fruit
   (c) seed
   (d) sporangium

___ (14) In complex land plants the diploid stage is resistant to adverse environmental conditions such as dwindling water supplies and cold weather. The diploid stage progresses through this sequence: _____.
   (a) gametophyte → male and female gametes
   (b) spores → sporophyte
   (c) zygote → sporophyte
   (d) zygote → gametophyte

___ (15) Fungi _____.
   (a) are producers
   (b) are generally saprophytic
   (c) usually have life cycles in which the diploid phase dominates
   (d) include *Fucus* and liverworts

CHAPTER TEST

INTEGRATING AND APPLYING KEY CONCEPTS

Explain why totally submerged aquatic plants that live in deep water never developed heterosporous life cycles.

41

ANIMAL DIVERSITY

General Objectives

1. Describe the major advances in body structure and function that made invertebrates and vertebrates increasingly large and complex.
2. Discuss the relationship between segmentation and the development of paired organs and paired appendages.
3. List the functions of a coelom and describe the role coelomic development played in animal evolution.
4. Be able to reproduce from memory a phylogenetic tree that expresses the relationships between the major groups of animals.

GENERAL CHARACTERISTICS OF ANIMALS
    Body Plans
    Representative Animal Phyla
SPONGES
CNIDARIANS
COMB JELLIES

Summary

More than 97 percent of all the species of animals in the world are backboneless invertebrates. Sometime between 2 billion and 1 billion years ago, the first animals arose; by the dawn of the Cambrian, there were already well-developed multicellular animals (*metazoans*). One of the early animal forms is thought to have led to the *sponges*, which are relatively simple vaselike animals that inhabit shallow seas. Seawater flows in through pores in the body wall, then out through the opening at the top; the interior is lined with microvilli-adorned *collar cells*, which trap and ingest microscopic organisms carried in on the water flowing through the various pores. Sponges may have evolved from colonial protistans that resemble collar cells, but because sponges have developed four distinct cell types, they are truly multicellular even though communication between cells and integration of activities is poorly developed. Sponges have no nerve cells, no muscles, and no gut.

  Cnidarians (jellyfish, corals, *hydra*, and sea anemones) have simple organs distributed in a radial symmetry. Stinging cells (*nematocysts*) aid in food capture and defense. Most cnidarian life cycles have a *planula* larval stage: a roving, ciliated, undifferentiated mass of cells that eventually settles down and becomes a sedentary, attached form called a *polyp*. For most cnidarians, the polyp stage eventually gives rise to free-swimming, bell-shaped forms celled *medusae*, which produce gametes in sex organs; the gametes later fuse, forming a zygote that divides mitotically to form the planula larva. Even though the body plan of both polyp and medusa consists essentially of two connected epithelial layers separated by a jelly layer that contains the few simple organs, cnidarians do have a true gut. The gut is little more than an epithelial cavity; enzymes from glands in the epithelium break down food that has been stuffed into the cavity. The circulatory system transports materials. Undigested residues are simply expelled through the mouth. These animals also show some integration of neural and muscular activity, having nerve nets, muscle tissue layers, and some sensory organs. Comb jellies (Ctenophora) are biradial forms that are predatory and often bioluminescent. Cilia arranged like teeth in combs beat in waves propelling the animal forward, mouth first.

Key Terms

metazoans	coelom, coelomate	amoeboid cells
vertebrates	hemocoel	spicules
invertebrates	acoelomate	gemmule
"primitive"	pseudocoelomate	cnidarians, Cnidaria
multicellular	pseudocoel	hydrozoans
asymmetrical	segmentation	sea anemones
radial symmetry	uncephalized	medusa
bilateral symmetry	cephalized	polyp
saclike gut	sponges, Porifera	nerve net
"complete" gut	collar cells	mesoglea

nematocysts
planula
epidermis
gastrodermis

colonial hydrozoans
Physalia
Obelia

comb jelly, Ctenophora
biradial symmetry
bioluminescent

Objectives

1. Distinguish *radial symmetry* from *bilateral symmetry*, and *acoelomate* from *pseudocoelomate*.
2. List two characteristics that distinguish sponges from other animal groups.
3. Describe each of the four cell types found in sponges. State whether or not you think sponges have tissues.
4. Describe the two cnidarian body types.
5. Explain how radial symmetry might be more advantageous to floating or sedentary animals than bilateral symmetry would.
6. State what nematocysts are used for and explain how they operate.
7. Design an illustration that summarizes a generalized cnidarian life cycle. Include these terms: *egg, sperm, zygote, planula, polyp, medusa.*
8. Tell how cnidarians obtain and digest food, and tell what they do with food they cannot digest.
9. Name several cnidarians.
10. Describe comb jellies in terms of body plan and preferred types of food.

Self-Quiz Questions

Fill-in-the Blanks

Ninety-seven percent of all animals on Earth are (1) _____. Multicellular animals are called (2) _____. (3) _____ form the most primitive major group of multicellular animals. They are nourished by microscopic organisms extracted from the water that flows in through pores in the body wall by sticky (4) _____ _____. (5) _____ between sponge cells and integration of activities are poorly developed. Sponges lack (6) _____ cells, muscles, and a gut. Cnidarians are (7) _____ symmetrical and have stinging cells called (8) _____, which aid in defense and food capture. Most cnidarian life cycles have a (9) _____ larval stage. Of the two body types, the (10) _____ is in the sexual stage, in which simple sex organs produce eggs or sperms. (11) _____ _____ are biradial predatory animals that appear to be made of jelly.

Labeling

Identify each indicated part of the accompanying illustration (p. 364).

(12) _____ _____

(14) _____ _____

(13) _____ _____

(15) _____

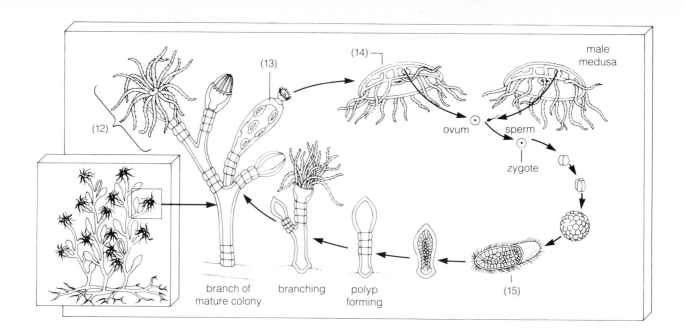

male medusa

(13)

(14)

(12)

ovum

sperm

zygote

branch of mature colony

branching

polyp forming

(15)

41-II
(pp. 649–655)

FLATWORMS
    Turbellarians
    Trematodes
    Cestodes
NEMERTEANS
NEMATODES
ROTIFERS

Summary

The most primitive *flatworms* resemble the planula stage of cnidarian life cycles. Perhaps, long ago, such a planuloid kept on crawling (rather than settling down and becoming attached); the leading end of such a mutant would have encountered food and danger more than any of its other body areas did; hence, nerve and sensory cell clustering at the head end would promote faster sensing and response. Predators of such crawling planuloids would be more likely to attack from above than below, so dorso-ventral differentiation would have been encouraged. Food and danger would be found as often on one side of a forward crawler as on the other; hence, the emergence of *bilateral symmetry*, which could have led to paired organs of the sort seen in many flatworms. Flatworms include free-living turbellarians, parasitic flukes, and parasitic tapeworms. All have flattened *acoelomate* bodies shaped like broad leaves or long ribbons, and they range in size from microscopic to eighteen meters long! Adult parasitic tapeworms attach to their hosts by suckers and hooks and are sustained by predigested nutrients from the host. Large free-living flatworms have a much-branched gut with only one opening, well-developed muscles, mesoderm, and simple tubes that pick up dissolved wastes from internal cells and expel them from the body much as kidneys do for other animals. Flatworms lack respiratory systems; gases easily enter and leave their thin bodies by diffusion.

Another major group of animals, the *round worms*, includes nematodes. This group encompasses hundreds of thousands of species, including some destructive parasites and many more free-living recyclers of nutrients. The body plan of nematodes consists of a tube (the gut) within another tube (the body wall).

The gut has a mouth *and* an anus, and food travels from mouth to anus. Reproductive cells in the *pseudocoel* between the gut and body wall give rise to gametes. Nematodes have only longitudinal muscles and no circular muscles; thus they thrash awkwardly about. A tough *cuticle* affords nematodes protection and flexibility.

Rotifers are tiny aquatic pseudocoelomate animals that feed on bacteria and single-celled algae with their crowns of cilia.

Key Terms

flatworms
Platyhelminthes
turbellarians
tapeworms
pharynx
protonephridium, -dia
flame cells
hermaphroditic system
trematodes

flukes
Schistosoma
schistosomiasis
cestodes
tapeworms
scolex
proglottids
host
roundworms (Nematoda)

cuticle
complete gut
pseudocoel,
    pseudocoelomates
hydrostatic skeleton
hookworm (*Necator*)
Trichinella spiralis
rotifers

Objectives

1. Outline how a planula could have given rise to a simple flatworm. Indicate the selective pressures that helped to promote bilateral symmetry.
2. List the three main types of flatworms.
3. State the evolutionary advances seen in flatworms as compared to cnidarians.
4. Describe the body plan of round worms, comparing its various systems with those of the flatworm body plan.
5. Describe a rotifer and tell where it lives and what it eats.

Self-Quiz Questions

Fill-in-the-Blanks

A mutant (1) _____ larva is believed to be the ancestor of the free-living flatworms. A shift from radial to bilateral symmetry could have led to (2) _____ _____ of the sort seen in many flatworms; for example, turbellarians have units called (3) _____ that regulate the volume and salt concentrations of their body fluid. Flatworms have no (4) _____ systems, and their (5) _____ system is saclike. Examples of parasitic flatworms are *Schistosoma*, a blood (6) _____ that causes schistosomiasis, and *Taenia saginata*, a tapeworm that attaches to the intestine with a (7) _____ and releases eggs that develop in proglottid segments. (8) _____ have complete digestive tracts, no circular muscles, and only a few longitudinal muscles. Between the gut and body wall of a nematode is a (9) _____, which contains (10) _____ organs and serves as both a circulatory system and a hydrostatic (11) _____. (12) _____ are tiny abundant aquatic pseudocoelomates that have ciliary crowns attracting to them their food of single-celled organisms.

TWO MAIN EVOLUTIONARY ROADS

MOLLUSKS

> **Chitons**
>
> **Gastropods**
>
> **Bivalves**
>
> **Cephalopods**

Summary

Two groups of animals are descended from ancestral flatworms. In *protostomes* (annelids, arthropods, and mollusks), the first opening into the gut that arises during embryonic development becomes the mouth; the anus forms later. In *deuterostomes* (echinoderms and chordates), the first opening to the gut becomes the anus, and the second becomes the mouth. The distinction between protostomes and deuterostomes seems trivial, but it helps us to identify two major lines of animal evolution that diverged long ago.

Mollusks include about 100,000 species of clams, chitons, scallops, nudibranchs, snails, slugs, squids, octopuses, and other diverse species. Unlike that of arthropods, the body wall of mollusks has stayed flexible, and its musculature has become well developed in some. The *mantle* secretes substances that form a hard shell. In chitons, bivalves, and snails, the shell is a protective shield from all but the most persistent, clever, and forceful predators. In other molluscan groups—notably nudibranchs and predatory squids and octopuses—the shell has been reduced to a tiny internal structure and the mantle of cephalopods has been converted into a highly muscularized, conical, jet-propulsive structure, making these animals the most magnificent swimmers of the invertebrate world. The nervous systems of squids and octopuses represent the peak of invertebrate complexity; in terms of size and complexity, their brains approach those of mammals that have the capacity for storing information and learning new behaviors. Like vertebrates, these animals have acute vision and refined motor control. Because of similar evolutionary pressures, the separately evolved vertebrate and cephalopod lines developed many similar neural and sensory structures, with similar functions.

Key Terms

protostomes	mantle cavity	nudibranch, sea slug
deuterostomes	radula	bivalves
gastrula	visceral mass	filter feeders
radial cleavage	hemocoel	siphons
spiral cleavage	open circulatory system	cephalopods
mesoderm	ctenidia	squids
mollusks, Mollusca	chitons	nautilus
foot	gastropods	cuttlefish
mantle	torsion	*Architeuthis*

Objectives

1. Define *protostome* and *deuterostome* and give examples of each group.
2. Explain why zoologists care to make what seems to be such a trivial distinction between the two groups.
3. Reproduce from memory the basic evolutionary relationships presented in Figure 41.4 in your main text.
4. Define *coelom* and list three benefits that the development of a coelom brings to an animal.
5. List eight examples of mollusks.
6. Define *mantle* and tell what role it plays in the molluscan body.

7. Explain why you think cephalopods came to have such well-developed sensory and motor systems and are able to learn.
8. Compare the nervous and circulatory systems of bivalves and cephalopods. Then compare the general types of food members of each group eat.

Self-Quiz Questions

Fill-in-the-Blanks

(1) _____ include echinoderms and vertebrates; in this group, the first opening to the gut becomes the (2) _____, and the second one to appear becomes the (3) _____. The situation is reversed in the (4) _____, which include annelids (such as (5) _____), arthropods (such as (6) _____ and crabs), and (7) _____ (such as abalones, limpets, squids, and chambered nautiluses). The most highly evolved invertebrates are generally considered to be the (8) _____, which include squids and octopuses; in terms of sheer size and complexity, the (9) _____ of these animals approach those of mammals. Like vertebrates, these animals have acute (10) _____ and refined (11) _____ control, which is well integrated with the activities of the nervous system. In less highly evolved mollusks, a structure known as the (12) _____ secretes one or more pieces of calcareous armor that protect these soft-bodied animals from predation. In the cephalopods, the mantle has become a conical cloak that surrounds the internal organs and the much-reduced shell; seawater moves in and out of the (13) _____ _____ in a jet-propulsive manner.

41-IV
(pp. 661–664)

ANNELIDS
Earthworms
Leeches
Polychaetes
Annelid Adaptations

Summary

Annelids, the true segmented worms such as earthworms, polychaetes, and leeches, differ from flatworms in having (1) a complete digestive system and (2) a *coelom* (a fluid-filled space between the gut and the body wall). In contrast, the flatworm gut is joined to the body wall by a solid mass of cells; hence, flatworms are *acoelomate* (without a coelom). Once organs become suspended in a coelom, they are bathed and cushioned in coelomic fluid and become insulated from the stresses of body movement.

A coelom permits increased size along with more activity, especially if, as in annelids, a *circulatory system* also is part of the body plan. Through various tubes, a fluid containing nutrients and wastes travels from the body's surface to its inner regions and back again. The circulating fluid (blood) provides a means for transporting materials between internal and external environments. In the annelid system, forceful contractions in muscularized blood vessels ("hearts") keep blood circulating in one direction. In some annelids, hemoglobin, which dramatically increases the blood's oxygen-carrying capacity, is one of the components of blood.

Segmentation, which is a repeating series of body parts, is well developed in annelids. Each annelid segment contains its own allotment of gut, coelom, body wall, muscles, ventral double nerve cord with a segmental ganglion ("brain"), circulatory tubes, nephridia, and bristles (setae) embedded in the body wall. The muscles are arranged in paired, antagonistic sets that move in a coordinated sequence in response to nerve signals relayed through the minibrains in each segment. Thus earthworms move forward with far more coordination than do the thrashing nematodes. Polychaetes have fleshy, paddle-shaped lobes called *parapodia*, which project from the body wall. Some annelids with muscle-endowed parapodia can swing them up, down, forward, and backward; thus in some annelids we see paired appendages and a mode of walking.

Key Terms

Annelida	polychaetes	nephridium, -dia
segmented worms	seta, setae	rudimentary brain
leeches	parapodium, -dia	segmental ganglia
oligochaetes	segmentation	

Objectives

1. Name the three groups of annelids and give a specific example from each group.
2. Explain how the development of a coelom and a circulatory system provided the potential for the development of larger and more massive animal bodies.
3. Define *segmentation* and explain how it is related to the development of muscular and nervous systems.

Self-Quiz Questions

(1) _____ include truly segmented worms such as earthworms,

(2) _____, and leeches; they differ from flatworms in having (first) a

complete digestive system with a mouth and (3) _____ and (second) a

(4) _____, which is a fluid-filled space between the gut and body wall. In

a circulatory system, (5) _____ provides a means for transporting

materials between internal and external environments. In some annelids,

(6) _____, a protein component of blood, dramatically increases the

blood's oxygen-carrying capacity. (7) _____, which is a repeating series of

body parts, is well developed in annelids. In each segment, there are swollen

regions of the (8) _____ _____ _____ _____ that control

local activity and a pair of (9) _____ that act as kidneys and bristles

embedded in the body wall. Many polychaetes (marine worms) have fleshy,

paddle-shaped lobes called (10) _____, which project from the body wall.

Labeling   Identify each indicated part of the accompanying illustration.

(11) _____   (14) _____   (16) _____

(12) _____   (15) _____ _____   (17) _____

(13) _____ _____

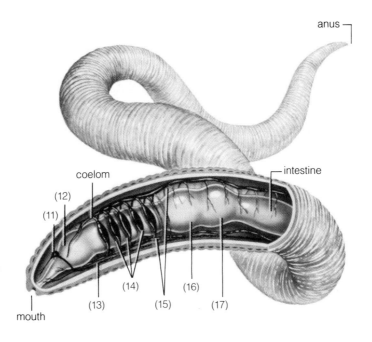

41-V
(pp. 664–671)

ARTHROPODS
    Arthropod Adaptations
    Chelicerates
    Crustaceans
    Insects and Their Kin

Summary

Arthropods (insects, crustaceans, arachnids, and their relatives) have developed a thickened *cuticle* and a hardened *exoskeleton*, both of which are superb barriers to evaporative water loss, give protection from some predation, and present surfaces to which antagonistically arranged muscles can attach. Between segments, the cuticle remains pliable and functions as a hinge. On land, the exoskeleton provides support for an animal body deprived of water's buoyancy. In the seas and in freshwater environments, the jointed-legged animals known as *crustaceans* have come to be well represented. The first land arthropods were the *arachnids*; their descendants (spiders, scorpions, ticks, and mites) still endure. *Centipedes* and *millipedes* arose later. *Insects* are thought to have evolved from centipedelike ancestors. The larger marine arthropods extract oxygen from water with *gills* (thin flaps of tissue richly supplied with blood vessels). Among insects, *tracheal systems*—which consist of branching tubes stiffened with *chitin*—supply each cell of the insect body with oxygen and afford insects the highest metabolic rates known. An *insect wing* is not a modified limb, as are the wings of bats and birds; rather, it consists of a modified flap of cuticle that lacks internal muscles.

Key Terms

trilobites
chelicerates
Onychophora
Peripatus
exoskeleton
chitin
molting
arthropod

specialized, specialization
tracheas
compound eye
book lungs
crustacean
carapace
mandibles
maxillae

antennae
copepods
barnacles
millipedes
centipedes
insect wings
Malpighian tubules

Objectives

1. Explain how the development of a thickened cuticle and a hardened exoskeleton affected the ways that arthropods lived.
2. State what ancestors are thought to have given rise to the arthropods and tell whether you think any other major groups of animals descended from the arthropods.
3. List the changes that the basic aquatic arthropod plan must undergo to adapt to life on land.
4. Contrast tracheal systems with gill systems and explain why land animals cannot use gill systems in obtaining oxygen from their environment.
5. Explain how insect wings differ from those of birds.
6. List six different groups of arthropods.

Self-Quiz Questions

Fill-in-the-Blanks

Arthropods developed a thickened (1) _____ and a hardened (2) _____. In freshwater and saltwater environments, arthropods known as (3) _____ came to be well represented. The first land arthropods were the (4) _____. Their descendants—(5) _____, scorpions, ticks, and mites—are still living. (6) _____ and millipedes arose later. (7) _____ are thought to have evolved from centipedelike ancestors. The larger aquatic arthropods extract oxygen from water with (8) _____, whereas most insects utilize (9) _____ systems, which provide the basis for the highest (10) _____ rates known. Arthropods are thought to have evolved from (11) _____.

Summary

Of the small number of *deuterostomes* that have survived to the present, only the echinoderms and the chordates are prominent. *Echinoderm* means spiny-skinned, in reference to the bristling spines possessed by some members of the group. In echinoderms, radial symmetry has been overlaid on an earlier bilateral heritage. Some biologists hypothesize that echinoderms are descended from deuterostomes that were evolving at a time when predators were increasing in both size and number. In response to selection by predators, the echinoderm ancestors came to be equipped with heavy, defensive armor of mineralized plates embedded in the body wall. Although heavy armor must have helped them avoid being eaten, it also may have forced them to settle to the sea bottom. Most echinoderms still go through a free-swimming, bilaterally symmetrical larval stage, but adults generally show the *radial symmetry* typical of bottom-dwelling organisms that cannot move quickly. The echinoderm way of moving about is based on a *water-vascular system* of canals and tube feet. Coordinated extension and retraction of many suckered tube feet enable echinoderms to move body parts in specific directions. Echinoderms include crinoids, sea stars, sea urchins, sand dollars, sea cucumbers, and brittle stars.

    Chordates are organisms that, at some time during their life span, develop (1) *gill slits*, (2) a *notochord*, (3) a *dorsal tubular nerve cord*, and some sort of (4) *endoskeleton*. Primitive chordates (such as tunicates and lancelets) tend to be more or less sessile as adults. They obtain food by passing plankton-laden water over sheets of mucus that line the gill slits; the entrapped plankton is then passed into the gut for digestion. Such a system is known as *filter feeding*. Sessile or sedentary animals generally produce many offspring, and they also produce motile larvae. Chordate larvae are bilaterally symmetrical, with a head end and a tail end. They have a nervous system with a dorsal nerve cord and a rod of stiffened tissue (the *notochord*) running the length of the body. Muscles attached to the notochord contract rhythmically, which causes the body to bend and move through the water. The notochord was the forerunner to the chordate's internal skeleton (endoskeleton).

    A strengthening of the notochord and its attached muscles encouraged the development of vertebrates: animals whose dorsal stiffening rod was segmented into a series of hard bones (vertebrae) arranged in a column that formed a backbone. The first vertebrates were the jawless fishes. As the brain region came to be protected by hard structures that eventually developed into a skull, various bone and muscle modifications occurred that led to the formation of jaws, which are characteristic of the *true fishes*. As the number of jawed fishes increased, there must have been a selective advantage in being able to detect food from a distance. Trends toward increasingly specialized sensory detectors and complex neural integration were established among the true fishes and were elaborated in the descendant forms that moved onto land. Gills were devices used mainly in filter feeding by primitive chordates, but as predatory activities increased, gills became more important in oxygenating the blood and supplying muscles with oxygen. Because gills can function only if they are kept moist, the first vertebrates that moved onto land during Devonian times relied increasingly on an internal swim bladder to extract oxygen from air.

In the fishes that seemingly gave rise to amphibians, blood traveled from gills to a two-chambered heart, which pumped it to the rest of the body. Amphibian (frogs, toads, salamanders) hearts are three-chambered. Reptiles, derived from amphibians, became even more successful on land because of their protective scaly skins, some physiological adaptations, and the amniotic egg, which allowed them to reproduce without going back to water (which most amphibians must do). Reptilian (snakes, crocodiles, lizards, turtles) hearts lie partway between three- and four-chambered hearts.

Birds and mammals evolved from separate groups of reptiles. Their success is partially related to their constant, warm body temperature and to the insulation provided by feathers or hair. These characteristics enable them to live in many environments. Good vision, a well-developed brain, appendages specialized for flight, swimming, running, grasping, and other activities are a few more features underlying the success of birds and mammals. Both birds and mammals have developed completely separate systemic and pulmonary circulations that work in harmony.

Key Terms

deuterostomes	Hemichordata, acorn	Agnatha
echinoderms,	worms	lamprey
Echinodermata	ostracoderms	scales
tube feet	gill slits	Chondrichthyes
water-vascular system	gill bars	Osteichthyes
ampulla	gill supports	Amphibia
chordates, Chordata	vertebrae	Reptilia
cartilage	skull	amniotic egg
notochord	jaws	Aves
urochordates	lungs	Mammalia
tunicates, sea squirts	nostrils	Prototheria
metamorphosis	heart	Metatheria
cephalochordates	coelocanths	marsupials
lancelets	lungfishes	Eutheria
Amphioxus	cerebral cortex	placental mammals

Objectives

1. Define *echinoderm* and list five groups of echinoderms.
2. Explain how the radial symmetry seen in adult echinoderms might have arisen.
3. Describe how locomotion occurs in echinoderms.
4. List three characteristics found only in chordates.
5. Describe the adaptations that sustain the sessile or sedentary life style seen in primitive chordates such as tunicates and lancelets.
6. State what sort of changes occurred in the primitive chordate body plan that could have promoted the emergence of vertebrates.
7. Describe the differences between primitive and advanced fishes, in terms of skeleton, jaws, special senses, and brain.

8. Describe the changes that enabled aquatic fishes to give rise to land dwellers.
9. State what kind of heart each of the four groups of four-limbed vertebrates has and list the principal skin structures that each produces.

Self-Quiz Questions

Fill-in-the-Blanks

Only two prominent groups of (1) _____ have survived to the present—the (2) _____ and the chordates. In echinoderms, (3) _____ symmetry has been overlaid on an earlier bilateral heritage; most echinoderms still go through a free-swimming (4) _____ symmetrical larval stage. Echinoderm locomotion is based on constant circulation of seawater through a (5) _____-_____ system of canals and (6) _____ _____. (7) _____ _____ have a rounded, globose body with bristling spines. (8) _____ have many long feather-duster arms; both mouth and anus lie within the circlet of arms. (9) _____ are among the most primitive of all living chordates; when they are tiny, they look and swim like (10) _____. A rod of stiffened tissue, the (11) _____, runs the length of the larval body; it was the forerunner of the chordate's (12) _____. Tunicates and (13) _____ practice (14) _____ _____; they draw in plankton-laden water through the mouth and pass it over sheets of mucus, which trap the particulate food before the water exits through the (15) _____ _____. The first vertebrates were the (16) _____ _____. The development of the formidable (17) _____ that are characteristic of the true fishes was surely an important evolutionary force. The first vertebrates that moved onto land during Devonian times relied more and more heavily on using the (18) _____ _____ to acquire oxygen from the air. Fish have (19) _____-chambered hearts. Chambers that receive blood are called (20) _____.

CHAPTER TEST

UNDERSTANDING AND INTERPRETING KEY CONCEPTS

___ (1) Which of the following is *not* true of sponges? They have no _____.
    (a) distinct cell types
    (b) nerve cells
    (c) muscles
    (d) gut

___ (2) Which of the following is *not* a protosome?
  (a) Earthworm
  (b) Crayfish or lobster
  (c) Sea star
  (d) Squid

___ (3) Bilateral symmetry is characteristic of _____.
  (a) cnidarians
  (b) sponges
  (c) jellyfish
  (d) flatworms

___ (4) Flukes and tapeworms are parasitic _____.
  (a) leeches
  (b) flatworms
  (c) nematodes
  (d) annelids

___ (5) Insects include _____.
  (a) spiders, mites, and ticks
  (b) centipedes and millipedes
  (c) termites, aphids, and beetles
  (d) All of the above

___ (6) Filter-feeding chordates rely on _____, which have cilia that create water currents and mucous sheets that capture nutrients suspended in the water.
  (a) notochords
  (b) differentially permeable membranes
  (c) filiform tongues
  (d) gill slits

___ (7) In true fishes, the gills serve primarily _____ function.
  (a) a gas-exchange
  (b) a feeding
  (c) a water-elimination
  (d) Both a and b

___ (8) The heart in amphibians _____.
  (a) pumps blood more rapidly than the heart of fish
  (b) is efficient enough for amphibians but would not be for birds and mammals
  (c) has three chambers (one ventricle and two atria)
  (d) All of the above

___ (9) Creeping behavior and a mouth located toward the "head" end of the body may have led, in some evolutionary lines, to _____.
  (a) development of a circulatory system with blood
  (b) sexual reproduction
  (c) feeding on nutrients suspended in the water (filter feeding)
  (d) concentration of sense organs in the head region

___ (10) Which of the following is associated with the shift from radial to bilateral body form?
(a) A circulatory system
(b) A one-way gut
(c) Paired organs
(d) The development of a water-vascular system

___ (11) The _____ body plan is characterized by simple gas-exchange mechanisms, two-way traffic through a relatively unspecialized gut, and a thin body with all cells fairly close to the gut.
(a) cnidarian
(b) nematode
(c) echinoderm
(d) flatworm

___ (12) The _____ have a tough cuticle, longitudinal muscles, and a complete digestive system, and they are facultative anaerobes.
(a) nematodes
(b) cnidarians
(c) flatworms
(d) echinoderms

___ (13) _____ insulates various internal organs from the stresses of body-wall movement and bathes them in a liquid through which nutrients and waste products can diffuse.
(a) A coelom
(b) Mesoderm
(c) A mantle
(d) A water-vascular system

___ (14) The annelid _____ may resemble the ancestral structure from which the vertebrate kidney evolved.
(a) trachea
(b) nephridium
(c) mantle
(d) parapodia

___ (15) The feeding behavior of true fishes selected for highly developed _____ .
(a) parapodia
(b) notochords
(c) sense organs
(d) gill slits

CHAPTER TEST

INTEGRATING AND APPLYING KEY CONCEPTS

Scan Table 41.1 (main text) to verify that most highly evolved animals have a complete gut, a closed blood-vascular system, both central and peripheral nervous systems, and are dioecious. Why do you suppose that having two sexes in separate individuals is considered to be more highly evolved than the monoecious condition utilized by earthworms? Would it not be more efficient if *all* individuals in a population could produce both kinds of gametes? Cross-fertilization would result in both individuals being able to produce offspring.

HUMAN ORIGINS
AND EVOLUTION

**General
Objectives**

1. Understand the general physical features and behavioral patterns attributed to early primates. Know their relationship to other mammals.
2. Trace primate evolutionary development through the Cenozoic era.
3. Understand the distinction between hominoid and hominid and distinguish between *Australopithecus* and *Homo*.

42-I
(pp. 691–697)

THE MAMMALIAN HERITAGE
PRIMATE CLASSIFICATION
TRENDS IN PRIMATE EVOLUTION
    From Quadrupeds to Bipedal Walkers
    Modification of Hands
    Enhanced Daytime Vision
    Changes in Dentition
    Brain Expansion and Elaboration
    Behavioral Evolution
    PRIMATE ORIGINS

Summary

When the world climate began to grow cooler and drier during the Cenozoic, birds and mammals, which have internal and external adaptations to help them maintain a relatively constant body temperature even when environmental temperatures rise and fall, rose to dominance. Primates, which include *prosimians* (lemurs, tarsiers) and *anthropoids* (apes, monkeys, and humans) apparently arose from insectivorous mammals like the tree shrews of Southeast Asia some 65 million years ago. Eyes adapted for sensing colors, shapes, and movements in a three-dimensional field, and body and limbs adapted for climbing trees were the key characters of primate evolution. The following traits later became developed in prosimians, monkeys, apes, and humans: (1) limbs and head that move freely in different planes; (2) digits that allow precise gripping of objects; (3) upright vertebral column and body posture; (4) enhanced daytime vision; (5) a nervous system that processes information quickly and then coordinates precise, rapid, flexible movements in response; and (6) strong social bonding with well-developed parental care and longer periods of learning.

The first prosimians date from approximately 54 million years ago. Ancestral anthropoids emerged sometime between 38–25 million years ago in Oligocene times. The anthropoid category is further subdivided into ceboids (New World monkeys), cercopithecoids (Old World monkeys) and *hominoids* (apes and hominids. *Hominids* include humans and their most recent ancestors. The dryopithecine apelike forms of the Miocene are thought to have given rise to the ancestors of humans because they had the Y-5 molar pattern, were larger brained than the other apes, and were more behaviorally flexible.

Key Terms

phylogenetic tree	Primates	convergent
Mammalia	arboreal	prehensile
axial endoskeleton	prosimians	dermal ridges
dentition	anthropoids	hominoids
incisors	New World monkeys	Y-5 molar pattern
canines	Old World monkeys	"cerebral Rubicon"
premolars	apes	culture
molars	humans	Paleocene
cusps	hominids	Eocene
infant dependency	platyrrhine	Oligocene
learning	catarrhine	Fayum Depression
placental mammals	brachiation	*Aegyptopithecus*
marsupial	quadrupedal	Miocene
monotremes	bipedal	dryopiths
flexibility	divergent	Miocene

Objectives

1. State when the earliest mammals evolved and when the shrewlike insectivores underwent adaptive radiation to give rise to the earliest primates.
2. Compare structural and behavioral features of the early primates with those of the cercopithecoids and hominids. Which anatomical features underwent the greatest changes along the evolutionary line to humans?
3. Name the places where prosimian survivors dwell today.
4. Name the two *key* characters of primate evolution.
5. Describe the environmental changes that occurred in Africa from about 35 million years ago up to about 10 million years ago and tell why these changes would have encouraged adaptive radiation.

Self-Quiz Questions

Fill-in-the-Blanks

The world climate began to grow cooler and drier during the (1) _____ era, when the birds and mammals rose to dominance. Primates, like all placental mammals, apparently arose from ancestral forms of the order (2) _____; representatives of this order living today are (3) _____, which are nighttime omnivores. Humans, apes, monkeys, and prosimians are all (4) _____; they have excellent (5) _____ perception as a result of their forward-directed eyes, and their fingers and toes are adapted for (6) _____ instead of running. One of the Fayum hominoids, (7) _____, lived 30 million years ago; it was the size of a house cat and had a skull resembling early hominoids, but it had a four-limbed gait. Hominoids of 25 million years ago were collectively known as (8) _____; they underwent (9) _____ _____ into three new ecological niches. The beginning of the hominoid line diverged from the dryopithecine-to-modern ape line sometime between 8 million and (10) _____ million years ago, during the Pliocene period.

THE HOMINIDS
Australopiths
Stone Tools and Early *Homo*
Homo erectus
Homo sapiens
SUMMARY

Summary

Geologic upheavals in the Great Rift Valley of East Africa during the late Cenozoic fostered environmental diversity, which in turn fostered distinct adaptations in the indigenous small, scattered bands of early apes. The first indisputable hominids (the australopiths, or "southern ape-men") are believed to have appeared in regions of increasing aridity during the Pliocene between 6 and 4 million years ago; they were foragers on the ground for seeds, fruits, nuts, and the remains of small animals. They also learned to hunt. Approximately 2 million years ago, there were also members of the genus *Homo*, of which we are the only living representatives. Since then, several distinctly different hominid populations appeared and disappeared, but by 30,000 years ago only one hominid species (*H. sapiens*) remained: man, the reasoner.

Key Terms

hominids	robust	*A. robustus*
plasticity	Lucy	*A. boisei*
australopiths	*Australopithecus afarensis*	*Homo sapiens sapiens*
gracile	*A. africanus*	specimen WT17000

Homo habilis	Homo erectus	Olduvai Gorge
Louis and Mary Leakey	Pleistocene	Neandertal
Laetoli		

Objectives

1. Consult Figures 42.1 (main text) and note the principal points of divergence during the evolution of humans from the primitive primates.
2. Beginning with the primates most closely related to humans, list the main groups of primates in order by decreasing closeness of relationship to humans.
3. Explain how you think the human species arose. Make sure your theory incorporates existing paleontological, biochemical, and morphological data.

Self-Quiz Questions

Fill-in-the-Blanks

About (1) _____ million years ago, the redistribution of land masses led to conspicuous changes in climate; there was a major cooling trend and a decline in (2) _____. Under these conditions, forests began to give way to (3) _____; however, in the interim, the environment was a mosaic of forests and (4) _____. Johanson's find, Lucy, was one of the earliest (5) _____, a collection of forms that combined ape and human features; they were fully two-legged ((6) _____), with essentially human bodies and ape-shaped heads. The oldest fossils of the genus *Homo* date from approximately (7) _____ million years ago. The (8) _____ were a distinct hominid population that appeared about 100,000 years ago in Europe and Asia; their cranial capacity was indistinguishable from our own and they had a complex culture. By 30,000 years ago, there was only one remaining hominid species: (9) _____.

CHAPTER TEST

UNDERSTANDING AND INTERPRETING KEY CONCEPTS

___ (1) Which of the following is *not* considered to have been a key character in early primate evolution?
   (a) Eyes adapted for discerning color and shape in a three-dimensional field
   (b) Body and limbs adapted for tree-climbing
   (c) Bipedalism and increased cranial capacity
   (d) Eyes adapted for discerning movement in a three-dimensional field

___ (2) Primitive primates generally live _____.
   (a) in tropical and subtropical forest canopies
   (b) in temperate savanna and grassland habitats
   (c) near rivers, lakes, and streams in the East African Rift Valley
   (d) in caves where there are abundant supplies of insects

___ (3) All ancestral placental mammals apparently arose from ancestral forms of the subclass _____.
  (a) Insectivora, which includes omnivorous shrews and moles
  (b) Carnivora, which includes dogs, cats, and seals
  (c) Rodentia, which includes mice and beavers
  (d) Marsupialia, which includes the koala ("teddy bear") and flying phalanger (like the flying squirrel)

___ (4) The hominid evolutionary line stems from a divergence (fork in a phylogenetic tree) from the age line that apparently occurred _____.
  (a) somewhere between 12 million and 6 million years ago
  (b) about 3 million years ago
  (c) during the Miocene epoch
  (d) less than 2 million years ago

___ (5) _____ was a dryopithecine with teeth and dental arches more like those of humans than of apes.
  (a) *Aegyptopithecus*
  (b) *Australopithecus*
  (c) *Homo erectus*
  (d) *Ramapithecus*

___ (6) Johanson's Lucy was a(n) _____.
  (a) dryopithecine
  (b) australopithecine
  (c) miocene
  (d) prosimian

___ (7) A hominid of Europe and Asia that became extinct nearly 30,000 years ago was _____.
  (a) a dryopithecine
  (b) *Australopithecus*
  (c) *Homo erectus*
  (d) Neandertal

Choose the one most appropriate answer for each.

(8) ___ anthropoids

(9) ___ australopiths

(10) ___ ceboids

(11) ___ Cenozoic

(12) ___ cercopithecoids

(13) ___ dryopiths

(14) ___ hominoids

(15) ___ metazoans

(16) ___ Miocene

(17) ___ Neandertals

A. A group that includes apes and humans
B. A population of *Homo sapiens* that lived from at least 100,000 to as recently as 30,000 years ago; tool users and artisans
C. A cultural period beginning with the earliest chipped stone tools about 750,000 years ago that lasted until about 15,000 years ago (deduction)
D. Organisms in a suborder that includes New World and Old World monkeys, apes, and humans
E. An era that began 65 to 63 million years ago; characterized by the evolution of birds, mammals, and flowering plants
F. A group that includes New World monkeys only
G. A term denoting multicellular animals

(18) ___ Paleolithic

(19) ___ Pliocene

(20) ___ prosimians

H. An epoch of the Cenozoic era lasting from 25 million to 17 million years ago; characterized by the appearance of primitive apes, whales, and grazing animals of the grasslands

I. Organisms in a suborder that includes tree shrews, tarsiers, lemurs, and others

J. An epoch of the Cenozoic era lasting from 7 to 2 million years ago; characterized by the appearance of distinctly modern plants and animals

K. A group that includes Old World monkeys only

L. Bipedal organisms from about 3.8 to 1 million years ago, with essentially human bodies and ape-shaped heads; brains no larger than those of chimpanzees

M. Transitional apelike forms that could climb about in trees and walk on the ground

---

CHAPTER TEST

INTEGRATING AND APPLYING KEY CONCEPTS

Suppose that someone told you that sometime between 12 million and 6 million years ago dryopiths were forced by predatory larger members of the cat family to flee the forests and take up residence in estuarine, riverine, and sea coastal habitats where they could take refuge in the nearby water to evade the tigers. Those that, through mutations, became naked, developed an upright stance, developed subcutaneous fat deposits as insulation, and developed a bridged nose had advantages in watery habitats that other dryopiths that remained inland never developed. As time went on, predation by the big cats and competition with other animals for available food caused most of the terrestrial dryopiths to become extinct, but the water-habitat varieties survived as scattered remnant populations, adapting to easily available shellfish and fish, wild rice and oats, various tubers, nuts, and fruits. It was in these aquatic habitats that the first food-getting tools (baskets, nets, and pebble tools) were developed, as well as the first words that signified different kinds of food.

How does such a story fit with current speculations and evidence of human origins? How could such a story be demonstrated to be true or false?

Crossword
Number Five

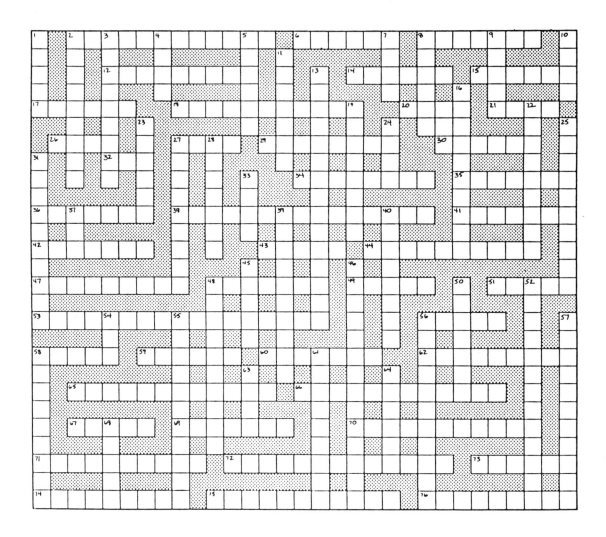

The terms in this puzzle are not necessarily biology terms.

Across

2. an oganism in which the second embryonic opening becomes the mouth
6. a bivalve mollusk
8. animal-like protistans
12. a snail or slug, for example
14. treeless arctic biome
15. a remnant or trace of an organism of a past geologic age
17. a tonguelike rasping organ in various mollusks
18. having only one type of spore
20. _____ worms construct their own "houses"
21. tiny arthropod
26. provides home for parasites
27. an encapsulated form of an organism

29. a severe kind of food poisoning
30. _____ algae, such as diatoms
32. geologic _____, a span of time
34. protistans with cilia
35. type of mammal
36. bivalve mollusks
38. primitive photosynthetic eukaryotes
41. free-swimming marine bivalve
42. club moss
43. a span of geologic time
44. makes all parts work in harmony
47. mutually rewarding dependence
48. same as #55 down
49. fungal sac
51. threadlike filament in fungi
53. earliest hominid type
56. a primitive seed-bearing gymnosperm
58. secreting organ
59. fluid-filled space
60. true segmented worm
62. same as #55 down
65. _____ earth is formed by certain protistans
66. an arrangement of repetitive compartments
67. a type of gastropod
69. _____ stars
70. the process of two different species evolving in response to each other
71. oxygen-bearing blood pigment
72. an asexual reproduction process in which an animal breaks into two or more pieces
73. sweet liquid of flowers or fruits
74. slow addition by deposition
75. disease caused by parasitic nematodes
76. substances that settle to the bottom

Down

1. necessary compound
2. loses leaves
3. _____ have hoofs
4. specific rodents
5. bell-shaped cnidarian
7. part of the respiratory system
8. receiving chamber in heart
9. rapid growth of plankton
10. organ that absorbs oxygen from water
11. type of angiosperm
13. what a glow-worm exhibits
16. one that breaks down organic matter
19. helical bacteria
22. elongated flexible grasping organs
23. parts of a circulatory system
24. a fungal disease of plants
25. division of brown algae
27. have a notochord sometime during their life histories
28. grassland interspersed with trees
31. smallest complete organism known

33. a fault
37. _____ fungi include *Penicillium* and the cup fungi
39. animals that can live on land or in water
40. sea squirt
45. _____ algae include *Ulva* and *Spirogyra*
46. club fungi
48. disease-causing bacteria
50. rodlike bacteria
52. photosynthetic plankton
54. _____ algae include the coralline algae and Irish "moss"
55. a time span on the geologic time scale
56. squids and octopuses
57. develop into sperm-bearing male gametophytes
58. bacterial venereal disease
61. not deciduous
63. fungal disease of plants
64. type of arthropod
68. aquatic plants

43

POPULATION ECOLOGY

General Objectives

1. Learn the language associated with the study of population ecology.
2. Understand the factors that affect population density, distribution, and dynamics.
3. Understand the meaning of the logistic growth equation and know how to calculate values for G by using the logistic growth equation. Understand the meaning of r_{max} and K.
4. Calculate a population growth rate (G); use values for natality, mortality, and number of individuals (N) that seem appropriate.
5. Use the equations for general population growth rate and the logistic growth equation.
6. Understand the significance and use of life tables; interpret survivorship curves.
7. Know the situation about the growth of human populations. Tell which factors have encouraged growth in some cultures and limited growth in others.

ECOLOGY DEFINED
POPULATION DENSITY AND DISTRIBUTION
    Distribution in Space
    Distribution Over Time
POPULATION DYNAMICS
    Variables Affecting Population Size

Summary

For humans, social, economic, and political considerations may influence the short-term distribution of resources on which any population depends. Biological principles predict the consequences of competition for scarce natural resources. *Ecology* is the study of the relations between organisms and the totality of the physical and biological factors affecting them or influenced by them. All evolutionary change is influenced by ecological interactions that operate on the *population, community, ecosystem,* and *biosphere* levels. Ecological studies may focus on one or more different levels: on populations of a single species, on an interacting community of many species, on the movement of matter and energy through a community in an ecosystem, or on global patterns in the biosphere.

    The number of individuals that are part of a population in a specified area determines the population density, and their arrangement in space defines their distribution. *Population density* may be no more than a head count in a total amount of defined space, or it may measure population density in space actually occupied for specific lengths of time. Individuals might be clumped together, spread out rather uniformly, or dispersed randomly in the environment. Distribution may change over time as animals migrate from place to place, often in response to environmental rhythms.

    Population density rises or falls with changes in number of births (*natality*), number of deaths (*mortality*), *immigration,* and *emigration. Population size (N)* = (births + immigration) − (death + emigration). When *N* remains constant over time, there is *zero population growth*.

Key Terms

ecology	herbivores	population density
population	carnivores	clumped distribution
habitat	parasites	random distribution
community	decomposers	uniform distribution
producers	detritivores	natality
autotrophic	ecosystem	mortality
photosynthetic autotrophs	biotic	immigration
chemosynthetic	abiotic	emigration
autotrophs	biosphere	zero population growth
consumers		

Objectives

1. Understand how the principles of ecology can influence human social, economic, and political considerations.
2. Explain how the kinds of interactions among species can shape the structure of a community.
3. Explain why too few head counts can yield an inaccurate estimate of true population density.
4. List the three ways that individuals can be distributed in space and provide an example of each.
5. Discuss the effects of the four parameters (natality, mortality, immigration, and emigration) on the equation for population size.

6. Define *zero population growth* and indicate how achieving it would affect the human population of the United States.

Self-Quiz Questions

Fill-in-the-Blanks

A group of individuals of the same species occupying a given area at a specific time is a(n) (1) _____. The (2) _____ maintains a system of energy use and materials cycling on a large scale. By sorting through soils or sediments, (3) _____ eat the fine particles of organic materials that remain from decomposition activities; (4) _____, nematodes, crabs, and a variety of other heterotrophs are examples of this kind of feeding role. (5) _____ extract blood, sap, and other nutrients from living hosts. Examples of nonrandom distribution are (a) uniform and (b) (6) _____. N = (natality + (7) _____) − (mortality + (8) _____).

43-II
(pp. 709–712)

POPULATION DYNAMICS (cont.)
Exponential Growth
Limits on Population Growth
Carrying Capacity and Logistic Growth

Summary

Any population that is not restricted in some way will grow in size at an increasingly accelerated rate. Population growth is the difference between the birth rate and the death rate, plus or minus any inward or outward migration. A convenient way to express the growth rate of a given population is known as its *doubling time* (how long it takes to double in size). What starts out as a gradual increase in numbers turns into explosively accelerated increases—a pattern of *exponential growth*. As new individuals are added to a population, the number of reproducing members increases, and the increase enlarges the potential reproductive base. When we plot the course of such exponential growth, we end up with a *J-shaped* curve. As long as the birth rate remains even slightly above the death rate, any population grows exponentially.

Any natural system keeps its populations in check. Initially, a bacterial population goes through an exponential growth phase. Next, population growth tapers off, only to give way to a plateau phase in which population size remains relatively stable. Then the population begins to decline from lack of essential resources and increases in toxic wastes.

When any essential resource is in short supply, it becomes a *limiting factor* on population growth. Only if the toxic medium were removed every so often and replaced with a fresh medium would things stabilize. Then we would end up with an *S-shaped curve*: an exponential growth phase that leads into a stable plateau phase. The effect of a limited carrying capacity on population growth is expressed by the *logistic growth equation*.

A population of any species can tolerate only a certain range of environmental conditions. Resource availability, prevailing conditions, and tolerance ranges interact to dictate where population growth must level off. For a given population, these factors define the environment's carrying capacity: the maximum density of individuals that can be indefinitely sustained in a given area, under a given set of environmental conditions—a brake on runaway growth.

Key Terms

r net reproduction per individual
exponential growth
J-shaped curve

doubling time
limiting factor
K, carrying capacity
logistic growth equation

r_{max}, maximum net reproduction per individual
S-shaped curve

Objectives

1. State how increasing the death rate of a population affects its doubling time.
2. Contrast the conditions that promote J-shaped curves with those that promote S-shaped curves in populations.
3. Explain why any population grows exponentially if the birth rate exceeds the death rate.
4. Define *limiting factors* and tell how they influence population curves.
5. Distinguish *tolerance range* from *carrying capacity*, explaining which is a feature of a population and which is a feature of the population's environment.

Self-Quiz Questions

Fill-in-the-Blanks

Any population that is not restricted in some way will show a pattern of (1) _____ growth because any increase in population size enlarges the (2) _____ base. When the course of such growth is plotted on a graph, a (3) _____-shaped curve is obtained. If (4) _____ factors (essential resources in short supply) act on a population, population growth tapers off. In the equation $G = r_{max} N (K - N / K)$, as the value of r_m increases and G and N remain constant, the value of G (5) [choose one] ☐ increases ☐ decreases ☐ cannot be determined by humans, even if they do know algebra. As the value of r_{max} increases and G and N remain constant, the value of K (the carrying capacity) (6) [choose one] ☐ increases ☐ decreases ☐ cannot be determined by humans, even if they do know algebra. (7) _____ _____ is a feature of the environment that is defined for one or more populations living in that environment and serves as a brake on runaway growth. The ability to produce the maximum possible number of new individuals is called the population's (8) _____. Sigmoid curves are characteristic of (9) _____ _____.

LIFE HISTORY PATTERNS
   Life Tables
   Patterns of Survivorship and Reproduction
   Evolution of Life History Patterns
LIMITS ON POPULATION GROWTH
   Density-Dependent Factors
   Density-Independent Factors
   Competition for Resources
   Predation, Parasitism, and Disease

Summary

Data summarized in life tables can be used to construct *survivorship curves*, which show trends in mortality and survivorship and reveal three basic kinds of population types. Type I has high survivorship until some age, then high mortality. Type II populations show a rather constant death rate at all ages. Type III populations experience high mortality early in life.

Populations differ with respect to when in the life cycle and how often they reproduce, as well as with respect to the number of offspring they produce. Reznick and Endler's studies of Trinidadian guppies have identified specific environmental variables that influence guppy life history patterns. A natural population is said to be stabilized when its population size typically fluctuates within a predictable range. Feedback mechanisms regulate birth and death rates; the mechanisms are related to population density. Some mechanisms are built into the species, and others are dictated by the environment. *Density-dependent factors* come into play whenever populations change in size; they depress population size when the density of individuals approaches the environment's carrying capacity, and they ease up when population density decreases. *Density-independent factors* work regardless of population size; they are environmental variables (such as periodic rains or temperature extremes) that may shift enough to push a population above or below its tolerance range for a given variable.

Members of the same population compete with each other for resources. Territoriality, social dominance, or simple defense of food sources are the most common forms of *interference competition*, in which certain individuals control access to the resource. A feeding frenzy in a shark population illustrates *exploitation competition*, in which all individuals have equal access to the resource.

Key Terms

age-specific patterns	timing of reproduction	intraspecific competition
demography	number of offspring	exploitation competition
life tables	density-dependent factor	interference competition
cohort	density-independent	territoriality
survivorship	factor	social dominance

Objectives

1. Explain how the construction of life tables and survivorship curves can be useful to humans in managing the distribution of scarce resources.
2. Define *density-dependent factors*, give two examples, and indicate how density-dependent factors act on populations.
3. Define *density-independent factors*, give two examples, and indicate how they affect populations.
4. Explain how timing of reproduction can affect the degree of intraspecific competition for available resources.
5. Define *intraspecific competition*, give two examples, and indicate how they can affect population growth and distribution.

Self-Quiz Questions

Fill-in-the-Blanks

(1) _____ _____ use information summarized from life tables, which show trends in mortality and life expectancy. Type (2) _____ populations have low survivorship early in life. Food availability is a density-(3) _____ factor that works to cut back population size when it approaches the environment's (4) _____ _____. Environmental disruptions such as forest fires and floods are density-(5) _____ factors that may push a population above or below its tolerance range for a given variable.

Matching

Choose the most appropriate letter(s) for each. A letter may be used more than once, and a blank may contain more than one letter.

(6) ___ cohort

(7) ___ density-independent factors

(8) ___ density-dependent factors

(9) ___ Reznick and Endler

A. Drought, floods, earthquakes
B. Reindeer in the Pribilof Islands in 1935
C. Food availability
D. Adrenal enlargement in wild rabbits in response to crowding
E. All of the 1987 human babies of New York City
F. Life history patterns of Trinidadian guppies

43-IV
(pp. 717–721)

HUMAN POPULATION GROWTH
Where We Began Sidestepping Controls
Age Structure and Reproductive Rates
PERSPECTIVE
SUMMARY

Summary

In 1988 the human population reached 5.1 billion. Between 5 and 20 million people each year now die from starvation or from malnutrition-related diseases, and the future is even more grim because the net population growth is increasing ever more rapidly. Net population growth for a given period is the difference between the total number of live births and the total number of deaths occurring in that time span. Values calculated on the basis of the number of births and deaths per 1,000 persons in the population at the midpoint of a given year are used in determining the percent annual growth rate, which can be extended to determine how long it will take the human population to double in size.

The world average growth rate in 1988 was 1.7 percent. In 1984 Kenya's growth rate was 4 percent; its population will double in eighteen years. Food production and supplies or other limiting factors are not likely to double even with the most intensive effort. In the past two centuries, the long-term human growth rate has undergone an astonishing acceleration that is due to three principal factors: Humans developed the capacity to expand steadily into new environments; the carrying capacity was increased; and a series of limiting factors was removed so that more of the available resources could be exploited.

We now have two options: Either we make a global affort to limit our numbers so that our population stabilizes according to the environment's carrying capacity, or we passively wait until the environment does it for us. Two important factors influence just how much we can expect to slow down the birth rate. *Age structure*—how individuals are distributed at each age level for a population—can give us an idea of the magnitude of the effort that will be needed to control birth rates. *More than a third of the world population consists of young people about to move into their reproductive years.* The *fertility rate* is determined by how many infants are born to each woman during her reproductive years. In industrialized countries, the average number is 2; in developing countries, it is 4.2. A world average of 2.5 children per family would stabilize the world population 70 to 100 years from now at about 6 billion people if the average were implemented now. A simple way to slow things down would be to encourage *delayed reproduction*—childbearing in the early thirties rather than in the mid-teens or early twenties. As long as birth rates even slightly exceed death rates, populations experience exponential growth.

Key Terms

age structure   replacement level   delayed reproduction
pre-reproductive base fertility rate

Objectives

1. Compare the doubling time of Kenya's human population with that of the United Kingdom's human population. Then compare both with that of the world's human population.
2. List three possible reasons why the human growth rate accelerated.
3. Define *age structure* and explain why this is the principal reason it would be 70 to 100 years before the world population would stabilize even if the world average became 2.5 children per family.

Self-Quiz Questions

Fill-in-the-Blanks

The (1) _____ _____ of a population shows how individuals are distributed in each age group. The number of infants born each year per 1,000 women between the ages of 15 and 44 is known as the (2) _____ _____. A world average of (3) _____ children per family is the estimated rate that would bring us to zero population growth; if that were achieved, it would still require at least (4) _____–_____ years before the human population stops growing because (5) _____ _____.

A simple way to slow things down would be to encourage (6) _____ _____. At the current rate of increase, Kenya's population will double in (7) _____ years, whereas the United Kingdom's will double in (8) _____ years.

___ (1) The total number of individuals of the same species that occupy a given area at a given time is _____.
  (a) the population density
  (b) the population growth
  (c) the population birth rate
  (d) the population size

___ (2) The average number of individuals of the same species per unit area at a given time is _____.
  (a) the population density
  (b) the population growth
  (c) the population birth rate
  (d) the population size

___ (3) A population that is growing exponentially in the absence of limiting factors can be illustrated accurately by a(n) _____.
  (a) S-shaped curve
  (b) J-shaped curve
  (c) curve that terminates in a plateau phase
  (d) tolerance curve

___ (4) If reproduction occurs early in the life cycle, _____.
  (a) higher population levels tend to result
  (b) it represents an extrinsic factor that limits population size
  (c) it represents a density-dependent factor that limits population size
  (d) All of the above

___ (5) A situation in which the birth rate plus immigration over the long term equal the death rate plus emigration is called _____.
  (a) an intrinsic limiting factor
  (b) exponential growth
  (c) saturation
  (d) zero population growth

___ (6) The rate of increase for a population (r) refers to the _____ the birth rate and death rate plus immigration and minus emigration.
  (a) sum of
  (b) product of
  (c) doubling time between
  (d) difference between

___ (7) _____ is a way to express the growth rate of a given population.
  (a) Doubling time
  (b) Population density
  (c) Population size
  (d) Carrying capacity

___ (8) Any population that is not restricted in some way will grow exponentially _____.
  (a) except in the case of bacteria
  (b) irrespective of doubling time
  (c) if the death rate is even slightly greater than the birth rate
  (d) All of the above

___ (9) Interaction between resource availability and a population's tolerance to prevailing environmental conditions defines _____.
  (a) the carrying capacity of the environment
  (b) exponential growth
  (c) the doubling time of a population
  (d) density-independent factors

___ (10) In natural communities, some feedback mechanisms operate whenever populations change in size; they are _____.
  (a) density-dependent factors
  (b) density-independent factors
  (c) always intrinsic to the individuals of the community
  (d) always extrinsic to the individuals of the community

___ (11) Which of the following is *not* an intrinsic factor that can influence population size?
  (a) Behavior
  (b) Metabolism
  (c) Predation
  (d) Fertility

CHAPTER TEST

INTEGRATING AND APPLYING KEY CONCEPTS

Assume that the world has reached zero population growth. The year is 2110, and there are 10.5 billion individuals of *Homo pollutans* on Earth. You have seen stories on the community television screen about how people used to live 120 years ago. List the ways that life has changed and comment on the events that no longer happen because of the enormous human population.

44

COMMUNITY INTERACTIONS

General Objectives

1. Define the following ecological terms: *habitat, niche, community, symbiotic, competition, predation, parasitism,* and *mutualism.*
2. List and distinguish among the several types of species interactions.
3. Discuss the positive aspects and the negative aspects of predation on prey populations.
4. Describe how communities are organized, how they develop, and how they diversify.

44-I
(pp. 723–724)

BASIC CONCEPTS IN COMMUNITY ECOLOGY
    Habitat and Niche
    Types of Species Interactions
MUTUALISM

Summary

To understand the character of populations, we must examine their interactions with organisms of other species in the same environment, as well as their interactions with conspecifics. A *community* may be defined as all populations of different species that occupy and are adapted to a given area; it differs from an *ecosystem* in that an ecosystem is a community plus its physical environment. A *habitat* is defined as the sort of place where an organism usually resides; it includes its geographic range and the array of organisms with which it is generally associated. A *niche* is defined by all of the environmental or ecological requirements and interactions that influence a particular population and define its role in the community.

In *symbiotic relationships*, one or both members of two species come to rely on the continuing presence of the other for survival. Such relationships vary in the degree of collaboration, the exclusiveness of the attachments, and the extent to which one species is helped or harmed by the presence of the other. In two-species interactions there can be a one-way or two-way flow of positive, negative, or neutral effects. Neutral interactions have no direct effect on either species.

Commensalism is a weak interspecific symbiotic attachment in which a guest species simply lives better in the presence of its host, and the host suffers no harm. *Predation, parasitism,* and *interspecific competition* are examples of symbiotic relationships that have a negative effect on one or both species.

Flowering plants and their pollinators provide examples of *mutualism*, in which positive benefits are exchanged between participating organisms. As time goes on, mutualism may become so mutually beneficial that the two participant species may coadapt structurally, functionally, and/or behaviorally because each acts as a powerful selective agent on the other; this is *coevolution*. Whenever selective advantages lead to specialized structures or behaviors that make each partner coadapted to and totally dependent on the other, the two species have entered an *obligate relationship*.

Key Terms

mycorrhizae	commensalism	predation
communities	mutualism	parasitism
habitat	symbiotic	facultative mutualism
niche	interspecific competition	obligate mutualism
neutral interaction		

Objectives

1. Define and distinguish between *community* and *ecosystem*.
2. Define and distinguish between *habitat* and *niche*.
3. Define and give one example each of *commensalism* and *mutualism*.

Self-Quiz Questions

Fill-in-the-Blanks

A(n) (1) _____ is defined as a community and its nonliving physical environment; a(n) (2) _____ is limited to all the different populations that occupy and are adapted to a defined area. The (3) _____ of a population is the sort of place where it is typically located; in contrast, the (4) _____ of a population is defined by its role in a community, including all the ecological requirements and interactions that influence that population in its community. Robins and fruit flies are (5) _____ with human populations. The flowering plants and their pollinators form (6) _____ relationships from which both participating populations benefit.

INTERSPECIFIC COMPETITION
  Exploitation Competition
  Interference Competition
PREDATION
  On "Predator" Versus "Parasite"
  Dynamics of Predator-Prey Interactions

Summary

When different species have some requirements or activities in common, their niches overlap. The greater the overlap, the greater the potential for competition between them.

In *competition* one population or individual exploits the same limited resources as another or interferes with another enough to keep it from gaining access to the resource. When two populations are competing for the same resource, one tends to exclude the other from the area of overlap. According to the theory of *competitive exclusion*, no two species can be exactly the same in their activities or in their ability to get and use resources. To a greater or lesser extent, one would have the advantage and the other one would be forced to modify its niche. In this view, no two species in a community can simultaneously occupy the same niche for very long. Population density and the abundance of resources influence whether or not exclusion occurs. Balancing the tendency to minimize niche overlap is a tendency for each species to expand its niche. Coexistence of two or more populations that compete in the same resource category may occur when each specializes within a narrow range of the resource gradient. Coexistence can also occur through resource partitioning, whereby populations share the same resource in different ways, in different areas, or at different times. Although requirements for different species might be similar, they cannot be identical, so interspecific competition is usually less intense than intraspecific competition.

Predators get their food from other organisms (their prey), but they do not live on or in the prey, and they may or may not kill it. *Parasites* also get their food from other living organisms (their hosts), but they live on or in the host for a significant part of their life cycle, and they may or may not kill it. Predators generally focus on prey that are large enough or numerous enough to make the hunt and the capture worth the energy outlay.

Predator populations often fluctuate in response to changes in the environment, changes in prey populations, and physiological changes in the predators themselves. Stable coexistence in both predator and prey populations tends to occur when predation keeps the prey population from exceeding its carrying capacity.

Key Terms

interspecific	competitive exclusion	carrying capacity
intraspecific	predators	functional response
exploitation competition	parasites	numerical response
interference competition	prey	stable coexistence
G. F. Gause	host	oscillation

Objectives

1. Explain how the concept of *competitive exclusion* is related to the concept of a *niche*.
2. State the two tendencies that act to define the extent to which a population influences a community.

3. Describe a study that demonstrates laboratory evidence in support of the competitive exclusion concept.
4. Describe a study that demonstrates interference competition.
5. Present two general ways in which two or more different populations can coexist in the same resource categories.
6. Define *predator* and tell how predators benefit from one another.

Self-Quiz Questions

Fill-in-the-Blanks

In (1) _____ one population or individual exploits the same limited resources as another or intervenes with another sufficiently to keep it from gaining access to the resource. According to the concept of (2) _____ _____, when two species are competing for the same resource, one tends to exclude the other from the area of niche overlap. To a greater or lesser extent, one would have the advantage; the other would be forced to modify its (3) _____. When one population denies another species access to a limited resource (usually by aggressive behavior) the tactic is called (4) _____ _____. When two or more populations share resources in different ways, in different areas, or at different times, coexistence is also possible through (5) _____ _____. Populations of Canadian lynx and snowshoe hare undergo (6) _____ _____, thus providing support for the Lotka-Volterra model of predator-prey interactions. Wolves and dogs establish (7) _____ to avoid direct competition.

44-III
(pp. 731–736)

PREY DEFENSES
    Camouflage
    Moment-of-Truth Defenses
    Warning Coloration and Mimicry
PARASITISM
    True Parasites
    Parasitoids
    Social Parasitism

Summary

Ecologically interacting populations can coevolve through reciprocal selection pressures. For example, predator and prey populations exert continual selection pressures on each other. *Mimicry* and *warning coloration* are two of the coevolutionary results of predator-prey systems. When two or more species of dangerous or noxious organisms resemble each other, the resemblance is Müllerian mimicry; no deception is involved, and there is no division into model and mimic as occurs in Batesian mimicry. *Camouflage* involves adaptations in form, patterning, color, or behavior that enable an organism to blend with its background and escape detection.

Chemical defenses such as warning odors, repellents, alarm substances, and poisons have frequently been developed among various prey species. Prey species avoid predation by adaptations for flight, hiding, fighting, or disguise.

Flowering plants and their pollinators provide examples of *mutualism*, in which positive benefits are exchanged between participating organisms. As time goes on, mutualism may become so mutually beneficial that the two participant species may coadapt structurally, functionally, and/or behaviorally because each acts as a powerful selective agent on the other; this is *coevolution*. Whenever selective advantages lead to specialized structures or behaviors that make each partner coadapted to and totally dependent on the other, the two species have entered an *obligate relationship*.

Parasitism is generally a one-way relationship in which a guest species benefits at the expense of a living host. Parasites spend less energy on survival by forcing their hosts to perform specialized tasks for them; as a result, their structure tends to be simple. *Parasitoids*, on the other hand, kill their hosts by completely consuming their soft tissues by the time they metamorphose into adults. *Social parasites* depend on the social behavior of another species to complete their life cycle; the American brown-headed cowbird depends on another species to shelter and nurture cowbird offspring.

Key Terms

coevolution	mimicry	aggressive mimicry
camouflage	aposematic	speed mimicry
terpenes	Müllerian mimics	parasitoids
pheromone	Batesian mimics	social parasitism

Objectives

1. Define *coevolution* and suggest why it might serve the interests of the populations concerned to coevolve.
2. Identify an outstanding vulnerability that coevolved populations share.
3. Distinguish *mimicry* from *camouflage*.
4. Name an animal that utilizes a startle display behavior when cornered.
5. Explain what energy outlay has to do with predation strategy. Contrast the strategy of the blue whale with that of the killer whale.
6. Explain how obligate relationships are established through coevolution.
7. Describe how prey populations avoid predation. Explain how the environment in which prey populations live helps shape the population's responses to predation.
8. Contrast Batesian mimicry and Müllerian mimicry.

Self-Quiz Questions

Fill-in-the-Blanks

(1) _____ often occurs through reciprocal selection pressures operating on two ecologically interacting populations. Prey populations attempt to avoid predation by adaptations for flight, hiding, fighting, and/or (2) _____.

(3) _____ refers to adaptations in form, patterning, color, or behavior that enable an organism to blend with its background and escape detection. In

(4) _____ mimicry, harmful species resemble each other; there is no model or mimic.

The same letter may be used more than once. Use only one letter per blank.

(5) ___ American bittern

(6) ___ blood flukes (schistosomes) and humans

(7) ___ blue whale and krill

(8) ___ bombardier beetle and grasshopper mouse

(9) ___ brown-headed cowbird, Kirtland's warbler, and botflies

(10) ___ Canadian lynx and snowshoe hare

(11) ___ fungal mycelia and plant root hairs

(12) ___ jackpine sawfly larvae and other parasitoid larvae

(13) ___ katydid

(14) ___ killer whale and seals and dolphins

(15) ___ *Lithiops*

(16) ___ short-eared owl

(17) ___ tiger and tall-stalked golden grasses

(18) ___ yucca moth and yucca plant

A. Parasitism
B. Mutualism
C. Predator-prey relationship
D. Camouflage
E. Startle display

44-IV
(pp. 736–745)

COMMUNITY ORGANIZATION, DEVELOPMENT, AND DIVERSITY
    Resource Partitioning
    Effects of Predation on Competition
    Species Introductions
    Succession
    Species Diversity: Island Patterns
    Mainland and Marine Patterns of Species Diversity
SUMMARY

Summary

Groups of functionally similar species living together in a community do different things. Through *resource partitioning*, groups of functionally similar species share the same resource in different ways, in different areas, or at different times. Weland and Bazzaz investigated the example of bristly foxtail grasses, mallows, and smartweeds living together in the same space and sharing the resources of water and soil nutrients in different ways.

Predation minimizes the intensity of competition among prey species. Introducing nonnative species into new regions may present problems. If appropriate means to control the population of the introduced species are not introduced simultaneously, the ecosystem may be disrupted.

The introduction of species that have evolved elsewhere into a region has sometimes been successful (honeybees, ring-necked pheasants) and sometimes disastrous. Water hyacinths from South America brought to Louisiana and Florida during the 1880s rapidly displaced the native plants and brought much of the river traffic to a halt. Dutch elm and Chestnut blight fungi have devastated millions of acres of U.S. forests.

Primary succession is a gradual process during which vacant land first becomes populated with pioneer species that later are sequentially replaced by other communities until a stable, complex climax community is achieved. The process may take hundreds of years, but sometimes it takes thousands; all the while, the total biomass slowly increases, offering more possibilities for the partitioning of niches.

Secondary succession may also occur when an established community is disrupted in whole or in part. Because most dominant plant species in a climax community may not grow unless an integrated community already exists, reestablishing a community on the exposed soil is not simply a matter of having seeds from nearby trees drift over and germinate. Only when a rich layer of decomposing matter has built up do the trees characteristic of the climax stage start to germinate. Some communities, such as those dominated by sequoia trees, depend on intermittent fires for maintaining long-term stability. Through interlocking food webs, materials and energy are cycled through communities, and overall community stability is established.

Key Terms

resource partitioning	secondary succession	distance effect
climax community	pioneer species	area effect
succession	perennials	species diversity
primary succession	cyclic replacement	

Objectives

1. Discuss (by giving examples) the effects of resource partitioning and predation on interspecific competition and niche relationships.
2. Explain how the introduction of nonnative species can disrupt succession. List five specific examples of species introductions into the United States that have had adverse results.
3. Describe the sequence of communities that might occur if the climax community nearest to you were burned to the ground.
4. Estimate *qualitatively* the differences in species diversity and abundance of organisms likely to exist on two islands: Island A has an area of 6,000 square miles, and Island B has an area of 60 square miles. Both islands lie at 10° N latitude and are equidistant from the same source area of colonizers.

Self-Quiz Questions

Fill-in-the-Blanks

(1) _____ minimizes the intensity of competition among prey species. Two different species with similar niche requirements can occupy that niche if they are able to (2) _____ that resource—that is, use it at different times, in different areas, or in different ways. (3) _____ _____ occurs after a disturbance wipes out one or more communities; in the years that follow, earlier community stages become reestablished. (4) _____ _____ is

the ratio of the total number of different species in an area to the total number of all individuals in the same area; its value (5) [choose one] ☐ increases ☐ decreases ☐ stays the same as one moves from Earth's poles toward the equator.

UNDERSTANDING AND INTERPRETING KEY CONCEPTS

____ (1) All of the populations of different species that occupy and are adapted to a given area are referred to as a(n) _____.
(a) biosphere
(b) community
(c) ecosystem
(d) niche

____ (2) The range of all factors that influence whether a species can obtain resources essential for survival and reproduction is called the _____ of a species.
(a) habitat
(b) realized niche
(c) carrying capacity
(d) ecosystem

____ (3) A one-way relationship in which one species benefits at the expense of another is called _____.
(a) commensalism
(b) competitive exclusion
(c) parasitism
(d) an obligatory relationship

____ (4) The weakest symbiotic attachment, in which one species simply lives in the presence of another species, is _____.
(a) commensalism
(b) competitive exclusion
(c) mutualism
(d) an obligate relationship

____ (5) A symbiotic relationship in which both species benefit is best described as _____.
(a) commensalism
(b) mutualism
(c) predation
(d) parasitism

____ (6) The relationship between the brown-headed cowbird and the Kirtland warbler is an example of _____.
(a) commensalism
(b) competitive exclusion
(c) mutualism
(d) parasitism

___ (7) During the process of community succession, _____.
   (a) the total biomass remains constant
   (b) there are increasing possibilities for niche partitioning
   (c) the pioneer community gives way quickly to the climax community, followed by a succession of more diverse arrays of organisms
   (d) All of the above

___ (8) In a community, friction between two competing populations might be minimized by _____.
   (a) partitioning the niches in time
   (b) partitioning the niches spatially
   (c) fights to the death
   (d) Both a and b

___ (9) Fruit flies probably have a(n) _____ relationship with humans.
   (a) parasitic
   (b) mutualistic
   (c) obligate
   (d) commensal

___ (10) The relationship between an insect and the plants it pollinates (for example, apple blossoms, dandelions, and honeysuckle) is best described as _____.
   (a) mutualistic
   (b) competitive exclusionist
   (c) parasitic
   (d) commensalistic

___ (11) The relationship between the yucca and the yucca moth that pollinates it is best described as _____.
   (a) camouflage
   (b) commensalism
   (c) competitive exclusion
   (d) an obligate relationship

___ (12) In 1934 G. Gause utilized two species of *Paramecium* in a study that described _____.
   (a) interspecific competition and competitive exclusion
   (b) resource partitioning
   (c) the establishment of territories
   (d) coevolved mutualism

CHAPTER TEST

INTEGRATING AND APPLYING KEY CONCEPTS

If you were Ruler of All People on Earth, how would you organize industry and human populations in an effort to solve our most pressing pollution problems?

Is there a *fundamental niche* that is occupied by humans? If you believe so, describe the minimal abiotic and biotic conditions that populations of humans require to live and reproduce. [Note that *thrive* and *be happy* are not criteria.] If you do not think so, state why.

These minimal niche conditions can be viewed as resource categories that must be protected by populations if they are to survive. Do you believe that the cold war between the United States and the Soviet Union primarily involves protection of minimal niche conditions, or do you believe that the cold war is based on other, more (or less) important factors?

(a) If the former, how do you think *minimal* niche conditions might be guaranteed for all humans willing and able to accept certain responsibilities as their contribution toward enabling this guarantee to be met?
(b) If the latter, identify what you think those factors are, and explain why you consider them more (or less) important than minimal niche conditions.

45

ECOSYSTEMS

General Objectives

1. Understand how materials and energy enter, pass through, and exit an ecosystem.
2. Describe an important study that determined the annual pattern of energy flow in an aquatic ecosystem.
3. Explain what studies in the Hubbard Brook watershed have taught us about the movement of substances (water, for example) through a forest ecosystem.
4. Understand the various trophic roles and levels.

45-I
(pp. 747–748)

STRUCTURE OF ECOSYSTEMS
    Trophic Levels
    Food Webs

Summary

An *ecosystem* is a community of organisms functioning together and interacting with their physical environment through (1) a flow of energy and (2) a cycling of materials. Ecosystems are open systems: They require inputs of energy and nutrients for sustenance, and they release energy and nutrient outputs. Inputs and outputs vary considerably, largely depending on the size of the ecosystem, the total metabolic rate of its producer organisms, the ratio of producers to consumers, and the age of the ecosystem. Feeding relationships among species determine the direction and extent of the materials used in and the energy flow through a community. In its simplest form, the trophic structure of community includes four levels of participants: *producers*, which most often secure the *energy and nutrients* used by the entire system; *consumers*, which include herbivores, carnivores, scavengers, parasites, and omnivores; *decomposers*, which generally are fungi and bacteria that break down organic debris and thereby help cycle nutrients and ions back to producers; and *detritivores* (invertebrates that feed on partially decomposed bits of organic matter).

The pattern of feeding relationships in a community is best expressed as a *complex food web* rather than as a simple, linear food chain. A *food web* is a network of crossing, interlinked food chains, encompassing primary producers and an array of consumers and decomposers.

Key Terms

system	energy input	"trophic group"
photosynthetic autotroph	nutrient input	food chain
"self-feeders"	mineral	food web
heterotrophs	energy output	primary producer
detritivores	nutrient output	primary consumer
ecosystem	"ecosystem approach"	secondary consumer
open system	trophic level	tertiary consumer

Objectives

1. Define the term *ecosystem* and state how autotrophic organisms are related to ecosystems.
2. List the principal trophic levels in an ecosystem of your choice; state the source of energy for each trophic level and give one or two examples of organisms associated with each trophic level.

Self-Quiz Questions

Fill-in-the-Blanks

Photosynthetic (1) _____ capture (2) _____ and concentrate (3) _____ for ecosystems. All organisms in a community that are the same number of energy transfers away from the initial energy input into their ecosystem constitute a (4) _____ _____. All (5) _____ are primary consumers. (6) _____ obtain their energy from the organic remains and products of other organisms. (7) _____ are invertebrates that feed on partially decomposed bits of organic matter.

45-II
(pp. 748–753)

ENERGY FLOW THROUGH ECOSYSTEMS
    Primary Productivity
    Major Pathways of Energy Flow
    Pyramids Representing the Trophic Structure

Summary

Primary productivity is the *rate* at which energy becomes stored in organic compounds through photosynthesis. *Gross primary productivity* is the total amount of solar energy converted into organic compounds during photosynthesis. *Net primary productivity* is the energy remaining—and stored as organic matter (biomass)—after aerobic respiration by autotrophs. This net production is the variable energy base for consumers of the ecosystem. A growing animal must take in about 10 kilocalories of photosynthetically derived energy to produce every 1 kilocalorie of stored energy (or potential food energy for predators). Energy storage becomes more efficient over time because energy must be expended on growth and maintenance. The faster energy can be transferred from one trophic level to the next, the higher the efficiency of transfer is within the community.

Energy flows through ecosystems by way of *grazing food webs* (based on the living tissues of photosynthesizers) and *detrital food webs* (based on organic waste products and remains of photosynthesizers and consumers). In most ecosystems, the largest portion of net primary production passes through detrital food webs. Energy generally leaves ecosystems as a result of metabolic activities (for example, respiration) that occur in both producers and consumers; such energy is not recycled.

Feeding relationships may change cyclically or permanently over time. The sizes of the organisms being counted in each trophic level may vary and must be accurately weighed. With such data, *pyramids of* either *numbers* or *biomass* can be constructed as crude forms of ecosystem analysis. An energy budget can be constructed for a community, and this information can be used to determine how much energy flows through the ecosystem in a given time period; H. T. Odum's 1957 study of an aquatic ecosystem in Florida provides a classic example of a simple energy budget. More complex ecosystems are more difficult to analyze. If energy budgets are worked out for each entire community, a general principle can be derived: A community cannot survive indefinitely if its members expend more energy than they take in. Only about 6–16 percent of the energy entering one trophic level becomes available to organisms at the next level.

Key Terms

primary productivity	detritus	"head count"
gross primary productivity	grazing food web	pyramid of biomass
net primary productivity	detrital food web	phytoplankton
net community productivity	"ecological pyramid"	pyramid of energy
	pyramid of numbers	

Objectives

1. Distinguish between *net* and *gross* primary production.
2. Contrast the ways in which energy and nutrients pass through an ecosystem. Explain why nutrients can be completely recycled but energy cannot.
3. Compare grazing food webs with detrital food webs. Present an example of each.
4. Explain why determining a pyramid of numbers or a pyramid of biomass for a particular ecosystem is, at best, an inadequate representation of energy flow through that ecosystem. Explain how one goes about preparing an ecosystem analysis such as the one prepared by H. T. Odum.

Self-Quiz Questions

Fill-in-the-Blanks

The energy remaining after respiration and contained in an organism's biomass is its (1) _____ _____ _____. (2) _____ typically is expressed as grams (dry weight) of organic matter per unit of area. In a (3) _____

_____ _____, decomposers feed on organic wastes, dead tissues, and

decomposed organic matter. Earthworms, millipedes, and fly and beetle larvae

are examples of (4) _____ feeders.

45-III
(pp. 753–759)

BIOGEOCHEMICAL CYCLES
   The Hydrologic Cycle
   Global Movement of Carbon—An Atmospheric Cycle
   Commentary: **Greenhouse Gases and a Global Warming Trend**

Summary

Biogeochemical cycles include the movement of water, nutrients, and other elements and compounds from the environment to organisms, and then back to the environment. An *ecosystem* is a community of organisms functioning together and interacting with their physical environment through (1) a flow of energy and (2) a cycling of materials. In nearly all ecosystems, photosynthetic autotrophs are the organisms that fix the energy and take up the nutrients used by other members of the community. Ecosystems are open systems, with energy and nutrient inputs and outputs. The following variables affect the degree of input and output: the ecosystem size and age, the overall rate of metabolism of its organisms, and the ratio of producer to consumer organisms. Water, carbon, and energy enter and leave ecosystems, whereas nutrients such as nitrogen are mostly recycled within ecosystems. Plants greatly influence the rate at which nutrients move through the ecosystem phase of biogeochemical cycles. Tree roots mine the soil efficiently, moving calcium into growing plant parts and making it available to food webs. Rain and weathering of rocks brought calcium replacements into the watershed. Stripping the land of vegetation disrupts nutrient retention for the entire ecosystem for a long time.

Undisturbed ecosystems have predictable rates of water and nutrient loss; the rates are altered when an ecosystem is disturbed, as by fires or by clear-cutting. Levels of carbon dioxide, for example, are increasing in the atmosphere through human burning of fossil fuels and human alteration of natural ecosystems.

Key Terms

biogeochemical cycles
hydrologic cycle
atmospheric cycles
sedimentary cycle
evaporation
precipitation
detention
transportation

nutrients
Hubbard Brook
   Experimental Forest
watershed
hectare
soil infiltration
runoff
"transpiration"

gauging weir
"holding stations"
"greenhouse gases"
chlorofluorocarbons ,
   CFCs
greenhouse effect
infrared

Objectives

1. Describe the hydrologic cycle, correctly using words such as *evaporation, precipitation, detention, transportation, infiltration, watershed,* and *transpiration.*
2. Describe the carbon cycle and explain how carbonate, carbon dioxide, organic materials, and the greenhouse effect affect your life.

Fill-in-the-Blank

In the (1) ＿＿＿＿ ＿＿＿＿, Earth's surface is warmed by sunlight and radiates (2) ＿＿＿＿ (infrared wavelengths) to the atmosphere and space. Greenhouse gases such as (3) ＿＿＿＿, which are used as refrigerants, solvents and plastic foams and (4) ＿＿＿＿ ＿＿＿＿, which is released from burning fossil fuels, deforestation, car exhaust and factory emissions, will together be responsible for about 75 percent of the global warming trend by 2020. The (5) ＿＿＿＿ ＿＿＿＿ studies demonstrated that stripping the land of vegetation disrupts nutrient retention by an entire ecosystem for a long time; in the watershed, (6) ＿＿＿＿ ＿＿＿＿ efficiently mined the soil for calcium, moving it up into the growing plant parts. (7) ＿＿＿＿ and weathering of rocks brought calcium replacements back into the watershed. In (8) ＿＿＿＿ cycles, the element does not have a gaseous phase.

45-IV
(pp. 760–763)

BIOGEOCHEMICAL CYCLES (cont.)
Global Movement of Nitrogen—An Atmospheric Cycle
Global Movement of Phosphorus—A Sedimentary Cycle
Transfer of Harmful Compounds Through Ecosystems
SUMMARY

Summary

All organisms must synthesize proteins, for which they need nitrogen in the form of ammonia (NH_3), ammonium ions (NH_4^+), nitrite (NO_2^-), or nitrate (NO_3^-); these forms of nitrogen are supplied by aspects of the nitrogen cycle known as *nitrogen fixation, nitrification,* and *ammonification. Denitrification* converts nitrate into gaseous nitrogen (N_2), thus closing the nitrogen cycle.

In the Hawaiian Islands, investigations of atmospheric gases revealed the integrated effects of the carbon balances of land and water ecosystems of a whole hemisphere. Carbon dioxide, chlorofluorocarbons, methane, ozone, and nitrous oxide are the principal greenhouse gases that absorb heat and warm Earth's surface layer.

Bacteria and blue-green algae are the microorganisms that can carry out nitrogen fixation. It can also be carried out by volcanic action and lightning. Enormous amounts of petrochemical energy are needed to synthesize fertilizer industrially; in many cases, modern agriculture has been pouring more energy into the soil in the form of fertilizer than it reaps in the form of food or gasahol.

Key Terms

gaseous nitrogen, N_2	ammonification	nitrogen scarcity
cycling processes	nitrification	leaching
nitrogen fixation	nitrite, NO_2^-	denitrification
ammonia, NH_3	nitrate, NO_3^-	nitrous oxide, N_2O
ammonium, NH_4^+	chemosynthesis	crop rotation

phosphorus cycle   chlorinated hydrocarbon   sylvatic plague
phosphate, $PO_4^{\equiv}$ biological magnification   ecosystem analysis
DDT   malaria

Objectives

1. Define the chemical events that occur during nitrogen fixation, nitrification, ammonification, and denitrification.
2. Explain why agricultural methods in the United States tend to put more energy into the soil in the form of fertilizers, pesticides, food processing, storage, and transport than is obtained from the soil in the form of energy stored in foods.
3. Describe how DDT damages ecosystems in general and how it damaged Borneo's rural habitats in particular.

Self-Quiz Questions

Fill-in-the-Blanks

(1) _____ can assimilate nitrogen from the air in the process known as

(2) _____ _____. In (3) _____, either ammonia or ammonium

ions are stripped of electrons, and (4) _____ (NO_2^-) is released as a

product of the reactions. Under some conditions, nitrate is converted into

(5) _____ _____ by denitrifying bacteria. DDT was sprayed in Borneo

to control (6) _____ responsible for transmitting the organisms that cause

(7) _____. Fill in the blank for this food web:

mosquitoes, flies, cockroaches → lizards

(8) _____ _____

fleas → rats

Because of its stability, DDT is a prime candidate for (9) _____ _____
—the increasing concentration of a nondegradable substance as it moves up
through trophic levels.

CHAPTER TEST

UNDERSTANDING AND INTERPRETING KEY CONCEPTS

___ (1) A network of interactions that involve the cycling of materials and the flow of energy between a community and its physical environment is a(n) _____.
   (a) population
   (b) community
   (c) ecosystem
   (d) biosphere

____ (2) An array of species arranged in an efficient use of materials and energy so that it is stable and self-perpetuating is called a(n) _____.
(a) ecosystem
(b) pioneer community
(c) succession
(d) climax community

____ (3) In the Antarctic, blue whales feed mainly on _____.
(a) petrels
(b) krill
(c) seals
(d) fish and small squids

____ (4) _____ is a process in which nitrogenous waste products or organic remains of organisms are decomposed by soil bacteria and fungi that use the amino acids being released for their own growth and release the excess as NH_3 or NH_4^+.
(a) Nitrification
(b) Ammonification
(c) Denitrification
(d) Nitrogen fixation

____ (5) In a natural community, the primary consumers are _____.
(a) herbivores
(b) carnivores
(c) scavengers
(d) decomposers

____ (6) A growing animal must take in _____ of photosynthetically derived food energy to produce every 1 kilocalorie of energy stored in its body.
(a) 1 kilocalorie
(b) 10 kilocalories
(c) 100 kilocalories
(d) 2,000 kilocalories

____ (7) Which of the following is a primary consumer?
(a) Cow
(b) Dog
(c) Hawk
(d) All of the above

____ (8) Of the 1,700,000 kilocalories of solar energy that entered an aquatic ecosystem in Silver Springs, Florida, H. T. Odum determined that about _____ percent was trapped during photosynthesis and used in generating plant biomass.
(a) 1
(b) 10
(c) 24
(d) 74

___ (9) Most dominant plant species in a climax community _____.
  (a) become quickly reestablished in a cleared area because they are adapted specifically to that geographic region
  (b) cannot grow or develop fully except as part of a certain integrated community structure
  (c) grow faster in areas exposed to sunlight by clear-cutting
  (d) Both a and c

___ (10) Chemosynthesizers utilize as their direct energy source _____.
  (a) energy released from degrading the organic remains and products of all other organisms
  (b) energy from sunlight
  (c) energy released during dehydration synthesis (condensation) reactions
  (d) energy released from the oxidation of inorganic substances

---

CHAPTER TEST

INTEGRATING AND APPLYING KEY CONCEPTS

In 1971 *Diet for a Small Planet* was published. Frances Moore Lappé, the author, felt that people in the United States of America wasted protein and ate too much meat. She said, "we have created a national consumption pattern in which the majority, who can pay, overconsume the most inefficient livestock products [cattle] well beyond their biological needs (even to the point of jeopardizing their health), while the minority, who can not pay, are inadequately fed, even to the point of malnutrition." Cases of *marasmus* (a nutritional disease caused by prolonged lack of food calories) and *kwashiorkor* (caused by severe, long-term protein deficiency) have been found in Nashville, Tennessee, and on an Indian reservation in Arizona, respectively. Lappé's partial solution to the problem was to encourage people to get as much of their protein as possible directly from plants and to supplement that with less meat from the more efficient converters of grain to protein (chickens, turkeys, and hogs) and from seafood and dairy products. Most of us realize that feeding the hungry people of the world is not just a matter of distributing the abundance that exists—that it is being prevented in part by political, economic, and cultural factors. Devise two full days of breakfasts, lunches, and dinners that would enable you to exploit the lowest acceptable trophic levels to sustain yourself healthfully.

46

THE BIOSPHERE

**General
Objectives**

1. Describe the ways in which climate affects the biomes of Earth and influences
   how organisms are shaped and how they behave.
2. Contrast life in lake ecosystems with that in oceans and estuaries.

**46-I
(pp. 765–770)**

COMPONENTS OF THE BIOSPHERE

GLOBAL PATTERNS OF CLIMATE
   Mediating Effects of the Earth's Atmosphere
   Air Currents
   Ocean Currents
   Effects of Topography
   Seasonal Variations in Climate

Summary

The world distribution of species results from climate, topography, species interactions, and geological history. The *biosphere* is composed of many ecosystems
that are linked through movements of materials and energy that span the globe.
Climate is the principal factor that determines the particular ecosystem that exists
at a specific place on Earth. "Climate" means prevailing weather conditions,
including temperature, humidity, wind velocity, degree of cloud cover, and rainfall. Four factors (the amount of incoming solar radiation, Earth's daily rotation

and annual migration around the sun, the arrangement of continents and oceans, and the elevation of land masses) predominate and interact to produce the prevailing winds and ocean currents in different regions. Atmospheric and oceanic circulation patterns influence the distribution of different types of ecosystems. In general, as one goes from the equator to the poles (and from sea level to the high elevations of mountains), specific types of vegetational cover typically occur. Landforms that interrupt or channel air and water movements influence regional climates. Cycles of biological activity often coincide with seasonal changes in climate.

Molecules in the atmosphere absorb heat (infrared radiation) and ultraviolet radiation; many of the potentially harsh effects of incoming solar radiation are moderated by the atmosphere. Heat energy derived from the sun warms the atmosphere—and that energy drives Earth's weather systems. Moist air tends to reduce temperature extremes as day changes to night.

Rain shadows are an example of how topography can influence the distribution of different types of organisms. A rain shadow is a region of diminished rainfall on the leeward side of high mountain ranges, and only plants and animals adapted to arid or semiarid conditions will be found on the leeward side. Weather and organisms of the region affect the composition and characteristics of soils and the distribution of primary producers, which constitute the basis of all food chains.

Key Terms

convergent evolution	"cold front"	gyres
biosphere	"warm front"	countercurrent
hydrosphere	easterlies	rain shadow
atmosphere	westerlies	"leeward"
climate	tradewinds	"windward"
ozone "layer"	doldrums	

Objectives

1. Describe how the biosphere is related to its three components: the narrow zone of water, the lower atmosphere, and the fraction of Earth's crust in which organisms live.
2. Name five abiotic factors that contribute to prevailing weather conditions.
3. Explain how certain components of Earth's atmosphere moderate some of the harsh effects of incoming solar radiation.
4. State why land masses heat and cool more rapidly than oceans do.
5. Explain what causes the prevailing air currents. Compare the prevailing air currents with the prevailing ocean currents and state whether they are similar.

Self-Quiz Questions

Matching

Choose the single most appropriate letter for each.

(1) ___ climate

(2) ___ doldrums

(3) ___ greenhouse effect

(4) ___ gyres

(5) ___ lithosphere

A. Absorbs ultraviolet wavelengths of incoming solar radiation
B. Located between 0° and ±30° N and S
C. Regions of rising, warmed air spreading northward and southward
D. Caused by lack of abundant precipitation and intervening mountains

(6) ___ westerly

(7) ___ ozone

(8) ___ rain shadow

(9) ___ tradewinds

(10) ___ upwellings

(11) ___ Benguela current

(12) ___ Canary current

(13) ___ California current

(14) ___ Gulf Stream

(15) ___ equatorial countercurrent

(16) ___ Humboldt current

(17) ___ Japan current

(18) ___ Labrador current

(19) ___ North Atlantic Drift

(20) ___ west wind drift

E. Regions of rising, nutrient-laden water
F. Prevailing weather conditions that are caused by four principal factors
G. Caused by heat radiated from plants and by soil retention by the atmosphere
H. A wind that blows in an eastward direction
I. Rocks, soils, sediments, and outer portions of crust
J. Circular movements of large water masses
K. A cold water mass that maintains the temperate rain forest associated with Washington and Oregon
L. A southward-flowing current that runs into the Gulf Stream and eddies
M. Feeds into the Gulf of Mexico
N. Southward off Spain and West Africa
O. An eastward-flowing warm-water current in the Pacific Ocean
P. Splits to give rise to a northeastward drift and a southward current
Q. Off the west coast of Peru
R. An eastward-flowing *cold*-water current of the southern hemisphere
S. Flows past the northern coasts of Great Britain and Sweden
T. Merges with the California current in the northern Pacific gyre

Fill-in-the-Blanks

From the base of the Rocky Mountains to the peaks, one generally encounters five zones: (21) _____ and (22) _____, adapted to dry, warm conditions; the (23) _____ belt, with trees that can tolerate moisture and cool temperatures; the (24) _____ belt, with conifers adapted to a still colder climate; the cold (25) _____ belt, where grasses dominate; and on the highest peaks covered with snow and ice, (26) _____ _____ _____. When descending dry air warms, it picks up moisture from the land and leaves it very dry; (27) _____ are created; most of these occur at about 30° latitudes.

46-II
(pp. 770–776)

THE WORLD'S BIOMES
  Desert
  Sclerophyllous Shrublands and Woodlands
  Grasslands

Summary

In terrestrial ecosystems, the composition of the underlying soil helps determine the kinds of producers that establish the basis of the food webs of the ecosystem.

Soils with large particle sizes (gravel and sand) tend to lose water and minerals quickly; only plants that can survive on little water and few minerals can exist in such places. Soils with a large clay component (which has smaller particle sizes) tend to hold water and be anaerobic; they can sustain only plants that can tolerate waterlogged, oxygen-poor soils. Loam (clay containing a large number of large particles to prevent packing) is a productive soil.

On land the distribution of biomes is influenced not only by the amount of sunlight but also by temperature and rainfall. The amount of incoming sunlight varies with latitude, slope exposure, the seasons, and recurring cloud cover or clear skies. The amount of cloud cover depends on interactions between Earth's topography and wind and ocean currents. Water droplets in the air moderate extremes in temperature. The land itself—its elevation, its mineral content, its ability to hold moisture or encourage runoff, its slope—helps dictate the kinds of life found there.

Desert biomes are arid regions that support little primary production. Many deserts occur at latitudes 30° from the equator; others have formed through rain shadow effects. Because there is little moisture in the air, many deserts undergo large daily temperature fluctuations. About 5 percent of the world's deserts bake during the day and are extremely cold during the night; the Sahara of Africa and Death Valley in the United States are such deserts. Cacti, yuccas, and various other plants adapted to endure long periods without water constitute the desert producers. When the brief spring rains arrive, these plants depend on extensive root systems just below the soil surface. Occasional oases indicate where rocks saturated with groundwater break through the surface at low elevation.

Sclerophyllous shrublands and woodlands such as chaparral, cold deserts, successional shrublands, and thickets generally occur in regions with long, hot, dry summers and cold, moist winters. The plants are highly flammable during the drier seasons and produce abundant seeds, many of which require heat and scarring before they can germinate. In these habitats, periodic fast-spreading fires clear away old growth and speed the recycling of nutrients. These biomes occur in semiarid coastal regions of continents between 30° and 40° latitudes.

Grasslands prevail where the land is flat, temperatures are moderate, and rainfall is limited by mountains that bar most of the storms moving in from the sea. Four to twelve centimeters of rain fall annually. In the United States, the dominant primary producers were short grasses (where rain was sparse) and tall trees (where there was a little more rainfall). Today, undisturbed grasslands are rare; tallgrass prairie has been converted to vast fields of corn and wheat, and shortgrass regions sustain herds of imported cattle. Indians and buffalo have been displaced from their ecosystem.

Tropical grasslands include the African *savanna* with its variable numbers of scattered trees or shrubs, as well as the *monsoon grasslands* of Southeast Asia.

Key Terms

W. Sclater and A. Wallace	Regolith, bedrock	sclerophyllous shrublands
biogeographic realms	sand	chaparral
biome	silt	sclerophyllous woodlands
A-horizon, topsoil	clay	shortgrass prairie
E-horizon, zone of	humus	Dust Bowl
leaching	loam	tallgrass prairie
B-horizon, subsoil	laterite	tropical grasslands
C-horizon, parent	desert	savanna
material	desertification	monsoon grasslands

Objectives

1. Describe the typical layered structure of soils and state your understanding of how each layer was formed from its original parent material—solid rock.
2. Arrange according to size (smallest to largest) these particles: sand, silt, and clay. State where loam fits into this series.
3. List the factors that influence the distribution of biomes on land. Then consider the prairie biome and indicate the factors that can bring about more specialized ecosystems within the prairie biome.
4. State the relationship between temperature and rainfall (on the one hand) and the abundance and diversity of producers (on the other hand) in the different biomes.
5. List the factors that encourage tallgrass and shortgrass prairie to form.
6. Construct a complex food web that would be characteristic of the American prairie as it existed before its conversion to agriculture.

Self-Quiz Questions

Fill-in-the-Blanks

(1) _____ is formed as the products of living and dead organisms (humus) are mixed with weathered and eroded loose rock. As rainwater percolates down through the A-horizon, minerals dissolve in it and are carried along with particles to the (2) _____. Tree roots penetrate far below the soil surface, breaking up rocks as they go and absorbing minerals in solution; the region of loose rock that extends to the underlying bedrock is the (3) _____ _____. Arid regions that support little primary production are known as (4) _____. Scrubby plants are dominant primary producers in (5) _____ _____, and cacti, yuccas, and other drought-resistant plants dominate (6) _____ _____. At one time, (7) _____ extended westward from the Mississippi River region to the Rocky Mountains, from Canada through Texas. (8) _____ _____ was converted into vast fields of corn and wheat, and (9) _____ _____ were converted into ranges for imported cattle. Overgrazing and ill-advised attempts to farm the region led to massive erosion, which crippled the primary productivity of much of this land.

46-III
(pp. 776–781)

THE WORLD'S BIOMES (cont.)
Forests
Tundra

Summary

Tropical rain forests emerge where warm temperatures combine with abundant and fairly uniform rainfall; they are stratified communities dominated by trees that spread their canopies far above the forest floor. At different levels beneath the high canopy in such forests are diverse plant species adapted to ever-diminishing amounts of sun. The complex food webs sustained by all the producer plants are astonishing. In such places, the kinds of organisms living in or on a single tree often exceed the kinds of organisms living in an entire forest farther from the equator. When the forest is cleared of plants and animals for agriculture, most of the nutrients are permanently cleared off with them; as a result, a tropical rain forest is one of the worst places to grow crops. Even with *slash-and-burn agriculture*, most of the nutrients are soon washed away because of the heavy rains and poor soils.

Deciduous forests are those in which most plants lose their leaves during part of the year, either as an adaptation to dry seasons (in tropical seasonal forests) or as protection against winter cold (temperate deciduous forests). The *monsoon forests* of India and Southeast Asia can survive years of drought, but human agriculture in these regions depends entirely on the timely arrival of the monsoons. The best examples of the remains of the *temperate deciduous forest* are in the Appalachian Mountains in the southern United States.

The *northern coniferous forest*, or *taiga*, exists where there are cold, long winters with fairly constant snow cover and summers warm enough to promote the dense growth of evergreen conifers (spruce, fir, pine, and hemlock), birch, and aspens.

The United States Forest Service manages all wildlife, watersheds, recreation, and lumbering in the National Forest system. Part of its program involves tree farming, which converts a diverse, self-sustaining ecosystem into a monocrop of fast-growing softwood trees whose harvest is carried out by clear-cutting. Ninety percent of our national forests are potentially available for clear-cutting, which increases the rate of soil erosion and leaves areas scarred and vulnerable for decades.

The *tundra* is a region where it is too cold for large trees to grow but not cold enough to be perpetually frozen over with snow and ice. A permanently frozen layer at least 500 meters thick (the *permafrost*) underlies the surface soil, and the main limiting factor is the low level of solar radiation. Species diversity is low, and nutrient cycling rates are low.

As diverse as all the world's biomes are, as simple or complex as their dominant communities and ecosystems may be, all have a self-perpetuating, self-contained stability.

Key Terms

canopy
evergreen broadleaf forest
tropical rain forest
bromeliads
slash-and-burn
 agriculture
deciduous broadleaf forest

monsoon forests
temperate deciduous
 forest
evergreen coniferous
 forest
boreal forest
taiga

montane coniferous forest
temperate rain forest
pine barrens
arctic tundra
permafrost
peat
alpine tundra

Objectives

1. List the factors that encourage the development of evergreen broadleaf forests.
2. Construct a food web that might typify a tropical rain forest.
3. Describe a monsoon forest and explain what causes a monsoon forest to be deciduous rather than to be a rain forest.
4. Describe a broadleaf deciduous forest and construct a typical food web for this biome.
5. List the factors that encourage the development of evergreen coniferous forests.
6. Describe the physical and biotic features of the tundra ecosystem. Distinguish between *alpine* and *arctic tundra*.

Self-Quiz Questions

Fill-in-the-Blanks

Warm temperatures and uniform, abundant rainfall combined encourage the growth of (1) _____ _____ _____, which contain (2) _____ communities dominated by trees spreading their (3) _____ far above the forest floor. Although this type of biome has a very high species (4) _____, it is one of the (5) [choose one] ☐ best ☐ worst places to grow crops. (6) _____-_____-_____ agriculture reduces the forest biomass to ashes, which are then tilled into the soil, but most of the nutrients are soon washed away because of the heavy rains and poor soils. In (7) _____ _____ forests, most plants lose their leaves during part of the year as an adaptation to the dry seasons that alternate with the wet monsoons. In (8) _____ _____ forests, most of the dominant trees drop their leaves—not in response to dry seasons, but as protection against (9) _____ _____. At lower elevations of the montane coniferous forest, (10) _____ _____ and Douglas fir predominate. (11) _____ is synonymous with boreal forest. Just beneath the surface of the (12) _____ is the permafrost. The main limiting factor in the tundra is the low level of (13) _____ _____.

46-IV
(pp. 782–785)

THE WATER PROVINCES
LAKE ECOSYSTEMS
Lake Zones
Seasonal Changes in Lakes
Trophic Nature of Lakes

Summary

Aquatic ecosystems can be classified according to salinity. *Freshwater ecosystems* with extremely low salinity are either lotic (running-water) or lentic (standing-water) habitats. Lentic ecosystems include lakes and ponds; lotic ecosystems include streams and rivers. Productivity in aquatic ecosystems varies with light penetration, temperature, concentrations of nutrients, topography of the basin, sediments, water depth, and pattern of water movement. All of these factors change with daily and seasonal variations in climate, which in turn affect the nature and distribution of aquatic life. Deep freshwater lakes have layers of water of differing temperatures, density, nutrients, oxygen, and plankton. These deep lakes generally undergo *spring* and *fall overturns* as they respond to changes in air temperature, and they are divided into *littoral*, *limnetic*, and *profundal zones*. Stratification prevents the movement of nutrients from the lake bottom to photosynthesizers in the upper waters and the movement of oxygen from the upper waters to the decomposers below. Thorough mixing occurs only during the spring and fall overturns.

Key Terms

lentic ecosystem
lotic ecosystem
littoral zone
limnetic zone
plankton
phytoplankton
zooplankton

profundal zone
epilimnion
thermocline
hypolimnion
fall overturn
spring overturn

thermocline
oligotrophic
eutrophic
basin
Lake Washington
W. Edmondson

Objectives

1. Describe the causes of temperature stratification in bodies of water.
2. Discuss how spring and fall overturns can occur in freshwater ecosystems.

Self-Quiz Questions

Fill-in-the-Blanks

(1) _____ ecosystems are inland bodies of standing fresh water; (2) _____ ecosystems include creeks, streams, and rivers. Water is densest at 4°C; at this temperature it sinks to the bottom of its basin, displacing the nutrient-rich bottom water upward and giving rise to spring and fall (3) _____. Portions of a lake where light penetrates to the bottom compose the (4) _____ zone, where rooted vegetation and decomposers are abundant. The (5) _____ zone includes areas of a lake *below* the depth where sufficient light penetration balances the respiration and photosynthesis of the primary producers; anaerobic bacterial decomposers are the principal organisms here. (6) _____ lakes are nutrient-poor. In bodies of water that are sufficiently deep, there is a depth at which the temperature of the water decreases rapidly with a small increase in depth; this region is known as the (7) _____.

46-V
(pp. 785–792)

MARINE ECOSYSTEMS
Types of Marine Environments
Estuaries
Life Along the Coasts
The Open Ocean
Commentary: El Niño and Oscillations in the World's Climates
SUMMARY

Summary

Marine ecosystems include high-salinity oceans, seas, and inland bodies of brackish water, but *estuaries* are regions such as bays, salt marshes, and channels where seawater mixes with fresh water. In the oceans, the distribution of producer organisms is governed by such variables as light, temperature, salinity, and avail-

The Biosphere **419**

able nutrients. Only in the epipelagic (photic) zone—the upper 100–400 meters of water—is light intense enough to drive photosynthesis. (The mesopelagic zone [200–1,000 meters below the ocean's surface] has little light but contains most of the inorganic nutrients [nitrates and phosphates].) Temperature and salinity in the bathypelagic zone (between 2,000 and 4,000 meters below the ocean's surface) do not vary much, mainly because major currents tend to circulate water masses on a global scale.

At the surface, especially in the shallow intertidal zone (between the high and low tide marks), temperature and salinity vary with the degree of latitude and with the amount of freshwater intrusion. The way materials are cycled back to the producers determines why marine ecosystems are distributed as they are. In shallow nearshore waters and coral reefs, decay occurs in the photic zone, and resources are recycled as much as they are on land. Over open oceans, wastes and decaying organisms generally drift far from the narrow zone of photosynthetic organisms that could thrive on them; thus, essential minerals such as nitrates and phosphates are constantly being removed from the waters of openwater ecosystems, and the region's productivity is generally low compared to estuaries, nearshore waters, and tropical reefs. Areas of upwelling and the deep-sea geothermal ecosystem of the Galápagos Rift are two notable exceptions to this general rule. Around the world, estuaries, coastal waters, and upwelling areas are being commercially fished at levels that are dangerous to their stability.

Key Terms

estuaries	ocean zonation	abyssal zone
intertidal zone	pelagic province	hadal zone
open ocean	neritic zone	hydrothermal vents
Spartina	oceanic zone	Galápagos Rift
detrital food web	benthic province	upwelling
vertical zonation	continental shelf	El Niño Southern
sandy shores	bathyal zone	Oscillation (ENSO)
muddy shores		

Objectives

1. List the variables that determine how producer organisms (and the consumer organisms that depend on them) are distributed in marine environments.
2. Describe the differences in the way materials are cycled (a) in open-ocean communities and (b) in the intertidal zone.
3. Explain how a hydrothermal vent ecosystem operates.

Self-Quiz Questions

Fill-in-the-Blanks

A region where freshwater mixes with saltwater is a(n) (1) _____.

Organisms that depend on currents and drifts to distribute them are

collectively referred to as (2) _____. Most marine ecosystems fall within

the shallow (3) _____ _____ (between the high and low tide marks)

and the (4) _____ _____ (between 2,000 and 4,000 meters beneath the

ocean's surface). The (5) _____ _____ is a geothermal ecosystem

2,500 meters beneath the ocean's surface, where sulfur-oxidizing bacteria are the primary producers. (6) _____ is a marsh grass that is the dominant primary producer in New England salt marshes.

UNDERSTANDING AND INTERPRETING CHAPTER CONCEPTS

___ (1) The Galápagos Rift, a geothermal ecosystem 2,500 meters beneath the ocean's surface, has _____ as its primary producers.
(a) blue-green algae
(b) protistans
(c) nitrogen-fixing organisms
(d) chemosynthetic organisms

___ (2) In a(n) _____, water draining from the land mixes with seawater carried in on tides.
(a) abyssal zone
(b) rift zone
(c) upwelling
(d) estuary

___ (3) A biome with grasses as primary producers and scattered trees adapted to prolonged dry spells is known as a _____.
(a) warm desert
(b) savanna
(c) tundra
(d) taiga

___ (4) Monocrops of fast-growing softwood trees _____.
(a) are replacing climax forests in parts of the national forest system
(b) are less vulnerable to insect predators than are constituents of a mixed deciduous climax forest
(c) use fewer nutrients than climax forests
(d) are desirable for their nitrogen-fixing ability

___ (5) _____ of our national forest land is potentially open to clear-cutting.
(a) Ten percent
(b) Fifty percent
(c) Ninety percent
(d) None; our national forests are protected as public domains.

___ (6) In tropical rain forests, _____.
(a) competition for available sunlight is intense
(b) diversity is limited because the tall forest canopy shuts out most of the incoming light
(c) conditions are extremely favorable for growing luxuriant crops
(d) All of the above

___ (7) In a lake, the open sunlit water with its suspended phytoplankton is referred to as its _____ zone.
 (a) epipelagic
 (b) limnetic
 (c) littoral
 (d) profundal

___ (8) Differences in _____ determine the distribution of producer organisms in marine ecosystems.
 (a) the intensity of incoming solar radiation
 (b) surface water salinity
 (c) the availability of nutrients
 (d) All of the above

___ (9) _____ is least influential in determining the distribution of biomes on land.
 (a) Light intensity
 (b) Rainfall
 (c) Salinity
 (d) Temperature

___ (10) A wind system that influences large climatic regions and reverses direction seasonally, producing dry and wet seasons, is referred to as a(n) _____.
 (a) geothermal ecosystem
 (b) upwelling
 (c) taiga
 (d) monsoon

___ (11) _____ *least* affects the amount of incoming light that strikes an area.
 (a) Latitude
 (b) Temperature
 (c) The amount of recurring cloud cover
 (d) The degree that a slope is exposed to incoming light

CHAPTER TEST

INTEGRATING AND APPLYING KEY CONCEPTS

If, at the end of Chapter 44, you said that the cold war was essentially a matter of the U.S. and the U.S.S.R.'s protecting minimal niche conditions (resource categories) that support living and reproduction by their respective populations, how do you view the capacity of each nation to bring about a nuclear winter by exploding just 1,000 of their thousands of large warheads? Can you suggest a better global way of guaranteeing the minimal resource categories of all nations on our beautiful Spaceship Earth? If so, outline the requirements of such a system and devise a way in which it could be established.

47

HUMAN IMPACT ON THE BIOSPHERE

**General
Objectives**

1. Understand the magnitude of pollution problems in the United States.
2. Examine the effects modern agriculture has wrought on desert, grassland, and tropical rain forest ecosystems.
3. Describe how our use of fossil fuels and nuclear energy affects ecosystems.

**47-I
(pp. 794–797)**

ENVIRONMENTAL EFFECTS OF HUMAN POPULATION GROWTH
CHANGES IN THE ATMOSPHERE
    Local Air Pollution
    Acid Deposition
    Damage to the Ozone Layer

Summary

In natural ecosystems, life's by-products are generally recycled and do not accumulate for long. *Pollutants* are by-products of our existence that are not recycled through natural disposal systems. Humans do not recycle most of their waste products; instead, they reduce, collect, concentrate, bury, and burn wastes or spread them out through other ecosystems. More and more frequently, as human populations increase, industrial and human wastes that were once put out of sight and out of mind come back in the form of illness and death.

    Into the thin layer of atmosphere we dump more than 700,000 metric tons of pollutants *each day* in the United States alone. Oxides of sulfur and nitrogen are among the most dangerous air pollutants. They corrode buildings. They ruin agricultural crops and forests. They cause humans to suffer everything from headaches and burning eyes to lung cancer, bronchitis, and emphysema. Pollutants that affect life everywhere are by-products of fossil-fuel burning in transportation vehicles, home-heating plants, power plants, and the countless furnaces of industry. Chlorofluorocarbons, oxides of sulfur, carbon dioxide, oxides of nitrogen, and photochemical oxidants are routinely pumped into the atmosphere. Under specific conditions of weather and topography, *thermal inversions* and *photochemical* and *industrial smogs* result. CFCs and other greenhouse gases destroy the protective ozone layer; they also cause the atmosphere to gain and retain too much heat. *Acid rain* kills fish in streams and lakes, as well as great expanses of forest.

Key Terms

greenhouse effect	photochemical smog	dry acid deposition
pollutants	nitric oxide	sulfuric acid
carbon dioxide	nitrogen dioxide	wet acid deposition
thermal inversion	hydrocarbons	acid rain
industrial smog	photochemical oxidants	ozone
oxides of sulfur	PANS, peroxyacyl nitrates	chlorofluorocarbons

Objectives

1. Identify the principal air pollutants, their sources, their effects, and the possible methods for controlling each pollutant.
2. Define *thermal inversion* and indicate its cause.
3. Distinguish *photochemical smog* from *industrial smog*.
4. Explain what acid rain does to an ecosystem. Contrast those effects with the action of CFCs.

Self-Quiz Questions

Fill-in-the-Blanks

(1) _____ _____ _____ attack marble, metals, mortar, rubber, and plastic; they also form droplets of (2) _____ _____ that create holes in nylon stockings. When a layer of dense, cool air gets trapped beneath a layer of warm air, the situation is known as a(n) (3) _____ _____. When (4) _____ and nitrogen dioxide are exposed to sunlight, they are converted into poisonous substances collectively called (5) _____ smog. (6) _____ _____ kills fish in northeastern U.S. lakes and streams, while greenhouse gases such as (7) _____ destroy the ozone layer and raise the (8) _____ _____ _____ for the Earth, melting the ice caps and flooding coastal lands.

CHANGES IN THE HYDROSPHERE
    Consequences of Large-Scale Irrigation
    Maintaining Water Quality
CHANGES IN THE LAND
    Solid Wastes
    Conversion of Marginal Lands for Agriculture
    Deforestation
    Commentary: Tropical Forests—Disappearing Biomes?
    Desertification

Summary

More and more of the water available for use by organisms is becoming polluted because it is used as a dumping ground for by-products of human existence. About 30 percent of the waste water in the United States goes through only *primary* treatment before the liquid effluent is discharged. Mechanical screens and sedimentation tanks are used to force coarse suspended solids out of the water to become a sludge. Chemicals such as aluminum sulfate are added to hasten the sedimentation process, and chlorine is added to kill disease-causing microorganisms. *Secondary* treatment depends on microbial action to degrade the sludge. *Tertiary* treatment involves advanced methods of precipitation of suspended solids and phosphate compounds, adsorption of dissolved organic compounds, reverse osmosis, stripping of ammonia to remove nitrogen from it, and disinfecting the water through chlorination or ultrasonic vibrations. Oxygen-demanding wastes, suspended solids, nitrates, phosphates, and dissolved salts such as heavy metals, pesticides, and radioactive isotopes are the items treated by tertiary treatment.

It is possible to live in the midst of a gradual trend toward unlivable conditions and not even be aware of what is going on. Solid wastes are dumped, burned, or buried at the tremendous rate of 4.5 billion metric tons per year in the United States. Very often, hazardous materials are dumped together with paper, glass, aluminum, steel, copper, and plastic objects that could be recycled or burned to generate heat. Garden and food wastes could be used to generate compost useful in rebuilding soil or to produce animal feed. Although an ecologically based system for handling solid waste is the most desirable solution, even a recycling system would be preferable to the dump, burn, or bury throwaway system in use now.

The conversion of marginal lands to agricultural uses involves massive applications of fertilizers and pesticides and ample irrigation to nurture high-yield crops. Energy from fossil fuels generally drives the farming machines. Subsistence farmers in developing countries usually cannot afford to farm marginal lands with these practices. *Deforestation* leads to loss of the fragile topsoil and disrupts the watershed. Overgrazing and improper irrigation practices have increased the rate of desertification. Salt buildup and waterlogging of soils stunt growth, decrease yields, and eventually kill crop plants.

Key Terms

saline, salination	precipitation	alkaloids
Ogallala aquifer	reverse osmosis	desertification
primary treatment	green revolution	hectare
sludge	slash-and-burn-agriculture	Dust Bowl
secondary treatment		
tertiary treatment		

Objectives

1. Define *primary, secondary,* and *tertiary waste-water treatment* and list some of the methods used in each of the three types of treatment.
2. Identify any products of waste-water treatment that could be of use to humans.
3. List the disadvantages and dangers of trying to maintain our present throw-away system for handling solid wastes.
4. Distinguish a recycling system for solid wastes from an ecologically based system. State the conditions under which an ecologically based system for handling solid wastes would benefit the culture more than a recycling system would.
5. Describe the basic strategy and layout of a generalized resource recovery system.
6. Contrast subsistence agriculture in parts of China with modern large-scale agricultural practices in the United States. State which you think disturbs the grassland ecosystem less.
7. Explain the repercussions of deforestation that are evident in soils, water quality, and genetic diversity in general.

Self-Quiz Questions

Fill-in-the-Blanks

In a generalized resource recovery system, (1) _____ could be used to extract steel and iron, and (2) _____ _____ could send plastic and paper to different recovery chambers. In (3) _____ waste-water treatment, mechanical screens and sedimentation tanks are used to force coarse suspended solids out of the water to become a sludge. (4) _____ treatment involves advanced methods of precipitation of suspended solids and phosphate compounds. (5) _____ on steep slopes leads to soil erosion and watershed disruption. The buildup of salt in irrigated soils is called (6) _____. Farmers take so much water from the (7) _____ _____ that the overdraft nearly equals the annual flow of the Colorado River.

47-III
(pp. 803–806)

A QUESTION OF ENERGY INPUTS
Fossil Fuels
Nuclear Energy
PERSPECTIVE

Summary

Paralleling the J-shaped curve of human population growth is a steep rise in energy consumption. In the highly industrialized United States, the average individual directly or indirectly uses more than 200,000 kilocalories a day.

Coastal forests and algae were buried and transformed, under the sediments and ooze and compression of time, into fossil fuels: coal, oil, and natural gas. Oil is due to be depleted during the next century. The coal reserves in the United States represent one-fourth of the world's known coal supply. We must certainly burn more coal over the next two to four decades, but the environmental costs will be great.

Nuclear-powered, electricity-generating stations in the United States produce 8 percent of the nation's electrical energy. The costs of constructing these plants are great, the net energy produced is low, and there is serious concern about short-term safety in and near the plants, as well as about the long-term feasibility of storing nuclear wastes.

Solar energy ranks with wind energy as the cleanest, most abundant, and safest of all potential energy sources now being considered as alternatives to fossil fuels.

Many sources of stress on the biosphere (human overpopulation, overconsumption, ecosystem oversimplification, massive industrial technology, lack of leadership, and me-first behavior) can be traced to indifference to the principles of energy flow and materials reuse. Like all living forms in the past, we are all potentially endangered species, kin to one another in vulnerability because of intricate, often invisible, threads of ecological interdependence stretched through the biosphere.

Key Terms

net energy	strip mining	fusion power
fossil fuels	meltdown	nuclear winter
oil shale	breeder reactor	cultural imperative
kerogen		

Objectives

1. State when the U.S. petroleum reserves are expected to run out if present consumption rates are maintained. Natural gas?
2. Explain why exploiting oil shale deposits may not be worth doing.
3. Compare the functions of a conventional nuclear power plant reactor with those of a breeder reactor and assess the principal benefits and risks of each.
4. Explain what a *meltdown* is.
5. Define *fusion power* and assess the prospects for its being used.
6. List five ways in which you personally could become involved in ensuring that institutions serve the public interest in a long-term, ecologically sound way.

Self-Quiz Questions

Fill-in-the-Blanks

The Chernobyl power station experienced a (1) _____ and a (2) _____ fire. Oil shale is buried rock that contains (3) _____, a hydrocarbon compound. Nuclear-powered electricity-generating plants produce (4) _____ percent of the U.S.'s electrical energy. Only (5) _____ has good long-term, intermediate, and short-term availability as an energy option for the United States. Unlike conventional nuclear-fission reactors, (6) _____ reactors could potentially explode like atomic bombs. A nuclear exchange involving about one-third of the existing American and Soviet arsenals would probably kill between 40 and 65 percent of the human population and most other forms of life; the detonations would create a (7) _____ _____, in which a huge dark cloud of soot and smoke would block out the sun.

UNDERSTANDING AND INTERPRETING KEY CONCEPTS

___ (1) Which of the following processes is *not* generally considered a component of tertiary waste-water treatment?
(a) Microbial action
(b) Precipitation of suspended solids
(c) Reverse osmosis
(d) Adsorption of dissolved organic compounds

___ (2) When fossil-fuel burning gives off particulates and sulfur oxides, we have _____.
(a) photochemical smog
(b) industrial smog
(c) a thermal inversion
(d) Both a and c

___ (3) _____ results when nitrogen dioxide and hydrocarbons react in the presence of sunlight.
(a) Photochemical smog
(b) Industrial smog
(c) A thermal inversion
(d) Both a and c

___ (4) Between _____ of the urban wastes are paper products.
(a) 10–20 percent
(b) 20–40 percent
(c) 50–65 percent
(d) 75–90 percent

___ (5) Breeder reactors would be _____ than other types of nuclear reactors.
(a) more energy efficient
(b) more widely used
(c) safer
(d) All of the above

___ (6) About _____ of the waste water in the United States is not even receiving primary treatment.
(a) 20 percent
(b) 35 percent
(c) 50 percent
(d) 65 percent

___ (7) Of *all waste water* in the United States, about _____ goes through primary treatment *only*.
(a) 20 percent
(b) 25 percent
(c) 30 percent
(d) 40 percent

___ (8) Primary treatment of waste water does *not* involve using which of the following?
   (a) Sedimentation tanks
   (b) Aeration with pure oxygen
   (c) Mechanical screens
   (d) Chemicals such as aluminum sulfate

___ (9) The wastes from a nuclear power reactor must be kept out of the environment for _____ years before they are safe.
   (a) 25
   (b) 250
   (c) 1 million
   (d) 250,000

___ (10) Operation Plowshare was _____.
   (a) an enormous program that promoted new efficient agricultural methods
   (b) a massive effort to harness the atom for peacetime use
   (c) also known as the Green Revolution
   (d) a program that bought up a great deal of land for the purpose of burying hazardous solid wastes

___ (11) What proportion of the world's human population does *not* have enough pure water?
   (a) One-fourth
   (b) One-half
   (c) Three-fourths
   (d) Two-thirds

___ (12) For every million gallons of water in the world, only about _____ gallons are in a form that can be used for human consumption or agriculture.
   (a) 6
   (b) 60
   (c) 600
   (d) 6,000

___ (13) Each day in the United States, _____ metric tons of pollutants are discharged into the atmosphere.
   (a) 1,000
   (b) 100,000
   (c) 700,000
   (d) 5,000,000

___ (14) The most abundant fossil fuel in the United States is _____.
   (a) carbon monoxide
   (b) oil
   (c) natural gas
   (d) coal

INTEGRATING AND APPLYING KEY CONCEPTS

If you were Ruler of All People on Earth, how would you encourage people to depopulate the cities and adopt a way of life by which they could supply their own resources from the land and dispose of their own waste products safely on their own land?

Explain why some biologists believe that the endangered species list now includes all species.

48

ANIMAL BEHAVIOR

General Objectives

1. Understand the components of behavior that have a genetic and/or hormonal basis.
2. Distinguish behavior that is primarily instinctive from behavior that is learned.
3. Know the aspects of behavior that have an adaptive value.

48-I
(pp. 808–812)

MECHANISMS OF BEHAVIOR
    Genetic Foundations of Behavior
    Hormonal Effects on Behavior
INSTINCTIVE BEHAVIOR
LEARNING
    Categories of Learning
    Imprinting
    Imprinting and Migration
    Song Learning

Summary

Neuromotor responses are determined by the central nervous system. Behavioral responses are created by integrating sensory, neural, endocrine, and effector components, all of which evolve by natural selection. Innate behavior is represented by genetically determined neuromotor responses that predictably run their course once some stimulus sets them in motion. Any stimulus that activates a neuromotor response (or even some part of it) is called an *innate releasing mechanism*. Perception as to what constitutes the correct releaser for a particular response also has a genetic basis; this and the response may undergo mutation. Any behavior is subject to selection pressures; if advantageous to survival and reproduction, the responses will probably be transmitted to the next generation;

disadvantageous neuromotor responses tend to be eliminated. Adaptive behaviors contribute to the reproductive success of the individual.

As long as a response contributes to adaptability, selection will tend to favor individuals in which that behavior is rigidly locked into the genetic program. But if some innate behavior could be modified and if the change made the individual more successful than other members of the population in finding food, escaping from predators, or producing offspring, the flexibility of response would tend to be perpetuated. If such changes are more than one-time motor responses to a new cluster of stimuli and if there is an enduring potential for adapting future responses to a new encounter with a cluster of past experiences, the modification is called *learning*. In constant or highly predictable environments, rigid genetically determined motor patterns are common; in changing and unpredictable environments, the capacity for behavioral flexibility and learning tends to prevail.

Classical conditioning, whereby a connection is made between a new stimulus and a familiar one, is a form of *associative learning*. It is adaptive because it enables animals to anticipate and recognize potentially dangerous situations in time to deal with them adequately and thus survive. Another form of associative learning is *instrumental conditioning*, in which a reinforcing stimulus (reward or punishment) appears after a behavioral response is given; the animals learn by trial and error. In the absence of the reinforcing stimulus, the learned behavior may soon become extinguished; this is a learning process called *extinction*. Exploration (curiosity) encourages *latent learning*. Some primates demonstrate *insight learning* (alternative responses to a situation are first evaluated mentally with reference to stored information in the brain, and an enlightened response is made). The limited time spans during which learning certain forms of behavior occurs are the sensitive periods of development; *imprinting*, in which an enduring preference for a particular object (or other animal or place) is formed, occurs during an animal's critical period. Stimulus enrichment during the first two years of human life is critically important in developing intelligence in humans.

Key Terms

animal behavior	sign stimulus	extinction
hormones	innate releasing	latent learning
behavioral primer	mechanism	insight learning
seasonal photoperiodicity	learning	imprinting
song system	associative learning	compass sense
instinct	conditioned reflex	navigational sense

Objectives

1. Define *behavior* and name four factors that produce it.
2. Explain what a neuromotor response is and state its relationship to innate behavior.
3. Explain how a researcher can determine whether a behavior is innate or learned.
4. Define *releaser* and give one example of how a releaser works.
5. Provide an example of selection pressures operating to eliminate a disadvantageous behavior.
6. Give one example of an innate behavioral response that has been modified in ways that are adaptive to changing times and circumstances.
7. Define *learning* and distinguish it from *innate behavior*.
8. State the generalization that expresses the relationship between predictability of environmental events and the capacity for behavioral change and learning.
9. Explain what is meant by classical Pavlovian conditioning and distinguish it from instrumental conditioning by describing an example of each process. Explain why both forms of conditioning are examples of associative learning.

10. Name some of the simplest animals in which learning has been demonstrated. Describe the experiment that demonstrated the associative learning process.
11. Define *insight* and give an example.

Self-Quiz Questions

Fill-in-the-Blanks

A (1) _____ _____ is a sequence of motor outputs determined by the central nervous system that results in a coordinated movement. It is genetically determined—that is, (2) _____—and predictably runs its course once some stimulus sets it in motion. Any stimulus that activates a neuromotor response is known as a(n) (3) _____ _____.

(4) _____ is a change in behavior that involves an enduring potential for adapting future responses as a result of past experience. In (5) _____ environments, rigid, genetically determined motor patterns are common. The work of Ivan Pavlov centered on a form of (6) _____ learning, classical conditioning, in which new responses called (7) _____ _____ were established in the behavior of dogs. Another form of associative learning, (8) _____ conditioning, uses a reinforcing stimulus (a reward or punishment) that appears *after* a response is given, and learning occurs by (9) _____-_____-_____. If the reinforcing stimulus is stopped, the learned behavior is discontinued—a learning process called (10) _____. (11) _____ _____ has been demonstrated to occur only in some primates; it is a trial-and-error process that goes on in the brain before an appropriate response is made. For many animals, learning certain forms of behavior occurs only during limited time spans called (12) _____ _____ of development; preferential behavior toward a stimulus acquired during such a time is called (13) _____.

48-II
(pp. 812–819)

THE ADAPTIVE VALUE OF BEHAVIOR
Foundations of Behavioral Evolution
Adaptation and Feeding Behavior
Anti-Predator Behavior
Reproductive Behavior
SUMMARY

Summary

Behavior tends to adapt an organism to its environment, enabling it to reproduce successfully. To determine the ways in which a behavior is adaptive for an organism, the researcher generally profits most by using an approach that assumes that selection is operating to the individual's benefit—not primarily for the good of the group.

Many prey organisms have developed *avoidance behavior* to a fine art: sophisticated escape responses based on ultrasensitive receptors; "freeze" behavior in combination with environmentally cryptic coloration; and behavior that threatens the predator in some way. Predators, on the other hand, have developed lures and camouflage with productive results, and some have employed the use of tools.

Among sexually reproducing animals, the other members of one's species create obstacles to reproductive success. Competition for available mates brings about sexual selection. Environmental factors that influence the distribution of females are important to male competition for access to mates. Presumably, females of some organisms select their mates. Two main types of benefits may result from the female's choice: superior genes in her offspring or superior material benefits for both herself and her offspring.

Key Terms

reproductive success
adaptive behavior
selfish behavior

altruistic behavior
natural selection
sexual selection

resource-defense behavior
female-defense behavior

Objectives

1. Contrast altruistic behavior with selfish behavior and cite examples of each in animal populations.
2. State whether Gary Larson's lemming cartoon supports or undermines group selection.
3. Describe Steven Arnold's studies that suggest that feeding behavior in certain California garter snakes has some genetic basis.
4. Contrast resource-defense behavior with female-defense behavior and cite specific examples.

Self-Quiz Questions

Matching

Match the example or definition with the term at the left.

(1) ___ altruistic behavior

(2) ___ female-defense behavior

(3) ___ reproductive success

(4) ___ resource-defense behavior

(5) ___ selfish behavior

(6) ___ sexual selection

A. Male bison, lions, elk, or bighorn sheep competing in combat to gain a ready-made harem
B. Some female dragonflies laying their eggs only in particular habitats on ponds or streams that male dragonflies defend
C. A male white-throated sparrow singing his "Sam-Peabody" song in his chosen breeding site, trying to attract a female
D. The captain of a ship shouting "Women and children first!" as the boat begins to sink and the lifeboats are launched
E. Survival and reproduction of offspring
F. A male cat eating a female cat's newborn kittens and then mating with her

UNDERSTANDING AND INTERPRETING KEY CONCEPTS

___ (1) Any stimulus that activates a motor score, or even some part of it, is called a _____.
 (a) satiety response
 (b) topographic cue
 (c) meteorologic cue
 (d) releaser

___ (2) Rhythms that are based on the daily rotation of the Earth about its long axis are _____.
 (a) examples of photoperiodism
 (b) governed by N-acetyltransferase
 (c) circadian
 (d) regulated by topographic cues

___ (3) Newly hatched goslings follow any large moving objects to which they are exposed shortly after hatching; this is an example of _____.
 (a) homing behavior
 (b) imprinting
 (c) piloting
 (d) migration

___ (4) The principal difference between classical Pavlovian conditioning and instrumental conditioning is that _____.
 (a) one uses a reinforcing stimulus and the other doesn't
 (b) one presents the reinforcing stimulus before the response and the other one does so after
 (c) one uses a bell as a reinforcing stimulus and the other uses an instrument as the reinforcing stimulus
 (d) a satiety center is involved in one and an aggression center is part of the other

___ (5) A submissive animal that exposes its throat or genitals to a dominant member of the same group is said to be engaging in _____ behavior.
 (a) appeasement
 (b) avoidance
 (c) ritualized
 (d) dispersive

___ (6) A cat explores all of the rooms of a new home even though such exploration is not rewarded. Later, when the cat begins to feel chilled, it goes directly to the warmest room. This is an example of _____.
 (a) classical conditioning
 (b) echolocation
 (c) latent learning
 (d) imprinting

___ (7) Motor activity and metabolic rates associated with biological clocks are coordinated by _____.
 (a) thyroxin secreted by the thyroid gland
 (b) melatonin secreted by the pineal gland
 (c) a magnetic sense that is attuned to variations in Earth's magnetic field
 (d) pheromones released by the dominant animal of the group

INTEGRATING AND APPLYING KEY CONCEPTS

Explain whether you think humans have any critical periods for establishing the ability to learn certain kinds of knowledge. State whether you think humans undergo imprinting. Do you think humans employ resource-defense behavior? If so, can you cite an example? Female-defense behavior? Example?

49

SOCIAL BEHAVIOR

**General
Objectives**

1. Describe how forms of communication organize social behavior.
2. List the costs and benefits of social life.
3. Explain the roles of self-sacrifice and altruism in social life.

**49-I
(pp. 820–825)**

COMMUNICATION AND SOCIAL BEHAVIOR
    Consider the Termite
    Communication Defined
CHANNELS OF COMMUNICATION
    Visual Signals
    Chemical Signals
    Tactile Signals
    Acoustical Signals

Summary

Communication signals are stimuli produced by one animal that change the behavior of another individual in ways that benefit both the signaler and the receiver. Visual, chemical, acoustical, and tactile signals provide a variety of types of information.

Key Terms

social behavior   bioluminescent   von Frisch
communication signal   pheromone   "waggle dance"

Objective

1. List four kinds of signals with which animals convey information.

**Self-Quiz
Questions**

Matching

Select the best linkage.

(1) ___ acoustical signal   A. Pheromone
                                B. Sam Peabody, Peabody, Peabody
(2) ___ chemical signal   C. Bioluminescent messages
(3) ___ tactile signal   D. Dance of the foraging honeybee
(4) ___ visual signal

49-II
(pp. 825–827)

COSTS AND BENEFITS OF SOCIAL LIFE
PREDATION AND SOCIALITY
    Dilution Effect
    Improved Vigilance
    Group Defense
    The Selfish Herd

Summary

Predation dominates the selection pressures favoring the evolution of social behavior. By banding together with other members of the same species for mutual defense, an individual's risk of being selected as prey is diluted, vigilance is improved, and the chances for survival of the population are increased.

    In the unconscious competition between bearers of selfish alleles and others with self-sacrificing tendencies, the selfish types should become more common as time progresses. More behavioral traits are convincingly explained by using Darwinian individual selection than by using good-of-the-group selection.

Key Terms

dilution effect   "selfish herd"

Objectives

1. Explain how being preyed upon improves a population's sociality.
2. List some factors that encourage large social groups such as herds and flocks to form.

**Self-Quiz
Questions**

Fill-in-the-Blanks

The reason for diversity in social life becomes clearer if (1) _____ are

considered, as well as benefits of social life, in terms of individual success with

(2) _____. (3) _____ is by far the dominant selection pressure favoring the evolution of social behavior. Some animals live in groups simply to use other individuals as living shields against (4) _____; such groups have been labeled (5) _____ _____.

Summary

Because of the genetic continuity between a parent and its offspring, it is possible for a parent to die in the defense of its young (and thus give up its own chances for future reproduction) and still have its genes spread through the population.

Natural selection favors species with reproductively selfish behavior, as exemplified by Caspian terns. Prolonged courtship rituals protect the reproductive concerns of both parents and help ensure successful rearing of the young. Male territorial displays and mate-guarding behavior help ensure that the fittest males, their mates, and their offspring will have access to the best environmental resources.

Submission finds expression in two ways: avoidance behavior and appeasement behavior (exaggerated displays of submission). In a natural environment, a dominant animal rarely takes advantage of such total vulnerability, so selection pressures must favor the behaviors that help keep the group together rather than the ones that would promote mutual destruction. Selection favors the dominant animal whose competitive show of strength ensures that it will get the choicest food and its pick of potential mates. A dominant animal advertises its higher status in formal displays, or ritualized behavior, which are exaggerated ordinary functional movements that send distinct signals to conspecifics. Initially, fighting establishes dominance by one individual and submission by others. A dominance hierarchy is established, and actual fighting dwindles as a result. Because less energy is wasted on aggression, more is available for the business of survival.

Among the social insects such as bees and termites, division of labor reaches the level present in a complex society; individuals have become so specialized that they cannot survive on their own. When some members become specialized in food gathering, others in defense, and others in reproduction, there is little energy wasted in the group as a whole, and selective agents must act upon the instinctive cooperative behavior of all individuals.

If natural selection is the basis for evolutionary change, then the behavioral as well as structural traits that are most adaptive—in other words, those that provide the edge in competition for resources—will be selected for. Social groups are based, to varying degrees, on mutual cooperation. Aggression is dispersive; cooperation is cohesive. Many writers have attempted to decipher the inherent, natural behavior of humans, and many tend to overestimate the role of heredity in determining behavior and status in human societies.

Key Terms

dominance hierarchy   cooperation   kin selection
altruistic behavior   self-sacrificing behavior

Objectives

1. Explain the factors that encourage large social groups such as schools, herds, and flocks to form.
2. Define *aggressive behavior* and state what it accomplishes for an animal.
3. Explain why an animal that became established at a low position in a dominance hierarchy would not simply leave the group and seek a more dominant position elsewhere.
4. State any long-term values that might accrue to an animal population from assigning dominant and submissive statuses to its members.
5. Contrast avoidance behavior with appeasement behavior.

Self-Quiz Questions

Fill-in-the-Blanks

Usually, large social units such as herds and flocks form in (1) _____ environments—the arctic tundra, savanna plains, open seas—where food is adequate but there are few places to (2) _____. Fighting results in establishing (3) _____ for one individual and submission by others. The (4) _____ animals end up being separated by a greater distance from the rest of the group, and a ranking of its members—a (5) _____ _____—serves to lessen actual fighting. Dominant animals advertise their status in formal displays, or (6) _____ _____. Submission finds new expression in avoidance behavior and (7) _____ behavior as a further show of deference. Among the (8) _____ _____, division of labor reaches the level that exists in a complex society; the (9) _____ becomes the reproductive "organ" for all, and the workers forage for food and repel invaders. (10) _____ are male members of the colony that have no sting; some mate with the queen. People who believe in the notion of (11) _____ _____ tend to overestimate the role of heredity in determining behavior and status in human societies.

CHAPTER TEST

UNDERSTANDING AND INTERPRETING KEY CONCEPTS

___ (1) A dominant wolf carries its head and tail erect and walks with a stiff, formal gait. These behaviors constitute what is referred to as _____ behaviors.
   (a) appeasement
   (b) courtship
   (c) ritualized
   (d) defensive

___ (2) An animal who arranges its schedule of activities to minimize competition for essential resources with a dominant member of the same group is said to be engaging in _____ behavior.
(a) appeasement
(b) avoidance
(c) ritualized
(d) territorial

___ (3) The studies of siblicide among the egrets demonstrated that _____.
(a) sibling aggression promotes reproductive success of the parents
(b) parents will often intervene when siblings fight with the runt of the litter
(c) parent egrets incubate the eggs so that they hatch synchronously
(d) the parents engage in complex rituals before mating

___ (4) The example used to demonstrate that Darwinian individual selection explained some behavioral traits better than "good of the group" selection was _____.
(a) the dilution effect in wildebeest and zebra populations
(b) siblicide among egrets
(c) courtship behavior in albatrosses
(d) the dispersal of Norwegian lemmings when population densities became extremely high

___ (5) The example used to demonstrate that competitive interactions lead to the formation of dominance hierarchies involved _____.
(a) albatrosses
(b) a honeybee colony
(c) baboon troops
(d) greylag geese

___ (6) Parental support of offspring is an example of _____.
(a) artificial selection
(b) kin selection
(c) natural selection
(d) negative selection
(e) stabilizing selection

___ (7) The top male in a baboon troop is called the _____ male.
(a) alpha
(b) beta
(c) ultra
(d) omega
(e) super

___ (8) In highly integrated insect societies, _____.
(a) natural selection favors individual behaviors that lead to greater diversity among members of the society
(b) there is scarcely any division of labor
(c) cooperative behavior predominates
(d) patterns of behavior are flexible, and learned behavior predominates
(e) All of the above

___ (9) Social behavior among insects depends on _____.
   (a) diversity
   (b) echolocation
   (c) polymorphism
   (d) genetic similarity
   (e) communication

___ (10) In the termite colony described in the text, there is _____.
   (a) one queen that reproduces parthenogenetically
   (b) one queen and a harem of males to fertilize her
   (c) one king and one queen
   (d) several queens operating at one time
   (e) a complete reproductive caste to ensure the continuance of the species

---

CHAPTER TEST

INTEGRATING AND APPLYING KEY CONCEPTS

If you were Ruler of All People on Earth, what mechanisms would you employ to distribute essential resources so as to minimize deadly competition and wars? Would a person who believed in biological determinism be likely to advocate the equal distribution of essential resources among all members of a social group?

Crossword
Number Six

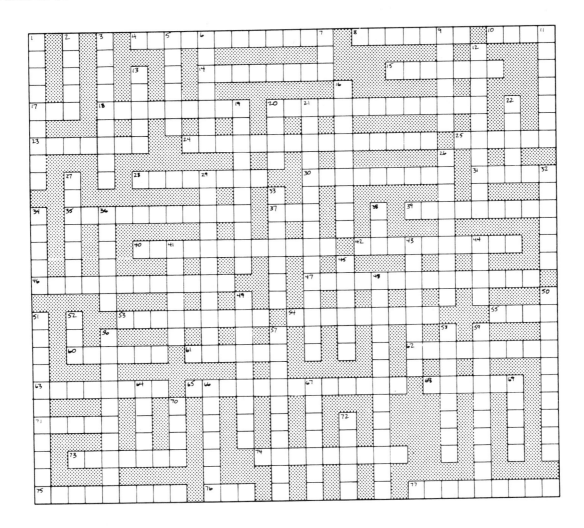

The terms in this puzzle are not necessarily biology terms.

Across

4. _____ canals detect acceleration or a change in the direction of movement
8. leads eggs from the ovary generally to the uterus
10. the ear _____ separates the outer ear from the middle ear
14. _____ eye
15. one of the major membranes that forms around the embryos of reptiles, birds, and mammals
17. nerve _____
18. has small clusters of living cells suspended in a firm, rubbery matrix, which they themselves secreted
20. an increasing concentration of nervous structures and coordinative functions in the "front" end, or head
23. fatty
24. detects energy associated with changes in pressure, position, or acceleration
25. the tough outer layer of the eyeball
28. hair _____

30. the adding together of graded potentials from different pathways to determine the course of further transmission
31. stimuli that activate sensory neurons and are transmitted to the central nervous system
35. _____ nerves dominate internal events in times of stress
37. rapid eye movements
39. the outermost protective epithelium of all animals
40. in the small spaces between cells
42. arranged in layers
46. _____ disks are components of cardiac muscle
47. having both sexes in one organism
53. capillary clusters in the kidneys
54. a receptor that distinguishes up from down
55. stimulates the adrenal cortex to secrete hormones
60. an organism in the phylum Porifera
61. inborn
62. release eggs that develop outside the mother's body
63. embryonic tissue that gives rise to bones and muscles
65. detects energy of visible and ultraviolet light
68. a spongelike and fatty tissue that fills the interior cavity of long bones or bone ends
71. relating to the eye
73. length-sensitive organ located within a skeletal muscle
74. the most anterior part of the brain
75. concerning the elimination of excess water, salts, and nitrogenous waste products
76. electroencephalogram
77. formed elements of the blood that are important in clot formation

Down

1. medulla _____
2. _____ glands in the skin help to control water and salt balance
3. having male individuals separate and distinct from female individuals
5. a fatty derivative that coats certain neurons
6. adenyl _____ is an enzyme important in producing cAMP
7. a photoreceptor cell concerned with perceiving gray and black shading in dim light and with perceiving movement
9. the neck of the uterus
11. the adrenal _____ secretes epinephrine
12. _____ tissues bind together and support other animal tissues
13. a group of fibers that connects the central nervous system and the organs or parts of the body
16. fundamental unit of muscle contraction
19. emit high-frequency sound waves followed by detection of waves that bounce back to their ears from objects in the surroundings
20. a photoreceptor cell that detects specific colors
21. _____ nerves generally dominate internal events when environmental conditions permit normal body functioning
22. a periodically recurring physiological state of rest characterized by lessened responsiveness to external stimuli
26. an agent that slows or stops an activity
27. alternative responses to a situation are first evaluated mentally so that accumulated experiences may suggest an appropriate response
29. neuron that connects sensory neurons with motor neurons

32. like cells united in form and function
33. connects the urinary bladder with the external environment
34. sound receptors in the human ear are hair cells in the organ of _____
36. the storage of individual bits of information somewhere in the brain
38. alimentary tract
41. shock
43. sticking to
44. contains circular and radial muscles in a contractile ring
45. connects a kidney to the urinary bladder
48. lens adjustments that bring about precise focusing onto the retina
49. relating to the sense of smell
50. the study of the function of a structure
51. synonym for platelet
52. reticular activating system
56. capable of long-range transmission of unaltered messages
57. _____ period: a time for insensitivity to stimulation
58. tubal _____, a form of sterilization
59. large white blood cell that phagocytizes
64. _____ window through which any vibrational energy remaining is dissipated
66. endocrine cell product
67. a movement of electrical energy
69. _____ blood cell
70. osseous tissue
72. an immature motile form of an organism that next becomes a juvenile

ANSWERS

CHAPTER 1   On the Unity and Diversity of Life

1-I (p. 2)	1. T 2. T 3. F 4. C	5. E 6. G 7. D 8. B	9. H 10. A 11. I

1-II (p. 4)	1. T 2. T 3. T 4. F 5. T 6. F 7. F	8. F 9. F 10. T 11. fat (or carbon- containing molecules) 12. ATP 13. Metabolism	14. egg 15. larva 16. pupa 17. adult 18. Homeostasis 19. Mutations

1-III (pp. 5–6)	1. F 2. T 3. T 4. T 5. F 6. T 7. F 8. A	9. E 10. H 11. F 12. G 13. B 14. C 15. D 16. J	17. K 18. I 19. producers 20. herbivores 21. carnivores 22. decomposers 23. nutrients (raw materials)

Understanding and Interpreting Key Concepts

(pp. 6–8)	1. d 2. a 3. b	4. c 5. c 6. d	7. d 8. b

CHAPTER 2   Methods and Organizing Concepts in Biology

2-I (pp. 10–11)	1. F 2. T 3. T 4. F 5. B	6. D 7. F 8. E 9. A 10. C	11. O 12. O 13. C 14. O 15. C

2-II (pp. 12–13)	1. Lamarck 2. Buffon 3. D 4. F 5. A 6. E	7. B 8. C 9. G 10. F 11. F 12. T	13. F 14. F 15. F 16. T 17. F 18. D

2-III (pp. 13–14)	1. T 2. F	3. T	4. F

2-IV (p. 14)	1. F	2. T	3. T

(pp. 14–16)

1. b	5. e	8. c
2. d	6. b	9. c
3. d	7. e	10. b
4. b		

CHAPTER 3   Chemical Foundations for Cells

3-I
(pp. 21–22)

1. T	8. C	15. J
2. T	9. N	16. E
3. T	10. K	17. F
4. F	11. L	18. B
5. F	12. G	19. H
6. T	13. A	20. I
7. D	14. M	

3-II
(p. 23)

1. F	2. T	3. F

3-III
(pp. 24–25)

1. T	4. F	7. See pages 40–41 of the main text for the answer.
2. F	5. T	
3. T	6. T	

3-IV
(pp. 26–27)

1. F	7. HCO_3^-	13. E
2. T	8. bicarbonate	14. B
3. T	9. salts	15. C
4. T	10. dissociate	16. A
5. K^+, Na^+, Ca^{++}, Mg^{++}, or Cl^-	11. hydrophobic	17. G
6. [same as above]	12. F	18. D

3-V
(p. 28)

1. A calorie	6. motion	10. F
2. hydrogen bonds	7. cohesion	11. F
3. heat of vaporization	8. solvent	12–13. See appropriate pages in main text.
4. evaporation	9. solutes	
5. absorbed		

(p. 29)

1. a	4. c	6. e
2. c	5. a	7. d
3. b		

CHAPTER 4   Carbon Compounds in Cells

4-I
(pp. 32–33)

1. hydroxyl	5. T	9. F
2. carboxyl	6. F	10. F
3. phosphate	7. T	11–14. See appropriate pages in main text.
4. amino	8. T	

4-II
(p. 34)

1. F	7. T	13. A, J
2. T	8. C, E	14. C, E
3. F	9. C, D	15. B, G
4. T	10. C, D	16. B, H
5. T	11. A, J	17. A, I
6. T	12. A, J	18. B, F

4-III
(pp. 35–36)

1. F	5. T	9. A
2. T	6. F	10. B
3. T	7. E	11. C
4. T	8. D	

4-IV
(p. 37)

1. F	3. T	5. T
2. T	4. F	

4-V
(p. 38)

1. T
2. F
3. T
4. F

5. T
6. T
7. T

8. B, D
9. C, E
10. A, F

Understanding and Interpreting Key Concepts

(pp. 38–39)

1. b
2. d
3. c

4. b
5. c

6. d
7. a

CHAPTER 5   Cell Structure and Function: An Overview

5-I
(p. 42)

1. T
2. F
3. T
4. E
5. A

6. D
7. C
8. B
9. light
10. light

11. Resolution
12. nanometer
13. transmission
14. scanning
15. nucleus

5-II
(p. 43)

1. D
2. D
3. B
4. F

5. C
6. E
7. Eukaryotes

8. prokaryotes
9. Animal
10. nucleoid

5-III
(pp. 45–46)

1. T
2. T
3. F
4. T
5. T
6. T

7. F
8. nucleolus
9. chromatin
10. chromosomes
11. E

12. F
13. C
14. A
15. B
16. D

5-IV
(p. 47)

1. F
2. T
3. F
4. F
5. T

6. C
7. A
8. D
9. B
10. G

11. infolded plasma
    membrane
12. chloroplast
13. $NADPH_2$ (ATP)
14. ATP ($NADPH_2$)
15. ATP

5-V
(p. 49)

1. microtubules
2. Microfilaments
3. cytoplasmic lattice
4. Microtubules
5. tubulins

6. H
7. I
8. C
9. D
10. A

11. E
12. E
13. B
14. F
15. G

5-VI
(pp. 50–51)

1. +
2. +
3. +
4. √
5.
6. √
7. √
8. √
9. √
10. √
11. +
12. +
13.
14. √
15.
16.
17. +
18.
19. √
20.
21.
22. √

23. √
24. √
25. √
26. √
27. √
28. √
29. √
30. √
31.
32. √
33. √
34. √
35. √
36.
37. √
38. √
39. √
40. √
41.
42. +
43. +
44. +

45. √
46. √
47. √
48. √
49. √
50. √
51.
52. √
53. √
54. √
55. √
56. T
57. T
58. T
59. F
60. ribosomes
61. Golgi complex
62. rough endoplasmic
    reticulum
63. nucleolus
64. nuclear pore
65. mitochondrion

66.	chloroplast	71.	ribosomes	76. mitochondrion
67.	central vacuole	72.	Golgi complex	77. lysosome
68.	plasma membrane	73.	nucleus	78. microfilaments
69.	cell wall	74.	nuclear envelope	79. cytoplasm
70.	flagellum	75.	centriole pair	

Understanding and Interpreting Key Concepts

(pp. 51–52)

1. b
2. d

3. a
4. a

5. d
6. b

CHAPTER 6   Membrane Structure and Function

6-I
(pp. 54–55)

1. Freeze-fracturing
2. freeze-etching
3. lipid bilayer
4. proteins
5. mosaic

6. Lipids
7. phosphate
8. carbohydrate
9. proteins

10. metabolism
11. tissues
12. F
13. T

6-II
(p. 56)

1. density
2. gradient
3. concentration
4. random collisions

5. greater
6. lesser (lower)
7. bulk flow
8. differentially permeable

9. Water
10. hypertonic
11. T
12. T

6-III
(p. 57)

1. T
2. F
3. F
4. T
5. F

6. Water (O_2, CO_2)
7. oxygen (H_2O, CO_2)
8. carbon dioxide (O_2, H_2O)

9. gases
10. Active transport systems

Understanding and Interpreting Key Concepts

(p. 58)

1. d
2. c
3. a

4. d
5. c

6. d
7. d

CHAPTER 7   Ground Rules of Metabolism

7-I
(p. 61)

1. F

2. T

3. T

7-II
(pp. 62–63)

1. D
2. F

3. C
4. A

5. B
6. E

7-III
(p. 64)

1. Enzymes
2. catalysts
3. equilibrium
4. substrate
5. active site
6. Temperature

7. pH
8. hydrogen
9. denaturation
10. active
11. product
12. active

13. active
14. T
15. F
16. T
17. F
18. T

7-IV
(p. 66)

1. cofactors
2. NAD^+
3. adenine
4. ribose
5. phosphate
6. water
7. enzyme
8. ADP
9. phosphate

10. energy
11. electron transport
12. electron
13. oxidation-reduction
14. electron
15. carrier
16. energy
17. phosphate

18. oxidized
19. $NADP^+$
20. NAD^+
21. FAD
22. cytochromes
23. F
24. T
25. T

Understanding and Interpreting Key Concepts

(pp. 66–67)

1. d
2. c
3. d

4. a
5. c

6. d
7. a

8-I
(p. 69)

1. Photosynthetic autotrophs
2. Chemosynthetic autotrophs
3. bacteria
4. bacteria
5. Heterotrophic
6. glycolysis (aerobic respiration)
7. aerobic respiration (glycolysis)

8-II
(p. 70)

1. thylakoid membranes
2. eukaryotic
3. stroma
4. pigments
5. chlorophyll
6. Carotenoids
7. photosystem
8. light
9. electron
10. acceptor molecule

8-III
(p. 72)

1. photosynthesis
2. carbon dioxide
3. water
4. O_2, oxygen
5. food molecules
6. sunlight
7. pigment molecules
8. ATP (NADPH)
9. NADPH (ATP)

8-IV
(p. 73)

1. ATP
2. NADPH
3. carbon dioxide
4. ribulose biphosphate
5. PGA
6. carbon dioxide fixation
7. six
8. PGA
9. ATP
10. NADPH
11. light-dependent reactions
12. PGAL
13. glucose
14. ribulose biphosphate
15. autotrophic
16. inorganic
17. oxidation
18. ammonium
19. sulfur (or iron)
20. nitrifying

Understanding and Interpreting Key Concepts

(pp. 74–75)

1. a
2. b
3. c
4. a
5. d
6. d
7. a
8. b
9. c

CHAPTER 9   Energy-Releasing Pathways

9-I
(pp. 77–78)

1. Autotrophic
2. Glucose
3. ATP
4. ATP (NADH)
5. NADH (ATP)
6. pyruvate (pyruvic acid)
7. O_2, oxygen
8. fermentative (anaerobic)
9. lactate
10. ethanol (CO_2)
11. carbon dioxide (ethyl alcohol)
12. aerobic
13. Krebs
14. electron transport phosphorylation
15. 36
16. ATP
17. aerobic respiration (cellular respiration)
18. carbon dioxide
19. electrons
20. NAD^+ (FAD)
21. FAD (NAD^+)
22. transport system (transport chain)
23. ATP
24. oxygen
25. mitochondria

9-II
(pp. 79–80)

1. Phosphorylating the sugars makes them unstable and quite reactive.
2. It is converted into PGAL.
3. It is used in the second part.
4. It is used to do work in the cell.
5. Its energy is used to convert pyruvic acid into fermentable products or is transferred to ATP.
6. They act as temporary acceptors for H^+ and e^- released during the oxidation of pyruvic acid into CO_2 and H_2O.
7. NADH and $FADH_2$
8. Electron transport phosphorylation
9. 0, 2, 2
10. 2, 6, 2
11. 0, 2
12. 3, 2
13. 4
14. $10 \times 3 = 30$
15. $2 \times 2 = 4$
16. Pyruvic acid is fed into one of several possible fermentation pathways.
17. 19 times as many
18. inner compartment of mitochondrion
19. inner membrane of mitochondrion
20. outer compartment of mitochondrion
21. outer membrane of mitochondrion
22. cytoplasm
23. ATP
24. O_2, oxygen
25. $FADH_2$
26. NADH
27. electron transport system

9-III
(pp. 81–82)

1. F
2. T
3. T
4. T
5. fatty acids
6. glycerol
7. glycolysis
8. amino acids
9. Krebs cycle
10. acetyl-CoA
11. pyruvate

Understanding and Interpreting Key Concepts

(pp. 83–84)

1. c
2. c
3. a

4. b
5. d
6. d

7. a
8. d

CHAPTER 10   Cell Reproduction

10-I
(pp. 90–91)

1. life cycle
2. reproduction
3. DNA
4. Genes
5. RNA molecules
6. RNA
7. enzymes
8. carbohydrates (lipids)
9. lipids (carbohydrates)
10. replication
11. DNA
12. microtubules

13. chromosomes
14. replicated
15. centromere
16. sister chromatids
17. T
18. T
19. T
20. F
21. T
22. F
23. interphase
24. mitosis

25. G_1
26. S
27. G_2
28. prophase
29. metaphase
30. anaphase
31. telophase
A. 27
B. 24
C. 26
D. 25
E. 23

10-II
(pp. 92–93)

1. F
2. F
3. T
4. F
5. T

6. I
7. F
8. H
9. B
10. A

11. G
12. J
13. C
14. D
15. E

Understanding and Interpreting Key Concepts

(pp. 93–94)

1. a
2. c
3. d

4. a
5. d

6. c
7. e

CHAPTER 11   A Closer Look at Meiosis

11-I
(pp. 96–97)

1. B
2. E
3. C
4. H

5. D
6. G
7. A
8. F

9. F
10. T
11. F
12. T

11-II
(p. 98)

1. 2
2. 5
3. 4
4. 1

5. 3
6. F
7. F

8. T
9. T
10. F

Understanding and Interpreting Key Concepts

(pp. 98–99)

1. a
2. a
3. d

4. b
5. b

6. b
7. a

CHAPTER 12   Observable Patterns of Inheritance

12-I
(p. 102)

1. traits
2. strains
3. pea
4. cross-fertilization

5. true-breeding
6. hybrids
7. recessive

8. blending
9. monohybrid
10. segregation

12-II
(pp. 103–104)

1. F   2. T
3. albino = pp
   normal pigmentation = pp or Pp
   Woman of normal pigmentation with an albino mother → Pp; received her recessive
   gene from her mother and her dominant gene (P) from her father. It is likely that half of

the couple's children will be albinos (pp) and half will have normal pigmentation but be heterozygous (Pp).

	P
P	Pp
P	pp

Pp × pp

kinds of gametes from Pp: P, P

only kind of gamete from pp: P

4. D   5. A   6. D   7. C

12-III
(p. 106)   1. F   2. F   3. F   4. F   5. T   6. F

12-IV
(p. 107)   1. T   2. F

Understanding and Interpreting Key Concepts

(pp. 107–108)
1. d
2. b
3. a
4. c

5. c
6. b
7. a

8. d
9. c
10. b

Integrating and Applying Key Concepts

(pp. 108–109)

1. Yes; the husband could not have supplied either of his daughter's recessive genes because his only X chromosome bears the N for normal iris.

$X^N Y$ father

$X^N X^n$ mother

N = normal iris
n = fissured iris

$X^n X^n$ daughter's genotype

The mother must also carry the recessive gene in order to be her daughter's mother.

2. 100% of offspring will be spotted because there are no dominant genes included.

ss × ss   parents

gametes possible

ss
offspring's genotype

s = spotted
S = solid color

3. black = B
solid color = S

red = b
white spots = s

Solid Red ♀ — Black and White ♂

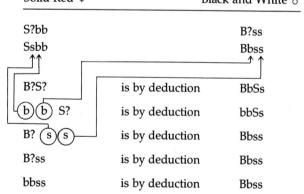

S?bb	B?ss
Ssbb	Bbss
B?S?	is by deduction	BbSs
(b)(b) S?	is by deduction	bbSs
B? (s)(s)	is by deduction	Bbss
B?ss	is by deduction	Bbss
bbss	is by deduction	Bbss

CHAPTER 13   Chromosomal Theory of Inheritance

13-I
(p. 111)
1. Flemming
2. Weismann
3. meiosis
4. Homologous
5. E
6. A
7. D
8. G
9. F
10. B
11. C

13-II
(pp. 112–113)
1. C
2. A
3. F
4. T

13-III
(pp. 114–115)
1. C
2. D
3. A
4. B
5. A
6. D
7. B

13-IV
(p. 116)
1. deletion
2. translocation
3. inversion
4. polyploidy
5. Colchicine
6. Nondisjunction
7. trisomy (triploidy)
8. crossing-over
9. tetraploid
10. a

Understanding and Interpreting Key Concepts

(pp. 116–118)
1. d
2. d
3. c
4. b
5. b
6. b
7. a
8. b
9. d

CHAPTER 14   Human Genetics

14-I
(p. 120)
1. C
2. A
3. B
4. D
5. E
6. E

14-II
(pp. 121–122)
1. B
2. E
3. E
4. C
5. A
6. A
7. F
8. E
9. B
10. D

Understanding and Interpreting Key Concepts

(pp. 122–123)
1. d
2. c
3. c
4. b
5. a
6. b
7. c
8. b
9. a

CHAPTER 15   The Rise of Molecular Genetics

15-I
(p. 125)
1. F
2. F
3. F
4. F
5. F
6. F
7. T

15-II
(pp. 126–127)
1. F
2. T
3. T
4. F
5. T
6. F
7. T

15-III
(p. 128)
1. F
2. F
3. T
4. T
5. C
6. deoxyribose
7. phosphate group
8. guanine
9. thymine
10. adenine
11. cytosine
12. nucleotide

Understanding and Interpreting Key Concepts

(pp. 128–131)
1. d
2. d
3. a
4. d
5. b
6. d
7. c
8. a
9. a

CHAPTER 16   Protein Synthesis

16-I
(pp. 133–134)
1. alkaptonuria
2. Garrod
3. enzyme
4. unit of heredity
5. enzyme
6. Beadle

7. Tatum	12. gene	16. RNA
8. pink bread mold	13. enzyme	17. ribose
9. haploid	14. RNA molecules	18. uracil
10. asexually	15. mRNA	19. adenine
11. sugar		

16-II
(p. 135)

1. transcription	5. complementary	8. introns
2. DNA	6. mRNA	9. exons
3. RNA polymerase	7. cytoplasm	10. poly-A tail
4. one of the strands		

16-III
(p. 136)

| 1. primary | 3. sixteen | 5. codon |
| 2. mRNA or three | 4. mRNA | 6. third |

16-IV
(pp. 137–138)

1. Messenger	4. transfer (t)	7. codon
2. translation	5. amino acid	8. anticodon
3. Ribosomal (r)	6. protein	

9. *AUG-UUC-UAU-UGU-AAU-AAA-GGA-UGG-CAG-UAG*
10. start-phe-tyr-cys-asp-lys-gly-try-gln-stop

16-V
(p. 139)

| 1. High-energy radiation | 3. Insertions | 5. Frameshift |
| 2. mutagenic chemicals | 4. deletions | |

Understanding and Interpreting Key Concepts

(pp. 139–141)

1. c	6. a	11. d
2. b	7. a	12. a
3. c	8. d	13. b
4. a	9. a	14. a
5. c	10. d	

CHAPTER 17 Control of Gene Expression

17-I
(pp. 143–144)

1. short	6. RNA polymerase	11. DNA (chromosomes)
2. development	7. operon	12. express
3. differentiation	8. repressor	13. nucleus (environment)
4. Repression	9. regulator	14. cells
5. repressor	10. promoter	

17-II
(p. 145)

1. differentiation	4. cyclic adenosine monophosphate	6. 2 (one less than the number of X chromosomes)
2. slug	5. environmental change	
3. fruiting body		

17-III
(pp. 146–147)

1. controls	5. lyonization	8. nucleosome
2. Transcriptional	6. Histones	9. T
3. transcript processing	7. nucleoprotein	10. T
4. Barr body		

17-IV
(p. 148)

1. F	5. B	8. A
2. T	6. E	9. B
3. T	7. D	10. C
4. F		

Understanding and Interpreting Key Concepts

(pp. 148–149)

1. d	5. a	9. D
2. c	6. a	10. A
3. b	7. E	11. C (B)
4. b	8. B (C)	

CHAPTER 18 Recombinant DNA and Genetic Engineering

18-I
(pp. 151–152)

1. F	5. C	8. B
2. F	6. I	9. E
3. D	7. A	10. H
4. G		

18-II   1. Restriction
           ucleases

4. Insulin
5. cDNA probe

6. transfection
7. Glycophytes

117 CHAUNCY ST. ESSEX ST. BOSTON, MA.

5. c
6. e
7. b

8. d
9. a
10. b

es, and Systems

         rm
         nbium
         d meristem

      6. Cortex

7. mesophyll (or parenchyma)
8. epidermis
9. Parenchyma
10. pith

11. Parenchyma
12. Collenchyma
13. pectin
14. Sclerenchyma
15. sclereids

19-II
(p. 159)

1. xylem
2. phloem
3. salts (ions)
4. vessel members
5. tracheids

6. pits
7. food (sucrose, sugar)
8. sieve cell members
9. sieve plates
10. Companion cells

11. epidermis
12. water absorption
13. water loss
14. microbial attack

19-III
(p. 161)

1. apical meristem
2. leaf primordia
3. Nodes
4. Vascular bundles
5. pith
6. cortex
7. nodes

8. Monocot
9. vascular bundles
10. sunlight
11. water
12. photosynthesis
13. Stomata
14. guard cells

15. Periderm
16. palisade mesophyll
17. spongy mesophyll
18. lower epidermis
19. vein
20. stoma (-ta)

19-IV
(pp. 162–163)

1. Taproot
2. fibrous
3. cap
4. apical meristem
5. Epidermis
6. cortex
7. vascular column

8. endodermis
9. vascular cambium
10. cork cambium
11. ground meristem
12. procambium
13. protoderm
14. root apical meristem

15. root cap
16. endodermis
17. pericycle
18. endodermis
19. pericycle
20. Casparian strip

19-V
(pp. 164–165)

1. vertical (ray)
2. ray (vertical)
3. ray
4. parenchyma

5. vascular cambium
6. vascular cambium
7. bark

8. periderm
9. phloem
10. xylem

Understanding and Interpreting Key Concepts

(pp. 165–167)

1. b
2. b
3. a
4. c

5. a
6. d
7. d
8. e

9. d
10. b
11. a
12. a

CHAPTER 20   Water, Solutes, and Plant Functioning

20-I
(pp. 169–170)

1. Water
2. vacuole
3. less expensive
4. nucleic acids
5. nitrate (NO_3^-)
6. Ammonium (NH_4^+)
7. Oxygen atoms
8. Hydrogen atoms

9. Turgor pressure
10. unwilted
11. B
12. A, I
13. B
14. B
15. B, H
16. A, D

17. B
18. B, F
19. A, G
20. A, C
21. A, E
22. A
23. B

20-II (p. 171)	1. Transpiration 2. cohesion 3. hydrogen bonding	4. potassium 5. active transport 6. heat	7. abscisic acid 8. potassium
20-III (p. 172)	1. active transport 2. xylem 3. Starch 4. sucrose	5. translocation 6. pressure flow 7. sieve-cell	8. water 9. pressure 10. roots

Understanding and Interpreting Key Concepts

(pp. 173–174)	1. c 2. b 3. b 4. a	5. b 6. a 7. d 8. b	9. b 10. d 11. b 12. c

CHAPTER 21   Plant Reproduction and Embryonic Development

21-I (p. 176)	1. sepals 2. petals 3. Stamens	4. anther 5. microspores 6. Carpels	7. ovary 8. ovule 9. megaspores
21-II (p. 177)	1. C 2. F 3. A	4. E 5. B	6. G 7. D
21-III (p. 179)	1. anthers 2. stigmas 3. pollen tube 4. sperm nuclei 5. sperms 6. zygote 7. endosperm 8. seeds 9. fruit	10. 100, 65 11. Wind 12. red 13. yellow 14. ultraviolet 15. seedling (sporophyte) 16. embryo 17. seed 18. sperm	19. tube nucleus 20. anther 21. megaspore mother cell 22. endosperm mother cell 23. egg 24. ovary wall 25. embryo sac 26. micropyle 27. ovule
21-IV (p. 181)	1. ovary 2. Endosperm 3. carpel	4. wind- 5. flowers 6. Grains	7. fleshy 8. aggregate 9. Multiple
21-V (pp. 181–182)	1. parthenogenesis 2. Nodes	3. bulbs	4. clones

Understanding and Interpreting Key Concepts

(pp. 182–183)	1. c 2. c 3. d 4. c 5. d	6. d 7. b 8. a 9. c 10. B	11. C 12. A 13. E 14. D

CHAPTER 22   Plant Growth and Development

22-I (p. 185)	1. imbibition 2. Germination 3. enlarge	4. meristematic 5. water 6. turgor pressure	7. polysaccharides 8. radicle
22-II (pp. 186–187)	1. hormones 2. Auxins (Gibberellins) 3. gibberellins (auxins) 4. Cytokinins 5. Abscisic acid	6. seed 7. Ethylene 8. IAA (auxin) 9. adventitious 10. Gibberellins	11. Florigen 12. Auxins 13. Abscisic acid 14. Ethylene 15. Cytokinins
22-III (p. 187)	1. B	2. D	3. C

3. organ	10. hormonal (chemical)	16. anterior
4. midsagittal	11. nervous	17. midsagittal plane
5. transverse	12. transverse plane	18. proximal
6. anterior	13. dorsal surface	19. inferior
7. frontal		

23-II
(pp. 198–199)

1. tissue
2. Somatic
3. epithelial
4. connective

5. muscular
6. nervous
7. T

8. F
9. F
10. F

23-III
(p. 200)

1. ground substance
2. Loose
3. collagen
4. elastin
5. Tendons (Ligaments)
6. ligaments (tendons)
7. Bone

8. cartilage
9. Haversian
10. Red marrow
11. plasma
12. Adipose
13. spongy bone

14. compact bone
15. outer membrane of dense connective tissue
16. Haversian system
17. concentric lamellae
18. osteocyte

23-IV
(pp. 201–202)

1. F
2. T

3. F
4. T

5. F
6. T

23-V
(p. 203)

1. epidermis
2. dermis
3. subcutaneous
4. adipose

5. liver
6. contraction
7. relaxation
8. nutrients

9. F
10. T
11. F

Understanding and Interpreting Key Concepts

(pp. 203–205)

1. e
2. b
3. c
4. a

5. d
6. d
7. d
8. c

9. c
10. d
11. a

CHAPTER 24 Information Flow and the Neuron

24-I
(pp. 207–208)

1. neurons
2. Neuroglial
3. dendrites
4. trigger zone

5. conducting zone
6. axon
7. terminals

8. Sensory
9. interneurons
10. motor

24-II
(p. 209)

1. differentially
2. concentration
3. against
4. potassium
5. sodium

6. negative
7. millivolts
8. resting membrane potential
9. sodium-potassium pump

10. message
11. graded potentials
12. summed (added together)

24-III	1. action potential	4. threshold	7. node of Ranvier
(pp. 210–211)	2. nerve impulse	5. refractory	8. saltatory
	3. all-or-nothing	6. myelin sheaths	

24-IV	1. synapse	3. Acetylcholine (or	4. excitatory
(p. 212)	2. transmitter substances	Dopamine or	5. inhibitory
		Epinephrine or	
		Norepinephrine)	

24-V	1. Integration	4. local circuits	7. stretch reflex
(p. 213)	2. summed	5. nerve pathway	8. muscle spindles
	3. synaptic integration	6. reflex	9. spinal cord

Understanding and Interpreting Key Concepts

(pp. 213–214)	1. a	4. d	6. a
	2. a	5. d	7. b
	3. c		

CHAPTER 25   Nervous Systems

25-I	1. radial	4. bilateral	7. brain
(pp. 216–217)	2. ganglia	5. nerves (muscles)	8. central
	3. Cephalization	6. muscles (nerves)	9. peripheral

25-II	1. somatic	6. vertebral column	10. D
(p. 218)	2. autonomic	7. cartilage	11. E
	3. sympathetic	8. cranial	12. C
	4. parasympathetic	9. B	13. A
	5. central nervous system		

25-III	1. spinal cord	5. medulla oblongata	8. hypothalamus
(p. 220)	2. white matter	6. cerebellum	9. endocrine
	3. gray matter	7. forebrain	10. thalamus
	4. synaptic (integrative, associative)		

25-IV	1. B	10. C	18. cerebellum
(pp. 221–222)	2. I	11. J	19. medulla oblongata
	3. H	12. G	20. pons
	4. A	13. E	21. pituitary gland
	5. M	14. memory	22. optic chiasm
	6. K	15. memory trace	23. hypothalamus
	7. L	16. short-term formative	24. thalamus
	8. D	17. long-term storage	25. corpus callosum
	9. F		

25-V	1. alpha rhythm	8. pain	15. D
(p. 224)	2. slow-wave sleep	9. transmitter substances	16. D
	3. EEG arousal	10. B	17. E
	4. REM sleep	11. E	18. C
	5. reticular formation (sleep centers)	12. E	19. B
	6. serotonin	13. E	20. B
	7. Endorphins (Enkephalins)	14. A	

Understanding and Interpreting Key Concepts

(pp. 225–226)	1. a	5. b	8. e
	2. d	6. c	9. c
	3. e	7. d	10. e
	4. b		

CHAPTER 26   Integration and Control: Endocrine Systems

26-I
(p. 228)
1. Transmitter substances
2. Neurosecretory
3. neurohormones
4. hormones
5. target
6. exocrine
7. same
8. ovaries
9. testes

26-II
(pp. 229–230)
1. hypothalamus
2. releasing hormones
3. pituitary
4. glandular (secretory)
5. nervous
6. portal
7. hypothalamus
8. ACTH
9. TSH
10. prolactin (PRL)

26-III
(p. 231)
1. adrenal cortex
2. cortisone, cortisol
3. ACTH
4. epinephrine, adrenalin
5. norepinephrine, noradrenalin
6. thyroid
7. inhibits
8. parathyroid glands
9. promotes

26-IV
(p. 232)
1. reproductive
2. Glucagon
3. insulin
4. thymus
5. pineal
6. pituitary
7. parathyroid
8. thyroid
9. thymus
10. adrenal
11. kidney
12. islets of Langerhans
13. ovary
14. placenta
15. testis

26-V
(p. 234)
1. H
2. F
3. K
4. L
5. M
6. D
7. E
8. A
9. G
10. B
11. I
12. O
13. C
14. J
15. N
16. cAMP

Understanding and Interpreting Key Concepts

(pp. 235–236)
1. a
2. e
3. d
4. b
5. d
6. c
7. b
8. a
9. c
10. a

CHAPTER 27   Sensory Systems

27-I
(p. 238)
1. receptors
2. stimulus
3. Chemoreceptors
4. mechanoreceptors
5. photoreceptors
6. thermoreceptors
7. olfactory (taste)

27-II
(pp. 239–240)
1. amplitude
2. frequency
3. higher
4. mechanoreceptors
5. middle
6. cochlea
7. organ of Corti
8. semicircular canals
9. middle earbones (malleus)
10. cochlea
11. auditory nerve
12. tympanic membrane/eardrum
13. oval window
14. basilar membrane
15. tectorial membrane

27-III
(pp. 241–242)
1. photons
2. Photoreception
3. Vision
4. Eyespots
5. Eyes
6. cornea
7. retina
8. ommatidia
9. focal point
10. Accommodation
11. Farsighted
12. Cone
13. fovea
14. vitreous body
15. cornea
16. iris
17. lens
18. aqueous humor
19. suspensory ligament
20. retina
21. fovea
22. optic nerve
23. blind spot/optic disk
24. sclera

Understanding and Interpreting Key Concepts

(pp. 242–244)
1. d
2. c
3. d
4. e
5. d
6. a
7. c
8. b
9. b
10. a

CHAPTER 29 Circulation

29-I (pp. 254–255)			
1. connective	5. iron	9. Stem cells	
2. pH	6. oxyhemoglobin	10. Neutrophils	
3. closed	7. erythropoietin	11. Platelets	
4. lymph vascular system	8. bone marrow		

29-II (p. 256)			
1. atrium	6. systemic	11. inferior vena cava	
2. ventricle	7. aorta	12. atrioventricular valve	
3. systole	8. left pulmonary veins	13. right pulmonary artery	
4. diastole	9. semilunar valve	14. superior vena cava	
5. pulmonary	10. left ventricle		

29-III (pp. 257–258)			
1. artery	5. interstitial	8. venules (veins)	
2. capillary	6. Valves	9. F	
3. endothelial	7. veins (venules)	10. F	
4. capillary bed			

29-IV (p. 259)			
1. stroke	5. medulla oblongata	9. collagen	
2. coronary occlusion	6. hemostasis	10. calcium	
3. Plaque	7. platelet plug formation	11. fibrin	
4. low	8. coagulation		

29-V (p. 260)			
1. tonsils	4. thoracic duct	7. Lymph	
2. right lymphatic duct	5. spleen	8. fats	
3. thymus	6. bone marrow	9. small intestine	

Understanding and Interpreting Key Concepts

(pp. 260–262)			
1. d	5. e	8. e (d)	
2. e	6. a	9. d	
3. c	7. b	10. a	
4. e			

CHAPTER 30 Immunity

30-I (p. 267)			
1. phagocytosis	3. Major histocompatibility complex	4. complement	
2. Antibodies			

30-II (p. 269)	1. clonal selection theory 2. nonself 3. lymphocyte 4. B cells (T cells) 5. T cells (B cells) 6. stem cells	7. markers 8. T cells 9. B cell 10. primary immune response	11. immunization 12. cancer 13. monoclonal (pure) antibodies 14. Target
30-III (pp. 270–271)	1. Allergy 2. Autoimmune disease 3. human immuno- deficiency virus	6. lymphokines 7. interferons 8. H	

	exchange mechanism	5. 700	6. swim bladders
31-II (p. 276)	1. diaphragm 2. rib cage 3. increases 4. drops 5. brain stem (medulla oblongata) 6. pleural sac 7. larynx	8. glottis 9. bronchi 10. bronchioles 11. alveoli 12. intercostal muscles 13. diaphragm 14. pharynx	15. epiglottis 16. vocal cords 17. trachea 18. bronchus 19. bronchioles 20. thoracic cavity 21. abdominal cavity
31-III (p. 278)	1. partial pressure 2. Passive diffusion 3. carbon dioxide 4. Hypoxia 5. bicarbonate	6. oxyhemoglobin (hemoglobin) 7. systemic (low-pressure) 8. carbonic anhydrase 9. reticular formation (medulla oblongata)	10. medulla oblongata (reticular formation) 11. Emphysema 12. lung cancer

Understanding and Interpreting Key Concepts

(pp. 279–280)	1. c 2. e 3. a 4. c	5. a 6. a 7. a	8. a 9. a 10. d

CHAPTER 32 Digestion and Organic Metabolism

32-I (p. 282)	1. Nutrition 2. ingestion	3. digestion 4. absorption	5. assimilated
32-II (pp. 284–285)	1. amylase 2. epiglottis 3. esophagus 4. peristalsis 5. stomach 6. small intestine 7. Pepsin 8. Bile 9. small intestine	10. Amylase (Lipase, Trypsin, or Chymotrypsin) 11. salivary glands 12. oral cavity 13. liver 14. stomach 15. gallbladder 16. small intestine 17. large intestine	18. anus 19. pancreas 20. esophagus 21. pharynx 22. T 23. F 24. F 25. T 26. F

32-III (p. 287)	1. D 2. calcium 3. A 4. D	5. K 6. niacin (vitamin B) 7. C	8. phosphorus 9. Iodine 10. Iron
32-IV (p. 288)	1. ammonia 2. urea 3. insulin 4. glucagon 5. glycogen	6. epinephrine (norepinephrine) 7. norepinephrine (epinephrine)	8. glycogen 9. glucose 10. glucose

Understanding and Interpreting Key Concepts

(pp. 288–290)	1. b 2. b 3. a 4. e	5. d 6. c 7. e 8. b	9. e 10. d 11. c

CHAPTER 33 Temperature Control and Fluid Regulation

33-I (p. 292)	1. hypothalamus 2. core	3. epinephrine 4. hypothermia	5. T 6. T

33-II (p. 294)	1. T 2. T 3. T 4. F 5. F 6. Filtration 7. permeability 8. pressure	9. filter 10. reabsorption 11. solutes 12. peritubular capillaries 13. Ureters 14. urinary bladder 15. urethra	16. glomerular capillaries 17. proximal tubule 18. Bowman's capsule 19. distal tubule 20. peritubular capillaries 21. collecting duct 22. loop of Henle

33-III (pp. 296–297)	1. Antidiuretic hormone 2. Countercurrent multiplication	3. hypothalamus 4. thirst	5. dialysis 6. T

Understanding and Interpreting Key Concepts

(pp. 297–298)	1. d 2. c 3. e 4. a	5. d 6. d 7. c	8. b 9. e 10. b

CHAPTER 34 Principles of Reproduction and Development

34-I (p. 300)	1. zygote 2. Cleavage 3. larva	4. regeneration 5. budding 6. Parthenogenesis	7. hermaphrodites 8. ovaries 9. testes

34-II (pp. 301–302)	1. F 2. T 3. Gametogenesis 4. egg 5. fertilization	6. Cleavage 7. blastula 8. Gastrulation 9. ovoviviparity 10. oviparity	11. RNA transcripts (maternal messages) 12. animal pole 13. gray crescent 14. body axis

34-III (p. 303)	1. F 2. cleavage 3. blastula 4. gastrulation	5. germ 6. gastrula 7. nervous 8. gut	9. skeleton 10. yolk 11. three

34-IV (pp. 304–305)	1. metamorphosis 2. indirect	3. identical twins 4. selective	5. T

34-V (pp. 306–307)	1. T 2. T 3. T	4. Morphogenesis 5. chemical gradients 6. epithelial sheets	7. embryonic induction 8. ooplasmic localization 9. controlled cell death

Understanding and Interpreting Key Concepts

(pp. 307–308)	1. c	4. a	7. d
	2. b	5. c	8. b
	3. b	6. e	9. e

CHAPTER 35   Human Reproduction and Development

35-I
(pp. 310–311)

1. seminiferous tubules
2. epididymis
3. testosterone
4. vas deferens
5. prostate
6. lytic (digestive) enzymes
7. midpiece
8. Prostaglandins
9. seminal vesicle
10. urinary bladder
11. prostate gland
12. bulbourethral glands
13. epididymis
14. seminiferous tubule

35-II
(pp. 312–313)

1. estrogen (progesterone)
2. progesterone (estrogen)
3. Oviducts
4. uterus
5. estrus
6. sphincter
7. follicle
8. ovulation
9. oviduct
10. corpus luteum
11. endometrium
12. Progesterone
13. LH
14. FSH
15. menstruation
16. oviduct
17. ovary
18. uterus
19. urinary bladder
20. urethra
21. clitoris
22. labium minor
23. labium major
24. vagina
25. endometrium

35-III
(pp. 314–315)

1. oviduct
2. implantation
3. placenta
4. second
5. fetus

35-IV
(p. 316)

1. 1,700,000
2. 200,000
3. 1,500,000
4. abstinence
5. Condoms
6. diaphragm
7. estrogens (progesterones)
8. progesterones (estrogens)
9. gonadotropins
10. tubal ligation

Understanding and Interpreting Key Concepts

(pp. 316–317)	1. d	5. c	9. c
	2. e	6. e	10. a
	3. a	7. d	11. b
	4. b	8. a	12. d

CHAPTER 36   Population Genetics, Natural Selection, and Speciation

36-I
(p. 320)

1. typological
2. population
3. Hardy-Weinberg rule
4. allele frequencies
5. genetic equilibrium
6. mutations
7. genetic drift
8. Gene flow
9. Find (b) first, then (c) and, finally, (a).
   (a) $2pq = 2 \times (0.9) \times (0.1) = 2 \times (0.09) = 0.18$
   $= 18\%$, which is the percentage of heterozygotes
   (b) $p^2 = \underline{0.81}$
   $p = \sqrt{0.81} = 0.9 = $ the frequency of the dominant allele
   (c) $p + q = 1$
   $q = 1 - 0.9 = 0.1 = $ the frequency of the recessive allele
10. (a) homozygous dominant $= p^2 \times 200 = (0.8)^2 \times 200 = 0.64 \times 200$
   $= 128 \text{ individuals}$
   (b) homozygous recessive $= q^2 \times 200 = (0.2)^2 \times 200 = (0.04) \times (200)$
   $= 8 \text{ individuals}$
   (c) heterozygotes $= 2pq \times 200 = 2 \times 0.8 \times 0.2 \times 200 = 0.32 \times 200$
   $= 64 \text{ individuals}$
   Check: $128 + 8 + 64 = 200$

36-II
(p. 322)

1. Directional selection
2. Disruptive selection
3. Horseshoe crabs
4. balanced polymorphism
5. polymorphism
6. natural selection

36-III
(pp. 323–324)

1. geographic isolation
2. genetic drift
3. divergence
4. Speciation
5. reproductive
6. Allopatric
7. polyploidy
8. polyploid

Understanding and Interpreting Key Concepts

(pp. 324–325)
1. c
2. a
3. d
4. d
5. c
6. b
7. a
8. e
9. b
10. c

CHAPTER 37   Phylogeny and Macroevolution

37-I
(p. 328)
1. gradualistic
2. punctuational
3. Proterozoic
4. prokaryotes
5. eukaryotes
6. Paleozoic
7. selection
8. speciation
9. cladistics
10. DNA hybridization

37-II
(p. 329)
1. Extinction
2. adaptive radiation
3. placental
4. Permian
5. asteroid/meteor
6. Cretaceous

Understanding and Interpreting Key Concepts

(pp. 329–330)
1. b
2. d
3. c
4. e
5. c
6. d
7. e
8. a
9. b
10. d

CHAPTER 38   Origins and the Evolution of Life

38-I
(p. 332)
1. 4.6
2. 3.8
3. clay crystals
4. amino acids
5. Microspheres
6. Liposomes

38-II
(pp. 333–334)
1. 2
2. oxygen
3. anaerobic
4. 1.4
5. aerobic (cellular)
6. stromatolites
7. anaerobic electron transport
8. symbiotic

38-III
(p. 335)
1. plates
2. mantle
3. Gondwana
4. trilobites
5. invertebrate
6. Jawless
7. amphibians
8. Carboniferous
9. Pangaea

38-IV
(pp. 336–337)
1. Mesozoic
2. gymnosperms
3. reptiles
4. archosaurs
5. endotherms
6. *Protoavis*
7. Gymnosperms (Reptiles)
8. reptiles (gymnosperms)
9. mammals
10. grasslands
11. mammals (birds)
12. flowering

Understanding and Interpreting Key Concepts

(pp. 337–338)
1. c
2. b
3. e
4. d
5. e
6. c
7. a
8. b
9. c
10. c

CHAPTER 39   Viruses, Monerans, and Protistans

39-I
(p. 340)
1. virions
2. nucleic acid
3. viral capsid (protein coat)
4. Viruses
5. pandemic
6. herpesvirus
7. viroids

39-II
(pp. 341–342)
1. Chemosynthetic
2. decomposers
3. parasites
4. obligate anaerobes
5. peptidoglycan
6. Archaebacteria
7. methanogens
8. Cyanobacteria
9. nitrogen fixation
10. endospores
11. *Thermoplasma*
12. photosynthetic cyanobacteria
13. bacterial (binary) fission

39-III (p. 343)	1. longitudinal fission 2. organelles	3. heterotrophic	4. Dinoflagellates (*Gonyaulax*)

39-IV (p. 344)	1. pseudopodia 2. Foraminiferans 3. *Plasmodium* 4. sporozoan (parasite) 5. gametes 6. freshwater	7. contractile vacuoles 8. gullet 9. food vacuoles 10. C, E, J 11. C, E, L 12. B, F, K	13. A 14. D, I 15. B, F 16. B, F, H 17. B, G (F)

39-V (p. 345)	1. *Prochloron* 2. *Chlamydomonas* 3. volvocines	4. plants 5. *Trichoplax adhaerens*	6. colony 7. multicellular

Understanding and Interpreting Key Concepts

(pp. 346–347)	1. d 2. a 3. c 4. b 5. b 6. a 7. c 8. c	9. c 10. a 11. B, I 12. A, E 13. A, H 14. B, N 15. B, D 16. B, I, O	17. B, F, G 18. C, M 19. A, H 20. A 21. B, L 22. B, J, K 23. A, H 24. B, F

CHAPTER 40 Fungi and Plants

40-I (p. 350)	1. chitin 2. spore 3. parasitic 4. mycelium	5. hyphae 6. zygote 7. asci	8. ascocarp 9. sac 10. club

40-II (p. 351)	1. Mycorrhizae 2. Lichens	3. Ringworm	4. *Penicillium*

40-III (p. 352)	1. passive 2. gametophyte 3. meiosis 4. spores	5. Red 6. pigments 7. Brown 8. kelps	9. holdfast 10. green 11. land

40-IV (p. 354)	1. vascular 2. sexually 3. mosses 4. gametophyte 5. sporophyte 6. homospores 7. gametophyte 8. gametes 9. diploid	10. sporophyte 11. sporophyte 12. gametophyte 13. vascular 14. ferns 15. rhizomes 16. sporophyte 17. zygote	18. sporophyte 19. sporangium 20. meiosis 21. spores 22. rhizoids 23. gametophyte 24. archegonium 25. antheridium

40-V (pp. 356–357)	1. heterosporous 2. female 3. pollen grains 4. wind 5. angiosperms 6. seeds 7. seed coat 8. gymnosperms	9. Cycads 10. Ginkgos 11. Conifers 12. Mesozoic 13. sporophyte 14. female gametophyte	15. male cones 16. female cones 17. meiosis 18. microspores 19. megaspores 20. pollen tube

Understanding and Interpreting Key Concepts

(pp. 357–360)	1. c 2. d 3. a 4. b 5. d	6. a 7. d 8. b 9. c 10. b	11. c 12. c 13. b 14. c 15. b

41-I
(p. 363)

1. invertebrates
2. metazoans
3. Sponges
4. collar cells
5. Communication
6. nerve
7. radially
8. nematocysts
9. planula
10. medusa
11. comb jellies
12. feeding polyp
13. reproductive polyp
14. female medusa
15. planula

41-II
(p. 365)

1. planuloid
2. paired organs
3. protonephridia
4. respiratory (circulatory)
5. digestive
6. fluke
7. scolex
8. Nematodes (Roundworms)
9. pseudocoel
10. reproductive
11. skeleton
12. Rotifers

41-III
(p. 367)

1. Deuterostomes
2. anus
3. mouth
4. protostomes
5. earthworms
6. insects (spiders)
7. mollusks
8. cephalopods
9. brains (eyes)
10. vision
11. motor (movement) (muscular)
12. mantle
13. mantle cavity

41-IV
(pp. 368–369)

1. Annelids
2. polychaetes (marine worms)
3. anus
4. coelom
5. blood
6. hemoglobin
7. Segmentation
8. double ventral nerve cord
9. nephridia
10. parapodia
11. brain
12. pharynx
13. nerve cord
14. hearts
15. blood vessel
16. crop
17. gizzard

41-V
(p. 370)

1. cuticle
2. exoskeleton
3. crustaceans
4. arachnids
5. spiders
6. Centipedes
7. Insects
8. gills
9. tracheal
10. metabolic
11. annelids

41-VI
(p. 373)

1. deuterostomes
2. echinoderms
3. radial
4. bilaterally
5. water-vascular
6. tube feet
7. Sea urchins
8. Crinoids
9. Tunicates (sea squirts)
10. tadpoles
11. notochord
12. spine (backbone)
13. lancelets
14. filter feeding
15. gill slits
16. jawless fishes
17. jaws
18. swim bladder
19. two
20. atria

Understanding and Interpreting Key Concepts

(pp. 373–375)

1. a
2. c
3. d
4. b
5. c
6. d
7. a
8. d
9. d
10. c
11. d
12. a
13. a
14. b
15. c

CHAPTER 42   Human Origins and Evolution

42-I
(p. 378)

1. Cenozoic
2. Insectivora
3. shrews
4. primates
5. depth
6. grasping (gripping)
7. *Aegyptopithecus*
8. dryopithecines
9. adaptive radiation
10. 5

42-II
(p. 379)

1. 25
2. rainfall (humidity)
3. grasslands (savanna)
4. savannas
5. australopithecines
6. bipedal
7. 2
8. Neandertalers
9. *Homo sapiens*

Understanding and Interpreting Key Concepts

(pp. 379–381)

1. c
2. a
3. d
4. a
5. d
6. b
7. d
8. D
9. L
10. F
11. E
12. K

13.	M	16.	H	19.	J
14.	A	17.	B	20.	I
15.	G	18.	C		

CHAPTER 43   Population Ecology

43-I
(p. 387)
1. population
2. biosphere
3. detritivores
4. earthworms (or any other detritivore)
5. Parasites
6. clumped
7. immigration
8. emigration

43-II
(p. 388)
1. exponential (logarithmic)
2. reproductive
3. J
4. limiting
5. increases
6. decreases
7. Carrying capacity
8. r_{max}
9. logistic growth

43-III
(p. 390)
1. Survivorship curves
2. III
3. dependent
4. carrying capacity
5. independent
6. E
7. A
8. C, D
9. F

43-IV
(p. 391)
1. age structure
2. birth (or fertility) rate
3. 2.5
4. 70–100
5. an immense number of children are still in the prereproductive base
6. delayed reproduction
7. 18
8. 700

Understanding and Interpreting Key Concepts

(pp. 392–393)
1. d
2. a
3. b
4. a
5. d
6. d
7. a
8. b
9. a
10. a
11. c

CHAPTER 44   Community Interactions

44-I
(p. 395)
1. ecosystem
2. community
3. habitat
4. niche
5. commensal
6. mutualistic

44-II
(p. 397)
1. competition
2. competitive exclusion
3. niche
4. interference competition
5. resource partitioning
6. cyclic fluctuation (or oscillations)
7. territories

44-III
(pp. 398–399)
1. Coevolution
2. disguise
3. Camouflage
4. Müllerian
5. D
6. A
7. C
8. C, E
9. A
10. C
11. B
12. A
13. D
14. C
15. D
16. E
17. D
18. D

44-IV
(pp. 400–401)
1. Predation
2. partition
3. Secondary succession
4. Species diversity
5. increases

Understanding and Interpreting Key Concepts

(pp. 401–402)
1. b
2. b
3. c
4. a
5. b
6. d
7. b
8. d
9. d
10. a
11. d
12. a

CHAPTER 45   Ecosystems

45-I
(p. 405)
1. autotrophs
2. sunlight (energy)
3. nutrients
4. trophic level
5. herbivores
6. Decomposers
7. Detritivores

45-II
(pp. 406–407)
1. net primary productivity
2. Biomass
3. detrital food web
4. detrital

45-III	1. greenhouse effect	3. chlorofluorocarbons	6. tree roots
(p. 408)	2. heat (infrared wavelengths)	4. carbon dioxide	7. Rain
		5. Hubbard Brook	8. sedimentary

45-IV	1. Bacteria	4. nitrite	7. malaria
(p. 409)	2. nitrogen fixation	5. nitrous oxide	8. house cat
	3. nitrification	6. mosquitoes	9. biological magnification

Understanding and Interpreting Key Concepts

(pp. 409–411)	1. c	5. a	8. a
	2. d	6. b	9. b
	3. b	7. a	10. d
	4. b		

CHAPTER 46   The Biosphere

46-I	1. F	11. M	20. R
(pp. 413–414)	2. C	12. N	21. grasslands
	3. G	13. K	22. shrubs
	4. J	14. P	23. montane
	5. I	15. O	24. subalpine
	6. H	16. Q	25. alpine
	7. A	17. T	26. lack of vegetation (or cover)
	8. D	18. L	27. deserts
	9. B	19. S	
	10. E		

46-II	1. Topsoil (or the A horizon)	4. desert	7. grasslands (prairie)
(p. 416)	2. subsoil (or B horizon)	5. chaparral shrublands	8. Tallgrass prairie
	3. loose rock (or C horizon)	6. hot (warm) deserts	9. shortgrass prairie

46-III	1. tropical rain forests	6. Slash-and-burn	10. ponderosa pine
(p. 418)	2. complex	7. tropical seasonal	11. Taiga
	3. canopy	8. temperate deciduous	12. tundra
	4. diversity	9. winter cold	13. solar radiation
	5. worst		

46-IV	1. Lentic	4. littoral	6. Oligotrophic
(p. 419)	2. lotic	5. profundal	7. thermocline
	3. overturns		

| 46-V | 1. estuary | 3. intertidal zone | 5. hydrothermal vent |
| (pp. 420–421) | 2. plankton | 4. abyssal zone | 6. *Spartina* |

Understanding and Interpreting Key Concepts

(pp. 421–422)	1. d	5. c	9. c
	2. d	6. a	10. d
	3. b	7. b	11. b
	4. a	8. d	

CHAPTER 47   Human Impact on the Biosphere

47-I	1. dry acid depositions	4. hydrocarbons	7. chlorofluorocarbons
(p. 424)	2. sulfuric acid	5. photochemical	8. average annual temperature
	3. thermal inversion	6. acid rain	

47-II	1. electromagnets	4. Tertiary	6. salination
(p. 426)	2. air blowers	5. Deforestation	7. Ogallala aquifer
	3. primary		

47-III	1. meltdown	4. 8	6. breeder
(p. 427)	2. graphite	5. coal	7. nuclear winter
	3. kerogen		

Understanding and Interpreting Key Concepts

(pp. 428–429)
1. a
2. b
3. a
4. c
5. a

6. c
7. c
8. b
9. c
10. b

11. c
12. a
13. c
14. d

CHAPTER 48 Animal Behavior

48-I
(p. 433)
1. neuromotor response
2. innate
3. sign stimulus
4. Learning
5. constant

6. associative
7. conditioned reflexes
8. instrumental
9. trial-and-error

10. extinction
11. insight learning
12. sensitive periods
13. imprinting

48-II
(p. 434)
1. D
2. A

3. E
4. B

5. F
6. C

Understanding and Interpreting Key Concepts

(p. 435)
1. d
2. c
3. b

4. b
5. a

6. c
7. b

CHAPTER 49 Social Behavior

49-I
(p. 438)
1. B
2. A

3. D

4. C

49-II
(pp. 438–439)
1. risks
2. reproduction

3. Predation
4. predators

5. selfish herds

49-III
(p. 440)
1. open
2. hide
3. dominance
4. dominant (stronger)

5. dominance hierarchy
   (social hierarchy)
6. ritualized behavior
7. appeasement

8. social insects
9. queen
10. Drones
11. biological determinism

Understanding and Interpreting Key Concepts

(pp. 440–442)
1. c
2. b
3. a
4. d

5. c
6. b
7. a

8. c
9. e
10. c

Crossword Number One

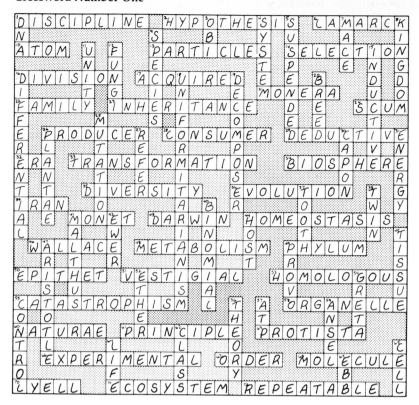

Crossword Number Two

Crossword Number Three

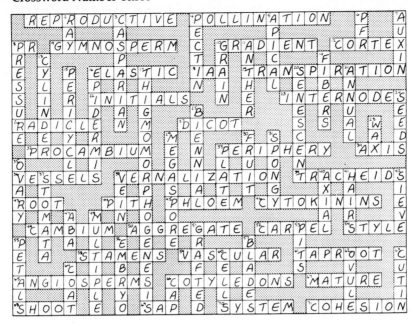

Crossword Number Four

Crossword Number Five

Across answers (as filled in grid):
- DEUTEROSTOME · MUSSEL · AMOEBAS
- GASTROPOD · TUNDRA · FOSSIL
- RADULA · HOMOSPOROUS · TUBE · MITE
- HOST · CYST · BOTULISM · GOLDEN
- CILIATES · PRIMATE
- OYSTERS · DINOFLAGELLATES · SCALLOP
- LYCOPOD · EPOCH · INTEGRATES
- SYMBIOSIS · PERMIAN · ASCUS · HYPHA
- AUSTRALOPITHECINE · CYCAD
- GLAND · COELOM · ANNELID · PALEOZOIC
- DIATOMACEOUS · SEGMENTATION
- SNAIL · BRITTLE · COEVOLUTION
- HEMOGLOBIN · FRAGMENTATION · NECTAR
- ACCRETION · TRICHINOSIS · SEDIMENTS

Crossword Number Six

Across answers (as filled in grid):
- SEMICIRCULAR · OVIDUCT · DRUM
- COMPOUND · CHORION
- CARTILAGE · CEPHALIZATION
- ADIPOSE · MECHANORECEPTOR · SCLERA
- FOLLICLE · SUMMATION · INPUT
- SYMPATHETIC · REM · EPIDERMIS
- INTERSTITIAL · STRATIFIED
- INTERCALATED · HERMAPHRODITIC
- GLOMERULI · STATOCYST · ACTH
- SPONGE · INNATE · OVIPAROUS
- MESODERM · PHOTORECEPTOR · MARROW
- OPTIC
- SPINDLE · FOREBRAIN
- EXCRETORY · EEG · PLATELETS